Evolution

The Four Billion Year War

Evolution

The Four Billion Year War

Michael Majerus, William Amos
and Gregory Hurst

Longman

Longman

Addison Wesley Longman Limited
Edinburgh Gate, Harlow,
Essex CM20 2JE, England
and Associated Companies throughout the world.

First published 1996

British Library Cataloguing in Publication Data
A catalogue entry for this title is available from the British Library

ISBN 0-582-21569-2

Library of Congress Cataloguing-in-Publication Data
A catalog entry for this title is available from the Library of Congress

Typeset by 16 in Times 10/12pt
Produced through Longman Malaysia, TCP

Contents

Preface

Until recently, most students of genetics and evolution could be split fairly readily into two categories: whole organism biologists and molecular biologists. Students equally interested in both were something of a rarity. Over the last decade that situation has been changing rapidly. Just as the old dichotomy of zoology and botany was eroded to a significant extent by the upsurge of interest in ecology, the split between molecular and whole organism biology is being bridged by more and more young (and a few older) scientists cognisant of the fact that to understand evolution, the unifying principle of biology, both the phenotype and its cellular parts must be understood. It is vital to combine knowledge of both the whole organism, how it works, behaves and interacts with other organisms and its environment, and knowledge of genetic material, how it behaves, is passed from generation to generation and is the root of the myriad of chemical reactions that interact to produce living organisms.

Most students of biology share a sense of wonder in the natural world, whether that wonder is for the beauty of a passion flower, the aggressive power of a killer whale, the precision of flight of a hummingbird or the intricacy of metabolic interactions that allow us to breath. We share that fascination, and in this book we attempt to convey not only some of the basic principles of evolution and some of the contemporary problems facing evolutionary geneticists, but also something of our own wonder in the natural world and our passion for investigating the processes that have shaped it, the pathways of evolution.

It is becoming increasingly apparent, as the intricacies of these evolutionary mechanisms are revealed, that one of the underlying themes in evolution is conflict. This is manifest in the phrase 'the struggle for existence', so often used to describe the process of natural selection identified by Darwin as one of the primary mechanisms of evolution. Biological conflicts can take many forms. The protagonists in conflict may be of different species (inter-specific competition), or any individuals of the same species (intra-specific competition), or individuals of opposite sexes (the battle of the sexes), or parents and their own progeny (parent–offspring conflicts), or different genes within the nucleus or in the cytoplasm of cells (intra-genomic conflict). We have used this theme as a binding thread throughout this book. Indeed, it was the realisation that the majority of the phenomena that we were dealing with involved differences of interests and antagonisms that led us to our title.

Assumed knowledge

The book is aimed primarily at final year undergraduate and postgraduate students, although we hope that it will also be a useful reference source both for undergraduates in earlier years, and for postdoctoral researchers and academics. We have assumed some knowledge of the basic principles of natural history, biology and genetics, and some very elementary knowledge of mathematics. We have not assumed any knowledge of molecular genetic techniques, or of population genetics.

Boxes

The main text of the book is divided into thirteen chapters. We have tried to make this text readable. Consequently, we have kept mathematical models and analyses to a minimum. Where we deemed full treatments were crucial to understanding we have placed them in boxes. The same is true for detailed explanations of some theoretical principles and for detailed descriptions of some case studies. At the end of each chapter, we have included a brief summary of the salient points of the chapter.

Further reading and references

We have included a short list of further reading for each chapter. The papers and books referred to are chosen to give further information, alternative views, background information or technical or mathematical back-up. The text is liberally, but not exhaustively, referenced. A full list of the publications referred to is included at the back of the book.

Glossary

In the text, certain crucial words or phrases are given in bold type. These terms are generally defined (according to our usage) when they first occur. However, because many of them crop up many times, we have also included a glossary of these terms at the end of the text.

Michael Majerus, Bill Amos and Gregory Hurst
University of Cambridge
21 July, 1995

Acknowledgements

There are various people that we would like to thank for their help in producing this book. First and foremost, we are grateful to Alexandra Seabrook for inviting us to write this book, and for her great patience during its preparation. We also acknowledge Bill Jenkins and other members of the production and marketing teams at Longman for all their work on this project.

Many parts of the text owe a considerable debt to colleagues and friends who have read particular sections or discussed particular ideas with us. These include Dr John Barrett, Dr Laurence Hurst, Dr Rufus Johnston, Dr Andrew Davis, Mark Ransford, John Sloggett, Dr Mark Tester, Mary Webberley, Emma Purvis, Andy Barron and Tamsin Majerus. Several referees made useful suggestions in respect of the original book proposal, and two anonymous readers made many cogent suggestions on parts of the first draft of the text. We thank them for their efforts on our behalf. Acknowledgement is also due to Tamsin Majerus for production of some of the draft figures, and for checking many parts of the manuscript for references and general consistency.

We also extend our thanks to all those undergraduate and postgraduate students who have worked with us over the years and whose interest and stimulation have so often provided a catalyst towards fuller understanding and novel perspectives.

We are also grateful to Dr Phil Clapham for the cover photograph, to Dr Kyle Summers for Plate 1 and to Mr Gerald Burgess for Plate 2a.

We are grateful to the following for permission to reproduce copyright material:

Figs 2.1a and 2.1b reprinted with permission from Bernardi, G. *et al.* (1985) *Science* **228**, 953–8. Copyright 1985 American Association for the Advancement of Science. Fig. 5.1 reprinted with permission from Burke T. *et al.* (1989) *Nature* **338**, 249–51. Copyright 1989 Macmillan Magazines Limited. Fig. 8.2 reprinted with permission from Hamilton, W. (1967) *Science* **156**, 477–88. Copyright 1967 American Association for the Advancement of Science.

Whilst every effort has been made to trace the owners of copyright material, in a few cases this has proved impossible, and we take this opportunity to offer our apologies to any copyright holders whose rights we may have unwittingly infringed.

Chapter 1

Introduction

Evolution is the real creator of originality and novelty in the universe.
(Bronowski, 1973)

The human species exists in the thinnest of temporal slices which caps a vast iceberg of evolution stretching back some four billion years. One of the great fascinations we have is trying to understand how we arrived at this point and why we cohabit the Earth with the organisms we do. Some try to piece together past events by studying fossils. Others study how the process of change continues today by looking at extant species and asking questions about the way in which they have come to fit their own particular ecologies. In all fields, novel methods of molecular analysis, particularly molecular genetic analysis, are causing a revolution in the way we think about evolution. This book examines how far this revolution has brought us and where it might be heading, focusing on the sorts of questions we want to ask and how molecular techniques melded to insights gained through more traditional techniques might provide answers.

Evolutionary questions

Right from its beginning, the course of evolution raises a host of important questions that have yet to be answered. To begin with, perhaps the greatest question of them all: how did life begin? How did the first complex molecules come to be formed in the primordial soup and then combine to create the first self-replicating assemblages? Which of the two genetic molecules, DNA or RNA, first carried hereditary information which allowed the forces of natural selection to take hold and begin to mould the first recognisable organisms? How did the link between DNA, RNA and proteins first arise?

Today we can divide living organisms into a number of large assemblages, which we call kingdoms. These include animals, plants and at least two major radiations of bacteria. Perhaps there were once several more. However, among those which survive we want to know which are the oldest, and how did the key divisions arise. If we could answer these questions we might learn more about the conditions which once prevailed on Earth, when the atmosphere first began to contain significant amounts of oxygen and which metabolites provided the earliest bacteria with their sources of energy.

Of particular interest is whatever happened a little over half a billion years ago. For some three billion years previously, an immense span of time, life had been exclusively single-celled. Then, during the Cambrian epoch, multicellular life appeared approximately 545 million years ago. Some 20 million years later, roughly 525–515 million years ago, there was a phenomenal explosion in the diversity of life forms. During this period, most of the major multicellular phyla came into being, from the sponges to sea squirts and lancelets, the first recognisable ancestors of the vertebrates. Within some 40 million years the first fish evolved. What happened about half a billion years ago to stimulate or allow the creation of this massive evolutionary radiation and was it really as sudden as many believe.

On Earth today there are probably somewhere in excess of 30 million species of organism. At a rough estimate, 65 per cent of these, maybe some 20 million species, belong to the class Insecta. At the same time, over 70 per cent of the Earth's surface is covered by sea. Yet there are no truly marine insects. Why is this, when life almost certainly began in the sea, and most other groups which came out onto land include radiations or individual species which have secondarily returned to the sea? Why are the insects different? On land they are masters of diversification and adaptation, yet none has managed to return to the sea?

Finally, but by no means least, extant life brims with products of evolution whose very existence challenges us to provide an explanation. One example will suffice. In alpine meadows along the sides of the Rhône Valley, in Switzerland, lives a small orange and black moth. It is a member of the tiger-moth family (Arctiidae), so the bright contrasting colours are probably adaptive, acting as a warning to potential predators that the moth is highly distasteful, if not toxic. The moth flies both during the day and at night. As it flies it makes a clicking noise that can be heard by humans up to 30 m away, and is perhaps similar to the sound we make when we click our fingernails together. The frequency of clicking is very regular.

At night, this moth, along with many other species, is attracted to street lights in the alpine villages. Bats are also attracted to the vicinity of the street lights, where they catch and feed upon the flying insects. It is very noticeable that one species of moth, the little arctiid, is never caught by the bats. One speculative explanation of the clicking sound made by this moth is that it interferes with the bats' sonar, so the bats are not able to locate these moths accurately. If this is the case, it must be assumed that the frequency of the clicks must be precise to jam the bats' sonar effectively. Two questions must be asked. Does the clicking sound act in this way, or is there an alternative explanation of the sound? And, if the clicking does serve this purpose, how could such a precise adaptation have evolved?

This journey through time gives a mere taste of the multitude of conundrums that face students of evolutionary biology. Many others could be cited. How did the first insect fly? How did the archer fish evolve to be able to spit a droplet of water accurately at an insect on foliage above the water, taking account of the refraction of light? How did the peacock get his tail? Where did viruses come from? Why do some plants have 200 times the amount of DNA of others? How and why did sex first evolve and why have some organisms reverted to an asexual existence? Why are about 40 per cent of all described species of organism beetles?

In essence questions related to these types of mystery may be split into two classes. The first class of questions deals with the design of different species of organism, whether extant or extinct. How do they work? Why do or did they have their own particular morphologies, physiologies and behaviours? The second class involves the ways in which the great diversity of species that the Earth has supported have arisen. What mechanisms cause new species to form? Why are there more insects than mammals? What are the evolutionary relationships between the lineages? Essentially what evolved from what, and why?

Unravelling the past

Questions within both classes may be addressed either by observation and analytical experimentation of changes within a species, or by comparison of the differences between species. Both have inherent shortcomings. The former of these approaches is strongly constrained by time. Because evolution is generally a slow process, most changes that occur within the working life of a human are small. However, it may be that the principles of change revealed by studying, say the evolution of insecticide resistance in mosquitoes, or the replacement of a pale form of a moth by a darker form in polluted environments, may usefully be applied to a broad range of evolutionary scenarios. The study of evolution in action by a mixture of laboratory and field work has been termed 'ecological genetics' (Ford, 1964).

The comparative approach has proved rewarding in many spheres. Indeed, it was an approach favoured by Darwin during his 20 year long accumulation of evidence to support his evolutionary theory prior to the publication of *The Origin of Species* (Darwin, 1859). However, deductions and conclusions based on comparative studies always face the criticism that they are inferences. The reasons for the design of the traits of organisms are inferred from differences between individuals in a population, between populations of a species and between species.

To build up a cogent picture of evolutionary progress, ideally two things are necessary. First, we need to be able to observe all aspects of the organisms under investigation: morphology, chemistry, physiology, behaviour and ecology. Second, we need to know the phylogenetic relationships between species, so that by knowing the order of divergence we can understand the evolutionary history of specific traits.

Can these aims be achieved? In some cases the answer must obviously be no. Many of the organisms that need to be investigated and understood if a full picture is to be built up are extinct. The origins of many of the traits we seek to explain are part of history. The dinosaurs have gone, the archer fish spits, the peacock has his tail, the leopard has its spots. With traits of extant organisms, at least we have the opportunity of investigating in detail how they do what they do, or how they develop what they develop. But we cannot study all aspects of organisms from the distant past now known to us only by their descendants, or as the imprints of their corpses left in the rocks. Of course some questions can be asked of these imprints. Fossils are now studied using a battery of sophisticated tools (electron microscopy, ultraviolet light, X-rays, three-dimensional graphics, image enhancers). New and more powerful technologies are continually being developed and methods of data collection, storage and analysis are moving forward in parallel. However, as yet there are precious few ways to study the biochemistry of long-gone species except by inference and assumption.

An alternative is to study evolution in progress in extant species. Done well, this may allow us to understand better the forces which gave rise to what we see today. This approach is clearly more accessible, but this does not mean it is easy. One classic example, the study of industrial melanism in the peppered moth, *Biston betularia*, illustrates the problems.

During the second half of the nineteenth century, a black form (f. *carbonaria*) of the peppered moth, which up until that time was usually white with black speckling, spread rapidly through many populations of this species in industrial regions of Britain. The great Victorian lepidopterist J. W. Tutt proposed an explanation of this spread. He proposed that the original, non-melanic form was well camouflaged when at rest on bark encrusted with foliose lichens, and that the dark *carbonaria* form had increased in frequency because the nature of many natural surfaces had changed. The major causes of this change were two industrial pollutants: sulphur dioxide, which kills foliose lichens, and soot fallout, which blackens the denuded surfaces. On these new and rather homogeneously coloured surfaces, Tutt (1896) proposed that *carbonaria* would be better camouflaged from avian predators than the strongly patterned typical form.

Tutt's hypothesis was not accepted for half a century, primarily because entomologists and ornithologists generally concurred in the view that birds are not major predators of cryptic day-resting moths. This view was based on anecdotal observations rather than controlled tests, yet it was not queried. Consequently, during the first half of this century, a variety of other hypotheses – from Heslop Harrison's ideas on the mutagenic effects of industrial pollutants (Harrison, 1928), to E. B. Ford's speculations on **heterozygote advantage** (Ford, 1937) – were put forward to explain the rise of melanic moths. Not until the 1950s did Bernard Kettlewell, encouraged by Ford and assisted by Niko

Tinbergen, investigate, in a scientific way, whether birds do eat cryptic day-resting moths, and whether they do so differentially with respect to morph. His now famous experiments in Birmingham and Dorsetshire woodlands demonstrated that birds do in fact do both (Kettlewell, 1955, 1956).

The case of the peppered moth has become one of the most cited examples of evolution being observed in action. It involves a change in a heritable trait and the main reason why the black form was more successful in industrial areas but not in unpolluted regions has been identified and is easily understood. Yet despite this case being one of the best known and comprehensively studied, much of the detail has still to be uncovered.

Problems in perception

That Victorian and Edwardian naturalists were so wrong in their belief that birds are not major predators of cryptic day-resting moths may be attributed to two factors. First, their adherence to a view without testing it. This was simply bad science. Second, and perhaps both more fundamental and problematic, they did not see with a bird's eye. They could not find cryptic moths on tree bark, so they assumed that birds would not be able to either. This difficulty is exemplified by an admission by Sir Cyril Clarke and his colleagues in 1985 (Clarke *et al.*, 1985): 'in 25 years we have found only two *B. betularia* on tree trunks or walls adjacent to our traps, and none elsewhere'.

As it happens, the reason for Sir Cyril Clarke's difficulties is now thought to be simply that peppered moths do not usually rest by day on tree trunks (Mikkola 1979; Howlett and Majerus 1987). However, many other species of moth do habitually rest on tree trunks. Most humans are simply not very good at finding them. In any case, it is difficult to understand why the peppered moth should mimic lichens so beautifully if there was really no danger in being seen.

The problem of differences in sensory perception between scientists and the species they study should not be underestimated. We have reasonably good sight, at least in respect of some wavelengths, and our sense of taste is passable. However, our senses of hearing, smell and touch are poor, and our ability to detect electrical impulses or magnetism are virtually non-existent. And these are just the senses that we recognise. What if there are senses that are so different from our own that we have little or no appreciation of them at all? How are we, with our highly differentiated and organised sensory system, to comprehend the sensory capabilities of a single plant cell, the stomatal guard cell, which has the ability to respond to all the basic types of stimuli that we respond to in our five senses?

While the difficulties in appreciating senses, or sensory information, that we are not naturally designed to receive should not be underestimated, we do have the technology to transform some of this information so that we can analyse it and work on it. For example, by using filters or special photographic film, we can visualise the ultraviolet and the infrared wavelengths of light outside our normal perception. In this way, we can study characters such as the honey-guides on flowers and the UV patterns on male butterflies that appear to be used by females in mate assessment. We can measure electrical output, ultrasonic stimuli and bats' sonar. Using flow chambers and choice experiments, we can assess reaction to scents and tastes. With electrical sensors we can measure the electrical discharges that fish such as electric eels use, both to perceive their surroundings, and to stun their prey. However, we cannot fully appreciate the way any other species perceives its world. Advances in technology can give us new insights into these worlds, but only within the constraints of the technology, our senses and our imagination.

A further problem in perception comes with our view of the organisms themselves. Animals have often been thought of as being simple and not capable of complex or subtle behaviours such as mate choice. However, recent evidence is beginning to shake this view. Most species do not obey simple rules; their behaviours are often highly evolved to be optimal in the face of a wide range of circumstances, both long-term and short-term, seen and unseen. Cuckoldry is not random, but instead relates to the quality and behaviour of the individuals concerned. Dispersal does not adhere to Walt Disney's fabrication about lemmings – a hell-for-leather, suicidal exodus. Instead, dispersal is an adaptive response which is modulated by a range of factors including age and sex, population density, relatedness, state of health and more.

Retrospectively, it seems a little optimisic to believe that a human who watches an animal, usually for less than half of each day, will be able to perceive, record and interpret correctly every facet of its behaviour. Evolution is rife with conflict, and cheating is often an important element. If an individual animal can successfully pull the wool over the eyes of a conspecific, possibly one living in the same burrow or nest, what are the chances that such a behaviour will be obvious to us? More importantly, we must ask whether such subtleties, when seen, will be interpreted correctly for what they are.

Opening the doors through molecular genetics

Perhaps the sphere of technological advance that offers most hope is that of genetics. To evolutionary biologists the genetic dimension is essential, for if we are to investigate and understand evolutionary mechanisms and pathways, our most precious materials for study are surely the hereditary materials DNA and RNA. It is these remarkable molecules that provide the evolutionary thread of life, the link between this and every previous generation.

During the twentieth century the field of evolutionary genetics has increased in importance as a sphere of biological endeavour. It draws from, and contributes to, all aspects of biological science. Theodosius Dobzhansky famously once said that 'Nothing in biology makes sense except in the light of evolution'. Later, Fransisco Ayala and John Kiger wrote that 'nothing in biology is understandable except in the light of genetics' (Ayala and Kiger, 1980). If both these propositions are true, then perhaps evolutionary genetics is the core of biological science.

New molecular techniques for cutting, marking, replicating, mapping and analysing DNA and RNA are now opening doors through which we can gather data reflecting a whole new dimension on interactions. DNA fingerprinting can be used to determine paternity, and so uncover where individuals either show fidelity or cheat in the mating game. Other techniques can be used to study relatedness and so help plot the passage of individual genes and chromosomes through time within a population. At a higher level still, methods now exist for studying how much mixing occurs between populations, allowing us to examine in detail how divisions occur, are maintained and may, eventually, lead to speciation. Above the level of the species, genetic studies offer ever better methods for phylogenetic reconstruction, relating species, genera and higher taxonomic groups to each other. In this way at least we can gain a map of what has happened, and reduce the extent to which we have to guess about affinities.

The extent to which these new methods are opening our eyes is becoming clear. Behavioural ecologists increasingly accept that, while in some cases they recorded an accurate picture, in others they were duped. Traditional taxonomists vacillate, on the one hand appreciating the power of molecular methods, and on the other being suspicious of the stories they sometimes tell. To begin with, there was mainly conflict. Fortunately, however, there is increasing reconciliation as the two disciplines unite towards a common cause, mainly through morphologists accepting slowly the use of molecular phylogenetics. Unfortunately, with a few notable exceptions, molecular biologists have tended not to reciprocate, preferring to place their faith entirely in DNA sequences.

Evolutionary biologists are perhaps now facing the most exciting period in the history of the subject. Over the last two or three decades, the rate of development of molecular tools allowing one to scrutinise hereditary variation has been accelerating rapidly. We have moved quickly from the development of electrophoresis and DNA hybridisation techniques in the 1960s, through restriction mapping and full nucleotide sequence analysis, DNA fingerprinting, the polymerase chain reaction (PCR) and mini- and microsatellite analysis, to the development of automated sequencing technologies which allow large amounts of data to be collected quickly. These molecular techniques provide evolutionary geneticists with extremely powerful tools. In parallel with these have come staggering advances in computer technology which facilitate greatly the collection and analysis of these large and intricate data sets. These tools thus allow us to amass the data to answer a myriad of evolutionary questions and problems which previously would have been intractable using traditional ethological, ecological and genetic techniques.

Consider for example a study of house sparrows nesting around a farm in Greece. Pairs formed, nests were built, eggs were laid and chicks were fledged. It was already known that sparrows indulge in extra-pair copulations. The aim of this study was to use DNA fingerprinting to determine the proportion of chicks that were the result of extra-pair matings. The basic result, that a significant proportion of the young were conceived on the 'wrong side of the nest-blanket', was perhaps not wholly surprising. However, the level of cuckoldry was much higher than had been expected from previous observations. One cock bird was of particular note. He appeared to be a most reproductively successful bird, working assiduously to feed ten broods of chicks over five years. The poignancy of the case is that the fingerprint analysis revealed that he had been cuckolded by his females on each of the ten occasions and was not the true father of any of the chicks he helped to raise.

The lack of reproductive success of this male sparrow could not have been revealed easily without the development of DNA fingerprinting techniques. The challenge, of course, is to apply these powerful molecular techniques to answer interesting questions about the evolution of whole organisms. The effort and funding that has been devoted to the development of these techniques demands a return in the form of answers to ecological, developmental, demographical and behavioural questions that impinge on the evolution of the diverse range of life on Earth.

Evolutionary conflicts

Darwin (1859) talked about natural selection in terms of a struggle for existence. He clearly endued organisms with a drive to survive and reproduce. In his conclusion to *The Origin of Species*, Darwin put it this way: 'Thus, from the war of nature, from famine and death, the most exalted object which we are capable of conceiving, namely, the production of the higher animals, directly follows.' Here Darwin clearly sees evolution as a process leading from the simple to the more complex.

One major source of evolutionary complexity is to be found in the conflicts of interest within and between organisms. Consider a simple instance of a bird whose best interest is in defending a territory, for by doing so it is able to increase its own fitness by monopolising a resource. However, at the same time this behaviour will lower the fitness of conspecifics who are barred from the resource. Such a situation typifies the selfishness implied in Darwin's notion of a struggle for existence, and creates a point of dynamic tension. Genes which increase a non-territory-holder's access to the resource, either by force or subterfuge, will tend to be favoured. At the same time, genes carried by territory-holders which help either to maintain or to increase the resource monopoly will also be favoured.

Conflicts of evolutionary interest take many forms. They may be between individuals of the same species, for resources or mates. Alternatively, they may occur between individuals in different species, for instance between predators and prey or between parasites and their hosts. Beyond this, it is becoming apparent that there may be a conflict of interests within an individual. Although all genes within an individual have the same 'goal' of maximising the reproductive success of their bearer, diploid individuals bear two copies of most genes, and only one of these copies is transmitted to each offspring. There can, therefore, be a conflict of interest between genes within the organism over which is transmitted via gametes.

All these conflicts have subtle differences in their evolutionary biology. They do, however, have one thing in common. Conflicts offer a potent evolutionary force capable of driving perpetual change. As the defender improves its defence, so the attacker will come under pressure to improve its attack. It is this concept that has made conflict a major focus of evolutionary genetics and of this book. Having said this, not all change is born out of conflict. Two hunters who hunt together may catch more then twice as many prey compared with if they hunt independently. Here, evolution will promote an alliance rather than conflict. Co-operation within a species and mutualism between species are both observed, and can be rationalised in terms of selection acting upon individuals and ultimately genes.

Finally, it is important not to forget that chance also plays an important role in evolution. Most genetic variability probably has little or no effect on fitness. Variants are free to drift up and down in frequency, like flotsam on the sea, some occasionally reaching the shore and becoming fixed. Although by definition these characters are individually insignificant, together they set the genetic context in which all evolutionary change must take place, facilitating some paths and blocking others.

All three examples further emphasise the link between genetics and evolution, and this is the true thrust of our book. If we are ever going to be able to make progress with understanding evolution, we must be able to quantify its unit parts. We must determine paternity, maternity and other relationships, assess fitnesses, identify selection pressures, measure levels of gene flow between populations and place species on accurate phylogenies. To this extent, advances in molecular genetics offer a light at the end of the tunnel; they provide a means to at least measure some of the most important parameters. This book examines the relationship between how, on the one hand, conflict acts to shape what we see, and how, on the other, genetics, particularly molecular genetics, can help us to dissect both ongoing and past evolution.

Chapter 2

Mutations – the clay of evolution

Introduction

Biological evolution is the cumulative change in the characteristics of organisms compared with their antecedents. Such changes range from the invisible, involving minute changes in **non-coding** hereditary material which has no apparent effect on **phenotype**, through the apparently insignificant, such as small changes in the number of ridges on a human fingerprint, to more overt changes, for example changes affecting the colour of a flower or conferring insecticide resistance on an insect. Evolutionary change may result from a number of different processes: **random genetic drift**, **natural**, **sexual** and **artificial selection**, and biases in the behaviour of the hereditary molecules **DNA (deoxyribonucleic acid)** and **RNA (ribonucleic acid)**. The mechanisms by which these changes occur form the substance of the majority of this book. However, ultimately the origin of all evolutionary change is rooted in **mutations**, heritable changes to the genetic material which give rise to the variation on which natural selection and other processes may act.

The term **mutation** is used to encompass any heritable change in the genetic constitution of the organism. Mutations therefore include changes as small as the substitution of one **nucleotide base** for another, or as big as the gain or loss of a complete set of **chromosomes**. Each type of mutation has its own characteristics, in terms of the frequency at which it occurs, the DNA sequences it tends to affect most and the consequences on the organism. In this chapter we examine the full range, attempting to identify which types of mutation are most important in the context of evolution.

Point mutations

A point mutation is any event where the DNA sequence changes by a single nucleotide. This could involve either gain (*insertion*), loss (*deletion*) or **substitution** of one for another. There are two primary sources: the process of DNA replication and the cell's own DNA repair systems. Substitutions occur more frequently than either deletions or insertions, at a rate of about one change every thousand million generations.

Point mutations do not affect all sequences or all organisms equally. For example, mammal and bird **mitochondrial DNA** mutates more rapidly than nuclear sequences (Brown et al., 1979, 1982), and the AIDS virus mutates more rapidly than either. We can identify four distinct levels which determine the mutation rate of any particular sequence: the species in which it is found, the precise region of the **genome** where it is located, the immediate surrounding sequence and the identity of the base itself. In addition, the sex of the individual may also influence the rate of mutation. In mammals at least, mutations may occur six times more frequently during male sperm formation than in female **gametogenesis**.

Influence of the species

There can be no doubt that organisms differ in the rate at which their genomes change (see for example Hillis et al., 1994; Schlötterer et al., 1991; Avise et al., 1992), part of which is likely to be due to differences in the rate at which mutations occur. These observations, along with theoretical considerations, form the basis of a fascinating debate about the extent to which mutation rate is itself an evolved character. Does selection favour ever improved fidelity of replication, or is the level of error we see an evolved optimum which could be selected either up or down? Certainly, if the system could be made error free, evolution could not occur, since variability would decay to zero, and without variability change cannot occur. In this sense, the argument is the same as one of those used to explain why animals and plants engage in sex. Sex (through **recombination**) and mutation (through the creation of novel sequences) both provide variability which can theoretically enhance a species' ability to adapt. In the long run, species would be expected to end up with a mutation rate which balances the deleterious effects of too much change against the benefits of long-term evolutionary plasticity.

One interesting suggestion, reviewed by John Cairns and colleagues (1988), is that the mutation rate of a gene may increase when under selection pressure to change. Several different experiments are described, exemplified by the following. A bacterial strain was created which can only grow on lactose in the presence of arabinose. The frequency of revertant mutations was then quantified either with or without exposure to lactose. In the absence of lactose, revertant bacteria were not detected, suggesting an actual mutation rate of less than 10^{-11}. However, in the presence of lactose, revertant rates of 10^{-8} were recorded. Somehow the bacteria appear to be able to increase the frequency of beneficial mutations.

Just how a gene can 'know' when to increase its mutation rate is unclear and lies beyond what can be envisaged given our current knowledge of the way in which cells function. Perhaps the bacteria are capable of testing the water by allowing a higher mutation rate but then repairing the vast majority of all mutations which do not prove beneficial. Who knows? As yet the phenomenon should not be taken as proven. Although the experiments are well enough controlled to make alternative explanations difficult, the mutations so far detected all involve reversions of artificially engineered defects, and it is possible that 'natural' mutations behave differently. Conversely, in support, one must bear in mind that the cellular machinery in species of the fruit fly *Drosophila* has the ability to recognise functional and non-functional *ribosomal RNA* genes (see Templeton *et al.*, 1989).

Influence of the genomic region: isochores

The way in which mutation rates vary between different regions of the genome is a little better understood, thanks largely to the work of Giorgio Bernardi (Bernardi *et al.*, 1985; Bernardi 1993). Bernardi has conducted pioneering experiments looking at the large-scale structure of the genome. The basic experiment is simple. Genomic DNA is cut up into very large pieces, then separated into fractions according to the proportion of Guanine (G) and Cytosine (C) nucleotides, as opposed to Adenine (A) and Thymine (T). One might expect to find some sort of normal distribution of GC composition, with most chunks of DNA lying near the overall mean for the genome. But this is not observed. Rather, in warm-blooded animals (mammals and birds), the fragments fall into discrete fractions, with either higher or lower GC composition (Fig. 2.1a). At first glance, it is not clear what this means. However, with a little thought, one realises that there is only one conclusion, i.e. that the genome is made up of large domains, some one million bases in length, which differ in mean GC content. Such regions were christened '**isochores**' by Bernardi.

Having split the genome into a number of well-defined fractions, the questions which next had to be asked were how precise are these domains and how do they relate to particular genes. In order to address these questions, molecular probes for a variety of genes were probed against a panel of isochore fractions. Virtually without exception, a wide variety of genes could be mapped to specific fractions, occurring in one isochore only. This result confirms that isochores are large and are not merely an artefact. Furthermore, **homologous** genes in divergent species appear in the identical or similar density isochores, indicating long-term stability in the relationship between genes and their isochores.

More interesting patterns emerge when comparisons are made between the GC content of a DNA sequence and that of the isochore in which it lies. It is found that there is a strong linear relationship between the GC content of the gene and that of its isochore. For intronic regions, the relationship appears to be that the GC composition is roughly 10 per cent higher than the isochore in which it lies. For **introns**, however, the relationship is exaggerated, such that

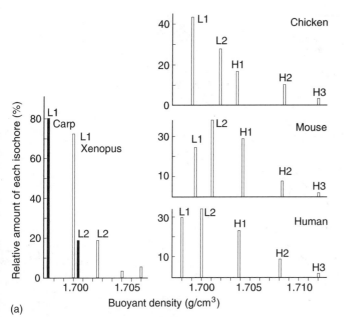

(a)

Fig. 2.1 (a) Isochores in warm-blooded and cold-blooded animals. Whole genomic DNA is fragmented into large pieces and then separated according to its GC content by centrifuging through a density gradient of caesium chloride. Instead of a continuous distribution, the DNA is seen to form discrete bands, or fractions, each corresponding to a different mean GC content. Warm-blooded animals, represented by the mouse, human and chicken, reveal five components, two light (low GC, named L1 and L2) and three heavy (high GC, labelled H1, H2 and H3). By contrast, cold-blooded animals (carp and frog) predominantly lack the heavy components.

those lying in high GC (heavy) isochores have extremely high GC content (Fig. 2.1b). It is as if isochores represent regions of mutational bias. For genes, there are constraints placed by the need to retain function. Changes in GC content thus reflect only differences in which one of two or more possible triplet codes (**codons**) is used to specify a particular amino acid. In contrast, most intronic regions probably lack function, and therefore may be free to drift to very high GC content.

The implications of a genomic structure based on isochore domains are profound. Any gene which moves from one isochore to another will experience a radical change in the rate and direction of mutations affecting it. In contrast, those genes which have remained for long periods in one domain may well have reached mutation saturation, such that further change is less likely. In other words, the rate of change may well depend less on an overall pattern for the whole genome, and more on the proportion of all possible mutations which produce a change in the direction favoured by the local isochore.

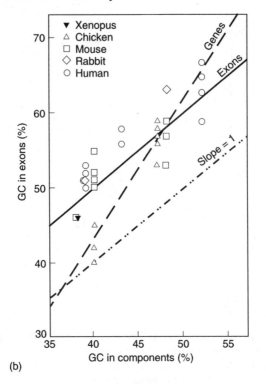

Fig. 2.1 (b) Comparing the GC content of particular genes with the GC content of the density component in which they lie reveals a linear relationship. Exonic (coding) sequences have on average around 10 per cent more GC. Whole genes show an exaggerated pattern in which high GC genes have a higher GC content and low GC genes have a lower GC content than their respective exons. By implication, non-coding introns are generally affected more than exons by whatever mutational bias generates isochore structure.

Much work is still required to find out the precise mechanisms which underlie isochore formation. We need to know how GC content differences come about and what determines their magnitude. We also need to disentangle cause and effect. To what extent are isochores created by the clustering of genes and other sequences with the same GC content, and to what extent is the GC content the result of isochore structure. Why, for instance, do cold-blooded vertebrates apparently lack the high GC content isochores found in warm-blooded species. One interesting possibility has been put forward recently by Adam Eyre-Walker, (Eyre-Walker, 1993). He suggests that GC-rich regions might be generated by the process of recombination, perhaps because the inevitable mismatch repair which occurs during recombination is biased in favour of Gs and Cs. It is certainly interesting to note, as he points out, that the long arm of the Y chromosome, which does not recombine, has one of the lowest GC contents of any region studied (Korenberg and Engels, 1978).

Influence of the immediate surroundings

The triplet code which specifies each amino acid is only a small part of the language of DNA. Control systems must direct the way in which these words are read by providing punctuation. There must be start signs and stop signs, read-now signs and read-me-only-once signs, signs to say continue, slow down, go back and more. Increasingly, we are finding that an important part of DNA's punctuation is held in its three-dimensional structure. Instead of the actual sequence of bases being important, it is the ability to form secondary structure constructed from hairpin loops that matters (Fig. 2.2).

Hairpin loops are formed by internal pairing within a strand of DNA. To a large extent, it appears that the order of bases is less important than the ability to maintain pairing without any gaps. Thus, any mutation which occurs within a stem structure will potentially disrupt the form and be selected against. There are two possibilities. In some cases, the cell's own repair mechanisms will spot the mismatch and 'mend' the pairing by making a compensatory change, either back to the original sequence or to a new pair. Alternatively, natural selection against impaired function may cause a second mutation which restores the

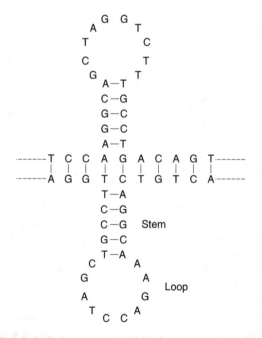

Fig. 2.2 A hairpin loop structure (also known as a stem–loop structure). Such structures occur widely, and include sections of ribosomal RNA molecules, transfer RNAs and, probably, a variety of genetic 'punctuation' marks. Apparently, it is often the maintenance of the length and integrity of the stem which is more important than the actual order of bases.

pairing to become fixed. If the mutation is 'repaired' then this will represent a new source of mutation which is related to the local sequence.

Stem–loop structures are probably common, and one of the better examples of compensatory changes comes from a study of the ribosomal RNA (rRNA) genes in *Drosophila*. In the 28S (large subunit) gene, a total of 70 compensatory changes were identified, all of which acted to preserve the conserved secondary structure of the rRNA molecule (Hancock and Dover, 1988, 1990). On a wider taxonomic scale, a comparison between homologous ribosomal RNA genes between kingdoms reveals a similar pattern in which stem–loop structures are preserved through billions of years of evolution, emphasising their role in the function of the molecule (Gray *et al.*, 1984).

A second example of a position effect comes from a nice database-searching study conducted by Adam Eyre-Walker. He compared published sequences for over 1000 bacterial genes, looking for patterns of codon usage (Eyre-Walker and Bulmer, 1993). His results showed a general codon bias in favour of those containing Cs and Gs. However, the pattern was not uniform. Instead, over the first 10–20 bases the bias is reversed, with As and Ts being favoured. These results show nicely how the cell's own translation machinery can influence the nature of the mutations which affect a particular region.

Influence of the base itself

Finally, the identity of the base itself can influence the rate and nature of the mutations which affect it. Early models of evolution assumed that each of the four different bases could mutate to any other with equal probability, the so-called Jukes–Cantor model (Jukes and Cantor, 1969). However, it soon became clear that this was a gross oversimplification; some base changes are much more likely than others. The most extreme case is provided by a study of the evolution of the AIDS virus (Gray *et al.*, 1984; Ou *et al.*, 1992; Hillis *et al.*, 1994). Here, estimates for the mutation rates were calculated for each possible change in turn, including all back mutations. Rates showed up to a 40-fold variation, with even forward and back mutations (for example, A changing to T versus T changing to A) differing by up to a factor of four. Clearly, 'mutation rate' has little meaning when quoted as an overall rate if one is interested in any particular nucleotide base.

The case of the AIDS virus evolution may be extreme. However, similar effects are common. The four nucleotide bases can be classified into two types, pyrimidines (cytosine and thymine) and purines (adenine and guanine). Mutations within a subgroup (i.e. A↔G or C↔T; called **transitions**) usually outnumber mutations between subgroups (all other changes; called **transversions**). The precise level of excess varies greatly between sequences, but appears to be a general feature of the mutation process.

In summary, the simple gain, loss or exchange of one base for another is not simple after all. The rate at which it occurs depends on many factors, including the organism, the starting base's genomic location, its neighbouring nucleotides and its identity. Until recently, it appeared that there was little chance that we would ever be able to make much progress with understanding the process of mutation. The events were just too rare to study in numbers. However, with the vast explosion in projects where kilobase upon kilobase of DNA is sequenced, we are now accumulating a suitable database from which to ask pertinent questions. Perhaps we will one day progress from trying to describe gross mutational biases to actually predicting accurately the mutation rates affecting individual bases.

Variable number of tandem repeats (VNTRs)

This mouthful was coined as a blanket term to encompass a heterogeneous class of highly mutable DNA sequences. The name is descriptive. It refers to any short sequence which occurs in blocks, all repeat units lying head to tail in a line. Such sequences tend to be non-coding, and we are currently ignorant of their function, should they have any. However, **VNTRs** are important. They provide the basis for the technique known as **DNA fingerprinting** (Jeffreys *et al.*, 1985a), cause a number of heritable diseases (Mandel, 1994), appear to affect the circadian rhythm of fruit flies (Hall and Rosbash, 1988), and offer a wealth of highly variable **genetic markers** for genome mapping studies and population analysis (Epplen *et al.*, 1991; Amos and Pemberton, 1992; Bruford and Wayne, 1993).

Minisatellites

Minisatellites are the sequences on which DNA fingerprinting is based. Their basic repeat length is variable but small, usually around 15–60 bp long (Fig. 2.3). Minisatellites were discovered serendipitously, by Polly Weller, while she was working with Alec Jeffreys in Leicester, comparing myoglobin genes

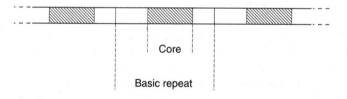

Fig. 2.3 Basic structure of a minisatellite. At each minisatellite locus there lies a block of short motifs, tandemly repeated many times. In a repeat unit, the central core sequence (shaded) tends to be short, of the order of 14 base pairs in length, and relatively conserved between different minisatellite loci. The flanking regions (pale) are less conserved and vary considerably in length between loci. In DNA fingerprinting, a probe made from a core sequence is used at low stringency to detect many loci. In single locus analysis, the entire repeat array is used as a probe at high stringency.

between humans and seals (Weller *et al.*, 1984). It was not long afterwards that their startling properties were recognised and described (Jeffreys *et al.*, 1985b). Minisatellite arrays tend to be highly mutable, gaining and losing repeat units with high frequency. Rates of up to 15 per cent (one sperm in six carries a mutation) have been observed (Vergnaud *et al.*, 1991). The key lies in the sequence itself. Comparing the sequences of several minisatellites from different chromosomal locations, Jeffreys identified a common 'core' sequence, a motif of around 12–16 bp which was, by and large, conserved between loci. When this core was run through a DNA sequence database, the closest match was to a bacterial recombination signal (Jeffreys *et al.*, 1985a).

It thus appears that minisatellites mutate because they comprise long chains of a single DNA phrase, and the phrase says 'rearrange me'. The important consequence is that the core sequence can be used as a molecular probe to detect other, similar sequences. In one experiment it is possible to examine patterns of variability at many different loci, all of which carry the core. The result is a ladder of bands resembling a supermarket bar code, which shows sufficient variability to identify individuals with confidence. Jeffreys christened this pattern a 'DNA fingerprint' (Jeffreys *et al.*, 1985a).

For the evolutionary biologist, the existence of minisatellites has many useful applications. For example, DNA fingerprinting can be used to identify individuals, to assign parentage and to identify closely related individuals (see Chapter 9). More pertinently here, the extreme mutation rates allow us to study the process of mutation. And Jeffreys has done just this. By using a DNA amplification process termed **polymerase chain reaction (PCR)** on DNA purified from sperm, Jeffreys and co-workers have managed to identify and characterise many different mutation events, and to provide estimates on the actual mutation rates (Jeffreys *et al.*, 1994). In addition, each mutation product was analysed for internal polymorphism among its constituent repeats and compared with both progenitor alleles, allowing the precise rearrangements which occurred during the mutation to be determined (see Box 2.1 and Jeffreys *et al.*, 1994; Monckton *et al.*, 1994).

From these experiments we have learned much about one particular form of mutation. The picture gained is both complex and incomplete. It transpires that most mutations occur at one end of the minisatellite array, rather than the other, and that gains in length tend to outnumber losses. Mutation rate does not vary with allele length, but some alleles show drastically reduced mutability, possibly due to the presence or absence of a neighbouring sequence. The mutation rate in sperm is often much greater than in the female germ line or somatic tissues (Vergnaud *et al.*, 1991). Although mutant alleles often appear to be composites, containing elements from both progenitor alleles, flanking markers are seldom exchanged, suggesting that genetic crossing-over is not involved. Instead, the process probably depends on complex gene conversion like events. Finally, a comparison between homologous loci in humans and a variety of non-human primates shows that microsatellite loci are capable of changing their appearance radically over short evolutionary periods (Gray and Jeffreys, 1991).

2.1 Complex mutation events in minisatellites

Most of what we know about how minisatellites mutate comes from elegant single-molecule analysis performed by Alec Jeffreys and colleagues in Leicester. To unravel precisely how repeat units are lost and gained, two primary techniques have been used, one to identify mutant molecules, the other to examine their molecular structure (Jeffreys *et al.*, 1994).

Small pool PCR

PCR has the potential to amplify a DNA fragment from as little as a single molecule. In theory, therefore, PCR can be used to pick out individual mutation events. When the mutations cause changes in length and occur frequently, as is the case for minisatellites, the technique of small pool PCR becomes applicable. Here, minisatellite alleles are amplified from small aliquots of DNA, each containing the equivalent of approximately 100 sperm. Naturally, the male's original alleles produce a strong signal. However, mutation products yield bands with novel lengths which, although some 100-fold less intense, are also seen (see Fig. a). Such mutant alleles can subsequently be gel isolated, reamplified and studied for their internal structure.

Internal mapping

Although each minisatellite locus is characterised by one basic type of repeat, limited sequence variability also occurs. Some human minisatellites consist of a mosaic of two or more different repeat sequences which differ from each other by a single base-pair substitution. When both the length of the allele and the order of repeats along its length are determined, virtually every allele in a human population is found to be unique (Jeffreys *et al.*, 1991). Furthermore, the internal structure provides a way by which minisatellite mutations can be studied. By determining the internal structure of a person's two alleles and then comparing these with the structure of mutation products obtained by small pool PCR, a detailed picture of the gain and loss of repeat units can be constructed.

Internal mapping is achieved by the use of two PCR primers, one which binds to one of the two repeat units and one that binds to the other. Consider what happens in a PCR reaction where one of these primers is present at low concentration. Primer molecules will be spoiled for choice, being able to bind to any of the repeat units they match. PCR will now generate a heterogeneous set of fragments, all of which end in

continued

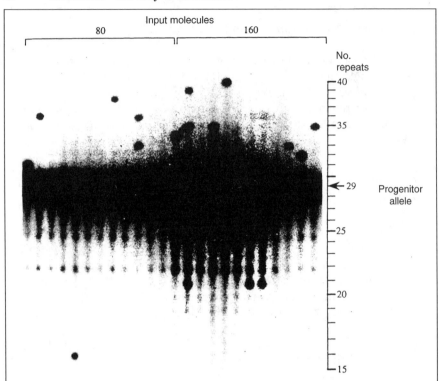

Fig. a Mutation products identified by small pool PCR. The intense central band is the progenitor allele. Mutant products appear as sporadic bands above and below (after Jeffreys *et al.*, 1994).

one type of repeat. The other primer will produce a complementary set. When both sets are run side by side on an electrophoretic gel the result is two complementary ladders, the rungs of which provide a digital read-out of internal structure of the minisatellite array. This same method can also be applied to samples containing two alleles. In this case the result is a form of 'trinary code': at any point on the ladder, a band in one track indicates only that both alleles carry the same variant repeat whereas a band in both tracks indicates that the two alleles carry different variant repeats (Fig. b). Note that in order to increase the quantity of PCR product, a secondary amplification is performed which makes use of a common 'tail' which was designed into both primers.

Jeffreys uses these analyses to formulate a tentative multistage model of how minisatellite mutations occur (Fig. c), in which strand breakage, slippage and gene conversion all play a part.

continued

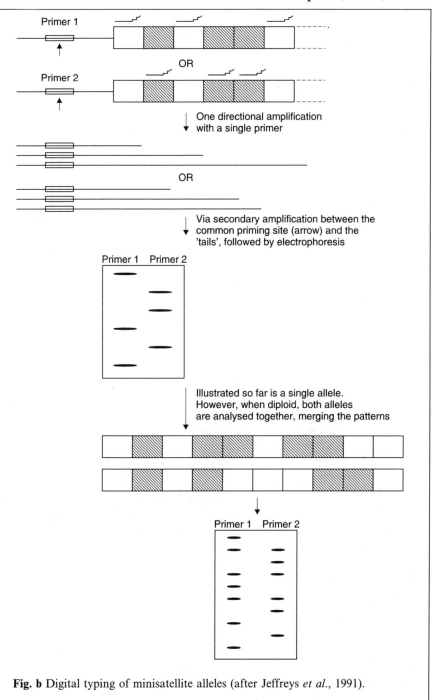

Fig. b Digital typing of minisatellite alleles (after Jeffreys *et al.*, 1991).

continued

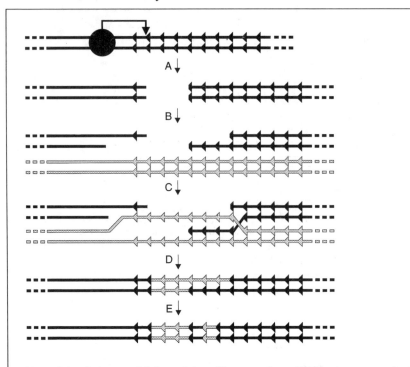

Fig. c Speculative model for minisatellite mutation. (i) Protein creates double strand break; (ii) an overhang arises, possibly through degradation by an exonuclease molecule; (iii) the overhang invades the homologous allele; (iv) the complex is resolved by a combination of DNA synthesis and strand breaks; (v) DNA repair enzymes correct mismatches, randomly converting mismatched repeats to one or other type (after Jeffreys *et al.*, 1994).

Clearly, we have learned a remarkable amount about how mutations occur and yet still have a long way to go. Importantly, the methods developed by the Leicester group will hopefully be applicable to the study of many other sequences, allowing us to discover many unknown parameters about the process of mutation.

Triplet repeat diseases

A number of human diseases result from the abnormal expansion of a block of triplet repeats within a gene (Willems, 1994; Shaw, 1995). These diseases include Huntington's disease, myotonic dystrophy, and fragile X. In each case the pattern is similar. Most of these diseases are manifest late in life, beyond the age of reproduction (fragile X syndrome may cause mental subnormality at an early age). However, disease severity and age of onset depend on the length of the triplet repeat block. The longer the block, the more severe the symptoms and the earlier the age of onset of the disease. Furthermore, the mutation rate among long disease alleles tends to be both extremely high and biased towards a gain in repeats. The result is a phenom-enon known as anticipation, where the children of affected people often inherit the disease in a more severe form, with earlier onset than their parents. Interestingly, anticipation occurs mainly through the male line due to the fact that large increases in repeat number occur much more frequently in sperm than in female gametes (Duyao *et al.*, 1993).

The process of mutation among triplet repeat diseases has, understandably, received a lot of attention. Once again, as with minisatellites, the picture which emerges is complex and incomplete (Snell *et al.*, 1993; Mandel, 1994; Richards and Sutherland, 1994). For example, David Rubinsztein and colleagues have modelled the evolution of allele lengths among normal individuals from both primates and a variety of human populations. They conclude that a mutational bias similar to that which affects sperm may apply also to shorter alleles (Rubinsztein *et al.*, 1994). Best fit to the data is achieved by including both length dependency and a bias towards an increase in length. If true, then these results make the worrying prediction that the future will see a gradual increase in mean allele length and hence in disease incidence.

Microsatellites

Triplet repeats are just one example of a much larger class of sequences known as **microsatellites**. The nomenclature is logical. Repeated DNA sequences were originally named satellites. This was because one method of purifying DNA used to be to centrifuge it through a density gradient of caesium chloride (CsCl). In so doing, the DNA would end up in the middle, and most impurities would either sink or float. However, the method was sensitive enough to sepa-rate different DNA molecules on the basis of their buoyant densities, which, in turn, are related to their GC content. Highly repeated sequences tend to form discrete 'satellite' bands away from the main bulk, hence the name. Minisatellites are short satellite sequences, even though they would not be identified by CsCl centrifugation. By extension, the logical name for tandemly repeated elements which have a basic repeat length of only 2–5 nucleotides is microsatellites.

Microsatellites occur abundantly throughout the genomes of all higher organisms, but are commoner among higher vertebrates (Tautz and Renz, 1984; Tautz, 1989). Mammals may carry several hundred thousand microsa-tellites in their genomes. Among invertebrates, the frequency per kilobase of genome may be lower by a factor of ten. Microsatellites are important because, like minisatellites, compared with other sequences in the genome they are highly mutable. Mutation rates of around one per 10 000 generations have been estimated (Weber and Wong, 1993; Banchs *et al.*, 1994). Most microsatellites are of the order of 20–40 bp in length.

Although microsatellites appear superficially similar to minisatellites, there are important differences, particularly in the primary cause of mutation. In minisatellites, the core sequence apparently interacts specifically with a protein, now known to be a single-stranded DNA binding protein (Collick and Jeffreys, 1990; Collick *et al.*, 1991). This protein appears to catalyse, or at least initiate events which lead to the gain or loss of repeat units. Microsatellites, on the other hand, are thought to mutate by a process called **slippage**. During replication, repeat units become misaligned, leading either to an increase or decrease in length (Fig. 2.4) (Schlötterer and Tautz, 1992).

The experiments of Christian Schlötterer and Diethard Tautz show how microsatellites can mutate in a test-tube by slippage. However, it is not yet clear how these results relate to what happens to microsatellites *in vivo*, in a chromosome. Preliminary data suggest once again that the process is not quite as simple as originally thought. These studies involve typing large known human pedigrees for large numbers of microsatellites and analysing the mutant alleles. What emerges is an intriguing pattern (Weber and Wong, 1993; Banchs *et al.*, 1994). First, as might be expected from the generally rather narrow distribution of allele lengths one finds, most mutations involve the gain or loss of only one or two repeat units. Larger mutations occur, but are rare. Microsatellites thus appear to conform to the so-called stepwise mutation model predicted by simulation experiments (Shriver *et al.*, 1993; Di Rienzo

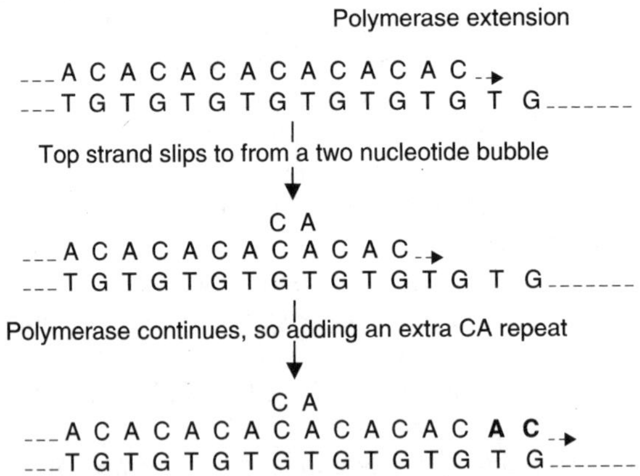

Polymerase extension

```
___A C A C A C A C A C A C_↳
___T G T G T G T G T G T G T G_____
              |
Top strand slips to from a two nucleotide bubble
              ↓
                C A
___A C A C A C A C A C_↳
___T G T G T G T G T G T G T G_____
              |
Polymerase continues, so adding an extra CA repeat
              ↓
                C A
___A C A C A C A C A C A C A C_↳
___T G T G T G T G T G T G T G_____
```

Fig. 2.4 Microsatellite mutation by slippage. Although the precise mechanism of microsatellite mutation is yet to be understood fully, it seems clear that some form of slippage event is involved, probably during DNA replication, an example of which is shown here. *In vitro* studies show that slippage is length-independent, suggesting that slippage bubbles are free to move back and forth along the microsatellite like a ruck in a carpet, becoming resolved only when they reach one or other end.

et al., 1994). Second, and much less expected, there is a distinct bias towards gains in length, just as is seen among minisatellites.

The data which show a mutational bias are rather limited. However, further evidence comes from a study in which the sizes of alleles are compared for homologous microsatellite loci between humans and chimpanzees (Rubinsztein *et al.*, 1995). In the majority of cases, human loci carry long alleles. Such a pattern is best explained if microsatellite loci show a general tendency to expand, but that the process has, for some reason, accelerated in the human lineage.

A second line of support comes from a series of experiments in which trinucleotide repeats were introduced, via plasmids, into bacteria (Kang *et al.*, 1995). The mutations they observed revealed a strong bias, the nature of which depended on the orientation of the repeat with respect to the origin of replication of the plasmid. In one orientation gains predominate over losses, in the other orientation the reverse holds. If the same or a similar pattern exists in higher organisms, then one can imagine microsatellites either declining to nothing or growing, depending on their position relative to the nearest origin of replication. Of these, we would see only the latter class. However, given the differences between plasmid DNA replication and the chromosomes of higher organisms, this is a big if.

The bias towards an increase in length seen in both minisatellites and microsatellites is fascinating and begs the question what happens next? There are two possibilities: either each locus expands until further change is prevented by some as yet unknown force, or expansion is followed by collapse. Neither of these possibilities is supported by direct evidence. However, a stabilisation predicts that the genomes of all organisms should gradually be filling up with long microsatellites which, surely, would have been found. The fact that long microsatellites are generally rare (Beckmann and Weber, 1992) therefore suggests that terminal collapse is the more likely option. There is thus the intriguing possibility that VNTR loci exhibit a distinct life cycle of initiation, expansion coupled with increasing variability and, finally, collapse.

Cryptic simplicity

The possibility of using microsatellites as genetic markers was originally proposed in three papers simultaneously (Litt and Luty, 1989; Tautz, 1989; Weber and May, 1989). Diethard Tautz's observation grew from a general interest in what he termed simple sequences, sequences with very simple arrangements of nucleotides (Tautz and Renz, 1984). His terminology, simple sequence repeats, has now been supplanted by the word microsatellites. However, Tautz's terminology lives on through his investigations of **cryptic simplicity** (Tautz *et al.*, 1986).

Cryptic simplicity is a form of variability in which particular two, three or four base-pair repeats are found more commonly than they would by chance, rather like a hybrid between a unique sequence and a microsatellite. Tautz has written a computer program to measure simplicity. This program works along a sequence, at each base defining a di-, tri- and tetranucleotide. It then searches a fixed distance either side and compares how frequently each of these motifs occurs with the number of times each should occur by chance. Where certain bases tend to occur together, the simplicity index rises. For example, a sequence containing only alternating AG motifs would give an extremely high simplicity value: instead of finding equal numbers AA, AG, GA and GG, we find only GA and AG. The end product is a simplicity profile, which looks rather like a cross-section of a mountain range, with peaks being regions of high simplicity.

Tautz used his program to look at patterns of simplicity in a variety of sequences. What he found was that genic regions, and other regions where we might expect to find constraints, have low simplicity. In contrast, non-conserved regions often show extremely high simplicity. In one beautiful example, cryptic simplicity appears to be speeding up the breakdown of a repeated structure. Genes coding the ribosomal RNA molecules are usually arranged in tandemly repeated blocks, with each gene separated by an apparently non-functional spacer. Often these spacer regions themselves contain tandemly repeated elements. Tautz and colleagues sequenced the spacer region from three closely related species of *Drosophila* and found that a strongly repeated structure present in *D. melanogaster* was beginning to break up in *D. hydei* and had almost disappeared in *D. virilis*. Breakdown of the repeat structure correlated closely with a massive increase in simplicity (Tautz *et al.*, 1987). The important implication is that cryptic simplicity, when it is allowed to increase, can result in a radical change in both the type of mutations occurring and the rate at which change occurs.

Telomeres and satellites

Variable numbers of tandem repeats are common. A genome may contain several hundreds, if not thousands, of minisatellites and hundreds of thousands of microsatellites. However, this is not the end of the list. At the ends of all chromosomes are so-called telomeric regions. **Telomeres** effectively cap the ends of the chromosomes and comprise a few tens of kilobases (kb) of tandemly repeated motifs around 6 bp in length. As with other VNTRs, telomeres appear to be highly mutable, probably through some kind of slippage replication mechanism (Kipling and Cooke, 1990). It is even possible that telomeres go through cycles of progressive expansion followed by sudden collapse, or maybe reach an equilibrium length in which expansion balances any processes which would otherwise cause degradation.

Under the heading of VNTRs we can also include rather longer sequences. Many true satellites have a tandemly repeated substructure, for example, the alphoid satellite family in humans. Furthermore, the ribosomal RNA genes are arranged in tandemly repeated blocks, and may be prone to certain types of mutation as a consequence. Unfortunately, as yet, processes which affect larger tandem repeats are poorly understood. This is because it is technically very difficult to study sequences which may be hundreds or thousands of kilobases in length. However, with the invention of pulsed field gel electrophoresis, a method for separating very large DNA fragments, this may change. Examples are discussed below in the section on repeated DNA sequences.

Jumping genes

A few simple operations make up most of the manipulations a genome can perform on its DNA. Two of these contribute to the phenomenon of **mobile elements**, otherwise called **jumping genes**. First, some viruses have evolved a lifestyle in which their genome alternates between circular and linear. Second, bacteria often carry **plasmids**, small circular passenger molecules which sometimes run into problems during replication by failing to separate and ending up as huge circles containing many basic plasmid molecules. In order to alleviate this problem, plasmids have evolved a site-specific crossing-over event which resolves the large molecule precisely into its component parts.

Combining these two functions, it is easy to see how a circular virus could enter a genome and later leave it unharmed, a behaviour which confers a number of advantages. Consider the bacterial virus, the *bacteriophage lambda*. This virus has two distinct phases, one virulent, when it enters cells, makes lots of copies of itself then lyses the cell to release its progeny, and the second rather more subversive. Instead of killing its host, the virus molecule inserts itself into the bacterial genome and waits, at the same time preserving its hiding place by preventing infection by a virulent form. The integrated virus is then inherited by the bacteria's progeny, as a living time bomb. At some future date, when the bacterial population becomes stressed, possibly as a result of high population densities, the virus will jump out and adopt the virulent pathway.

An integration pathway has obvious advantages. As the virulent form gives explosive population growth and destroys its host, there is a great risk that a local pocket of bacteria, along with the phage *lambda*, would be wiped out. In contrast, an integrated form is like an amazing biological insurance policy. It prevents local annihilation by the virulent form, it allows concurrent dispersal by whatever means the bacteria uses and it provides a far better launching pad from which to invade. The dormant virus only turns virulent when the host is stressed, probably when it was going to die anyway.

It is probable that most, if not all organisms carry, or have carried, some form of integrated viruses. Such viruses may no longer cause death, many having evolved a more benevolent relationship with their host. Consider evolution from the point of view of a virus which has lain dormant for long enough to have lost the ability to cause disease. What sort of properties will selection now favour. As always, the answer is any characteristic which leaves a greater number of progeny (see Chapter 3). Thus, if changes occur such that the virus does not excise to become virulent, but merely transfers a copy of itself to somewhere else in the genome, it will spread. The molecular wherewithal is in place and the path seems logical. No surprise then that the genomes of higher organisms in particular are littered with families of mobile elements, variously sized DNA sequences whose primary feature is an ability to transpose, that is to jump, from one site to another.

Jumping genes are not completely harmless. Whenever a jump is made, it is possible that the element lands in the middle of a gene, causing disruption of function. Indeed, many experimental observations of jumping genes involve monitoring colour pattern changes which occur when an element enters or leaves a pigment gene. Examples include the classic experiments of Barbara McClintock (1951) on maize and more recent studies by Enrico Cohen and his group on colour pattern variation in the snapdragon flower, *Antirrhinum* (Coen and Carpenter, 1986; Almeida *et al.*, 1989). As we shall see later, the frequency of jumping is an evolving strategy, reflecting the outcome of a battle between the interests of the virus, which lie in increasing in frequency, and those of the host, which aim to minimise disturbance to its genome.

In evolutionary terms, jumping genes have a number of important features. Through their coevolution with the host genome, they can lead to hybrid sterility and, therefore, promote speciation (see Chapter 10). Possibly, jumping elements can give rides to genes or control elements, moving them to new locations within the genome. From the point of view of the evolutionary biologist, they provide an amazing tool for tinkering with the genome to examine its multitude of functions. This is particularly true for the way in which P-elements are being used to study the fruit fly *Drosophila*.

The P-elements are probably the best studied of all mobile elements and are described in detail in Chapter 10 (for a review, see Kidwell, 1994). First, by controlling the jumping of a single element, Lynn Cooley and colleagues developed a method for producing 'libraries' of fly strains, each containing a single P-element inserted randomly in the genome (Cooley *et al.*, 1988). Wherever the P-element lands in a gene, it is likely to disrupt the gene's function. Consequently, many of the fly strains have abnormal phenotypes associated with individual genes. Critically, because each strain contains just one P-element, the location of the P-element can then be used to clone and characterise the gene, thereby linking gene and **genotype.**

This method thus uses P-elements simultaneously to mutate and tag random genes. An extension of this is called enhancer trapping, and involves the elegant use of P-elements to look at gene activity (Wilson *et al.*, 1989). It is known that the activity of many genes is regulated by neighbouring motifs known as enhancers, the molecular equivalent to volume controls. Enhancer trapping involves constructing a P-element containing β-galactosidase, a gene whose protein product can be made to give an intense blue staining. This gene only shows significant activity when it lies close to an enhancer element. The system works as follows. Single P-elements are allowed to jump randomly into a fly's genome. If one ends up close to an enhancer element, the β-galactosidase will be controlled in parallel with the enhancer's target gene, producing a blue colour in all the cells where it is switched on. For example, a P-element which lands near to a gene involved in eye formation will give rise to a fly with blue eyes. Enhancer trapping is particularly powerful as a way to find and study the expression of developmental genes.

Evolution among repeated sequences

All genomes are complex, but those of higher organisms are particularly so. Early work suggested that sequences could be classified into those that were single copy, those that were repeated tens or maybe hundreds of times, and those that were present in large numbers of copies. Very highly repeated sequences were sometimes recognised by the way they formed distinct bands away from the main genomic DNA when spun through a CsCl gradient. As mentioned above, such bands were referred to as satellite bands, and the highly repeated sequences they contain became known as satellite DNAs.

Repeated sequences include a motley range of members. A few are actual genes, such as those coding for the ribosomal RNA molecules, and appear to be repeated because their product is either needed rapidly, or in bulk. Other classes appear to be primarily selfish; they reach high copy number by promoting their own duplication. These may include mobile elements, human *Alu* sequences and the like. The function of true satellites, those present at extremely high copy number, is yet to be identified with certainty. These sequences often occur in large blocks, where the basic repeat units are arranged, head to tail, in vast tandem arrays. Such blocks are often found around **centromeres** or at the ends of chromosomes.

It seems reasonable to assume that any one repeat unit will evolve independently, quietly accumulating mutations without reference to its neighbours. In this way, all repeat units would be expected to evolve divergently away from one another. With time, one would expect to find that the individual repeat units would have become so dissimilar from one another that they no longer constitute an identifiable family. Surprisingly, this is not what we find.

Whales and dolphins, like all other higher organisms, possess satellite sequences. One whale satellite is spectacular, comprising almost 10 per cent of the entire genome. Being around 1.5 kb in length, this equates to around 100 000 copies. As with any other tandemly repeated sequence, any restriction enzyme which cuts just once within the basic repeat will release single repeat units. Comparing dolphins (delphinids) and other whales in this way, we find that the basic repeat length is around 200 bp shorter. Think about what this means. Since the divergence of the delphinid lineage, perhaps 20 million years ago, 100 000 repeat units have all suffered the same deletion. No one can argue sensibly that each of these deletion events occurred independently.

The whale satellite provides a spectacular example of a process which has been observed in many other repeated sequences. The process is called **concerted evolution**, and relates to the general observation that repeated sequences tend to show greater divergence between species than they do within species. It appears that any new mutation affecting a single repeat either gets eliminated, or somehow manages to spread to all other repeats, both within the individual and within the population/species. This unusual phenomenon generally receives little attention, but one person, Gabriel Dover, has taken it to heart.

Dover (1982) advocates recognising the processes which underlie concerted evolution as a new force, to be ranked equally alongside selection and neutral drift. He terms this force **molecular drive** and suggests that the changes it causes will continue to accumulate until blocked by natural selection. At this point, continued drive might eventually force a compensatory change. By analogy, molecular drive is seen as a blind force like the sea, beating continually against a sea wall, the organism's fitness. The sea will not go away, therefore the wall will have to be repaired or moved. As an example, one might think of the promoter sequences of the ribosomal RNA genes. If molecular drive begins to cause the spread of a variant sequence which is recognised less than perfectly by the polymerase molecule, the resulting selection pressure would favour compensatory changes in the polymerase molecule. As yet, although most people now recognise that concerted evolution is an inescapable fact, the mechanisms remain obscure, and the actual importance of molecular drive remains unclear.

Concerted evolution probably occurs as a result of the many different turnover processes which affect the genome. By turnover, we mean mechanisms which can cause repeat units to be gained, lost or duplicated. Among these, **unequal exchange**, **transposition**, **RNA mediated events** and **gene conversion** are some of the more likely perpetrators which Dover has identified as being involved in concerted evolution. The problem with most of these is that they appear to act too slowly and with little direction. Where we might expect a large gene family to appear like a large population, with neutral alleles drifting up and down in frequency until they become fixed, we usually find end states. New variants are either rare or nearly fixed, but seldom lie at intermediate frequencies.

The primary problem faced by anyone wishing to resolve the molecular puzzle posed by concerted evolution is the very fact that the sequences concerned are highly repeated. Most analytical methods used to examine DNA polymorphism are aimed at single sequences, and yield confusing signals when multiple copies are present. It is like standing in the doorway of a busy pub and trying to work out what any one person is saying. Some words can be caught, but much information gets lost in the general mêlée.

Recently, elegant technical advances have been used to learn more about molecular drive. A good example is Warburton and Willard's analysis of α-satellite sequences which lie near the centromere of human chromosome 17 (Warburton and Willard, 1990). The α-satellite block contains a large number of tandemly arranged 171 bp repeats. A proportion of repeats contain a cutting site for the restriction enzyme *Eco Rl*, and these can be used to identify blocks of between four and sixteen repeats (Fig. 2.5a). A second restriction enzyme, *Dra 1*, cuts very rarely, and can be used to divide the satellite block into eight pieces, each containing several thousand monomers.

Warburton and Willard used two different types of electrophoresis in order to find out whether different types of repeat unit were clumped or dispersed

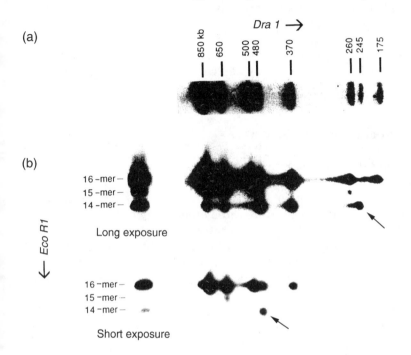

Fig. 2.5 Analysis of the long-range organisation of human α-satellite repeats on chromosome 17. The block of α-satellite repeats which lies near the centromere on human chromosome 17 contains some 20 000 copies of basic 171 bp long monomer. However, at least two higher levels of structure are apparent. First, larger repeats containing between four and sixteen monomers can be identified by digestion with the restriction enzyme *Eco Rl*, which cuts only a small proportion of monomers. Second, the entire block can be divided into eight very large blocks by digestion with a second enzyme, *Dra 1*. These two levels are illustrated in Fig. 2.5(a). To find out whether the *Eco Rl* repeats are clustered or dispersed evenly throughout the array, Warburton and Willard (1990) took advantage of two-dimensional electrophoresis. In the first direction genomic DNA is digested with *Dra 1* and separated on a pulsed field gel (a form of electrophoresis capable of resolving extremely large pieces). The resulting fragments are then digested within the gel slice with *Eco Rl* before being subjected to standard electrophoresis at right angles. The result is shown in Fig. 2.5(b) (after Warburton and Willard, 1990), and reveals a strongly non-random distribution, indicating a clustering of similar repeats. For example, 14-mer repeats are disproportionately abundant in the 480 and 254 Kb *Dra 1* blocks (arrows).

evenly within the block. First, genomic DNA was cut with *Dra 1*, and electro-phoresed using pulsed field gel electrophoresis. This method can resolve very large chunks of DNA, up to a megabase (1 000 000 bp) or so in length, and yields a profile containing eight bands, one for each major block of the α-satellite. Ingeniously, this DNA is then digested with the second enzyme while still in the gel, prior to standard electrophoresis at right angles to the first run. Consequently, each of the major *Dra 1* blocks is broken down into its *Eco R1* sub-blocks (Fig. 2.5b). Interestingly, the relative proportions of each of the *Eco R1* sub-blocks was found to vary greatly between *Dra 1* sections, indicating that similar repeat variants are very much clustered. By repeating the analysis with other enzyme combinations, they were able to show that the α-satellite block comprises many different, individually homogeneous domains.

Changes in chromosome structure

Deletions and duplications

The phenomena of duplication and deletion are not, of course, confined to single bases. Much larger segments of chromosomes may be lost by deletion or repeated by duplication. One process that may concurrently produce both types of change is unequal crossing-over. That is to say crossing-over, which usually occurs at precisely homologous points on the two chromosomes of a pair, may occasionally be imprecise (Fig. 2.6). This is most likely to occur when

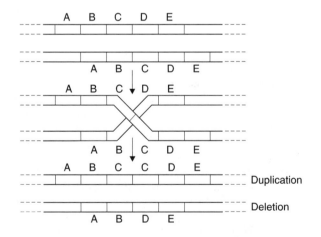

Fig. 2.6 The origin of deletions and duplications through unequal crossing-over. The chromosomes shown pair with one displaced slightly from the other, with the result that a cross-over in the mis-paired region results in gene C being deleted from one chromosome and duplicated in the other.

similar DNA sequences occur in neighbouring regions of a chromosome. This means that the production of an initial duplication may increase the likelihood of further repeats occurring in the same region, due to the obvious sequence similarity in a small domain.

What evidence is there that this process occurs? One case concerns the alpha-globin genes of which, in humans, there are two closely linked copies on chromosome 16. Thus most individuals have a total of four copies. However, in some populations (e.g. parts of south-east Asia) there is a high incidence (1 per cent) of individuals with only two copies (one on each chromosome). This leads to an anaemic disorder known as alpha-thalassemia, in which the production of alpha-globin chains is considerably reduced. The deletion of one of the two alpha-globin genes could be the result of random loss, or it could have resulted from unequal crossing-over. Were the latter to be the cause, chromosomes bearing three alpha-globin genes should also be produced. Restriction mapping analysis confirmed this expectation. A similar example is seen in the variation in the number of facets in the eyes of 'wild-type', bar-eyed and ultra-bar *Drosophila melanogaster*, bar being a duplication, and ultra-bar a triplication of a small region of the X chromosome (Bridges, 1936).

Deletions are often severely deleterious or lethal as they may involve the loss of crucial functions. In such cases, the deletion products of unequal crossing-over are likely to be rapidly purged from the population by selection. Some duplications are also deleterious, but others may survive as loss of function does not necessarily result from their production. The existence of duplicate genes, multi-gene families and possibly some **super-genes**, bears witness to the durability of some duplications.

Once duplicated, evolution can follow one of two paths. Either the copy is not needed and is eventually lost as a functional unit, or it may assume a new function. Evidence for gene loss comes from the existence of **pseudogenes**, degenerate copies of functional genes which no longer produce an active protein. Pseudogenes can be likened to shipwrecks. Once a duplicate gene has been 'holed' by a mutation which prevents its normal functioning, the organism will become reliant on the functional copy. The shipwreck will then no longer be maintained by natural selection and will therefore 'rot' by mutation. Interestingly, some pseudogenes have been found which lack **introns**. By implication, at least some gene duplication occurs by processed messenger RNA molecules becoming transcribed back into DNA and thence back into the genome.

The second fate of gene duplication is the one which has greater implications for evolution. Logic dictates that the only way by which the number and diversity of genes can increase is through a process of duplication and functional divergence (Ohno, 1970). Examples abound. The globin gene family includes a number of variations on a theme, all of which show strong similarities. Myoglobin is a single polypeptide molecule which binds oxygen. Haemoglobin performs a similar function, but is composed of four myoglobin-like molecules which come together to form a single functional protein.

This molecule itself is coded for by two separate genes, one specifying the α chain, the other specifying the β chain. In the foetus, a further form is found, the γ chain, which has a higher affinity for oxygen, allowing the foetus to take what it needs from its mother. All these genes are closely related and must have originated by duplication followed by change of function.

Inversions and translocations

Point mutations, duplications and deletions all involve changes (loss, gain or substitution) in the nucleotides present in a piece, or pieces, of DNA. However, these are not the only types of change in the structure of chromosomes. Some mutations result simply from the reordering of nucleotide sequences which occurs when chromosomes break and are then repaired. If the broken segments rejoin in a different order, then either an **inversion** or a **translocation** will result, and each has its own evolutionary implications.

Inversions, for example, are 180° reversals of chromosome segments, producing alterations in base sequences and consequently changes in **genetic linkage** groups. Two classes of inversion are recognised. If the inverted segment includes the centromere, the inversion is said to be **pericentric** (Fig. 2.7a); if not, it is **paracentric** (Fig.2.7b). In individuals **homozygous** for an inversion, **meiosis** proceeds normally. However, for those **heterozygous** for an inversion, **synapsis** formation during meiosis requires the formation of loops containing the inverted segments if homologous chromosomes are to pair fully. A cross-over event in such a loop has serious consequences. If the inversion is pericentric the result will be the production of two normal gametes (the non-cross-over products), one carrying the inversion, the other the non-inverted sequence, and two abnormal gametes (the cross-over products) each lacking some part of the inverted segment and having another part in duplicate. If the inversion is paracentric, again the result will be two normal (non-cross-over) gametes and two abnormal gametes. The abnormal gametes result because cross-over in the inversion loop gives rise to both a bridge chromosome, containing two centromeres, and a fragment containing none. Migration of the linked centromeres in the bridge to opposite poles of the meiotic cell, during anaphase I, will cause the bridge to rupture, while the fragment, having no centromeric attachment to the **meiotic spindle**, will be left stranded on the **metaphase plate**.

The result for both types of inversion is that recombination in heterozygotes is effectively suppressed, or considerably reduced. In addition, there will, be a reduction of up to 50 per cent in the production of viable gametes, the precise figure depending on the size of the inversion and the likelihood of a cross-over event occurring within the inversion loop. Individuals heterozygous for small inversions containing just a few genes may not suffer significant fertility loss, but, because the occasional gametic products of cross-over that do occur are still likely to be inviable, the genes in the inversion will act as a single conserved

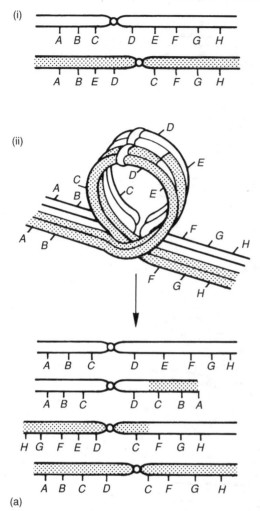

(a)

Fig. 2.7 (a) The result of a crossing-over in a heterozygote for a pericentric inversion. (i) Homologous chromosomes showing a pericentric inversion involving genes C, D and E. (ii) A cross-over between two non-sister chromotids results in two daughter chromosomes with the normal genic complement (top and bottom), and two abnormal chromosomes with some genes duplicated and others deleted. (After Ayala and Kiger, 1980.)

linkage group protected from recombination. Such inversions will therefore act as a super-gene.

A translocation is the name given to a piece of DNA that has moved its location either within or between chromosomes. Often translocations are reciprocal, involving the exchange of chromosomal segments between non-homologous chromosomes. Unidirectional translocations (i.e. those not involving reciprocal exchange) are sometimes termed **transpositions**. Heterozygotes

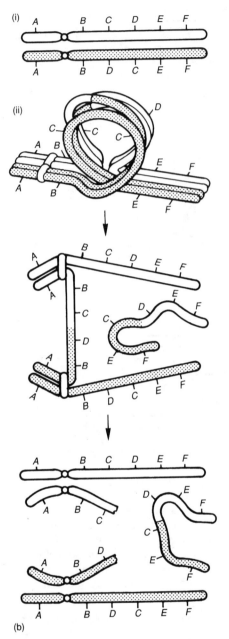

Fig. 2.7 (continued) (b) The result of a crossing-over in a heterozygote for a paracentric inversion. (i) Homologous chromosomes showing a paracentric inversion for genes C and D. (ii) A cross-over between two non-sister chromatids results in two daughter chromosomes with the normal genic complement (top and bottom), and two abnormal chromosomes. One has two centromeres (a bridge) which split to yield gametes with deletions. The other fragment, which lacks a centromere, becomes stranded on the metaphase plate and is lost. (After Ayala and Kiger, 1980.)

for both reciprocal and non-reciprocal translocations can cause problems during gametogenesis (Fig 2.8). In the case of large transpositions, chromosome pairing during meiosis will lead to the production of unpaired loops at both the old and new position of the translocation. These unpaired loops are liable to breakage, with a consequent loss of genetic material from some gametes, the broken segments that do not contain a centromere being left on the metaphase plate. For individuals heterozygous for reciprocal translocations between non-homologous chromosomes, the meiotic difficulties are more acute. Pairing of homologous chromosomal segments leads to an association between four chromosomes, usually in a cross-shaped configuration. Recombination is not then a problem, but following disjunction, the only gametes that are likely to

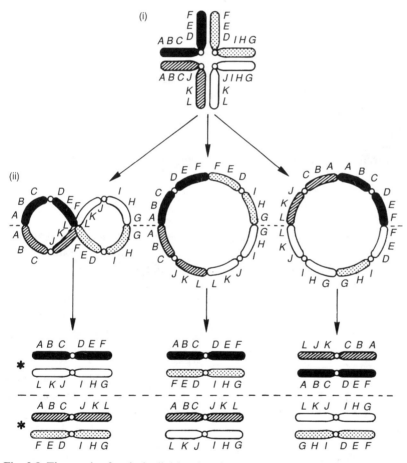

Fig. 2.8 The result of meiotic division in a heterozygote for a reciprocal translocation. (i) Homologous sections of the two chromosomes involved pair to form a cross-shaped configuration during prophase. (ii) At metaphase I, three different configurations on the metaphase plate are possible. These will give rise to six different types of gamete, of which only two (marked *) have the complete genic complement. (Adapted from Ayala and Kiger, 1980.)

be viable are those containing either both or neither of the translocated segments. These gametes will be balanced, while those containing a translocation on just one of the two chromosomes, in any of the four possible combinations, will have both duplications and deletions. Pollen grains with deletions and duplications formed by plants heterozygous for translocations are frequently aborted. This is not so often the case with animal gametes, but the zygotes formed by unbalanced gametes are usually inviable or have a markedly reduced fitness. In practice, up to two-thirds of the gametes produced by such heterozygotes will be inviable, or will produce unfit **zygotes**. The evolutionary significance of translocations is unclear, although the potential for heterozygotes to show lowered fitness might sometimes play a role in speciation (see Chapter 10).

Changes in chromosome number

Thus far we have concentrated on mutations which affect either the sequences or the nature of the nucleotides in a genome. We now turn to mutational mechanisms that alter the chromosome number. Changes in the number of chromosomes in a normal cell may be divided into two types: those that give rise to variation in the number of individual chromosomes, and those that produce variation in the number of sets of chromosomes.

Chromosomal fusion and fission

In the nineteenth century it became possible to see chromosomes microscopically. Tremendous amounts of time and effort were devoted to counting and characterising the chromosomal complements of all types of organism. It soon became evident that chromosome number varied between species, but was more or less constant within species. Further, closely related species of animal usually had fairly similar chromosome numbers. For example, humans normally have 23 pairs of chromosomes, while chimpanzees, gorillas and orang-utans have 24 pairs. How could such a change in chromosome number arise. William R. Robertson was the first to hypothesise that the fusion of two non-homologous chromosomes, with the loss of a centromere, could be the source of such a reduction. Conversely, the splitting, or fission of a chromosome into two would increase chromosome number by one, as long as a new centromere was formed so that both portions of the split chromosomes had a spindle attachment. **Chromosomal fusions** and **fissions** are thus sometimes termed **Robertsonian changes**.

Generally, chromosomal fusions and fissions do not lead to any significant loss or gain of hereditary material. The difference between hominid and pongid chromosome numbers appears to be due to a Robertsonian fusion. Where humans have one large **metacentric chromosome** (chromosome 2), the pongids have two smaller **acrocentric chromosomes**. Banding pattern analysis shows strong homology between these acrocentrics and the two arms of the human metacentric, strongly supporting the contention that human chromosome 2 is the product of the fusion of the two pongid acroeentrics.

Chromosomal fusions are thought to be much commoner than fissions, possibly because the latter requires the spontaneous creation of a new centromere. Comparative **karyotype** analysis has established the existence of fusions in a number of taxa, including the drosophilid flies (with a haploid chromosome number varying from three to six), and the salmonid fishes (around 60 to 100). In some instances Robertsonian changes appear to have a considerable effect on phenotype. However, in others they produce very little noticeable effect. For example, the normal haploid number of the house mouse, *Mus musculus*, is 20. Yet a population in the Italian Apennines, which shows no significant phenotypic abnormality, has a haploid number of eleven. This reduction is presumed to be the result of a series of fusions. More interesting still is the case of the dog whelk, *Nucella lapillus*, some populations of which are polymorphic for chromosome number.

Aneuploidy

Aneuploids are organisms in which one or a few of a normal chromosome set are either lacking or present in excess. The terms nullisomic, monosomic, trisomic, tetrasomic, etc., refer to the presence of zero, one, three, four, etc., copies of a chromosome. Aneuploidy is of unclear evolutionary significance. In animals it is often deleterious. For example, in humans the presence of an extra copy of chromosome 21 is responsible for the phenotypic condition known as Down's syndrome or mongolism, while trisomics and monosomics for the sex chromosomes (XXY – Klinefelter's syndrome; XXX – triple X; XO – Turner's syndrome) are affected by a range of serious physical and mental abnormalities. For example, Klinefelter's syndrome individuals are phenotypically male, but have under-developed gonads, little body hair and disproportionately long limbs and are mentally retarded, while Turner's syndrome individuals are phenotypic females with under-developed ovaries, short stature and a webbed neck and they are mentally retarded. The deleterious nature of aneuploidy in animals is probably responsible for the fact that, even in well-scrutinised species, trisomy and monosomy has never been recorded with respect to many chromosome types: aneuploid zygotes of many types may simply be lethal.

In plants, aneuploidy is more frequent, and appears to be less deleterious. For example, in the Jimson weed (*Datura stramonium*) in which trisomy has been studied extensively (Blakeslee, 1934), all 12 possible trisomics have been recorded to occur naturally. Each has a different set of effects on the phenotype such that experienced researchers can recognise different types of trisomic by the examination of external features such as seed capsules. Trial crosses using the different trisomics produce abnormal transmission ratios due to the loss of chromosomes during meiosis, the lower germination rate of trisomic mutants, compared to normal zygotes, and the slower pollen tube growth of pollen carrying an extra chromosome. Potentially, aneuploidy could lead to permanent changes in chromosome number, although it is not known how frequently this occurs (see Khush (1973) for review).

Haploidy and polyploidy

Most eukaryotic organisms are diploid, having two sets of homologous chromosomes in their somatic cells. In **haploid** (sometimes called monoploid) and **polyploid** individuals the number of sets of chromosomes in somatic cells is one or more than two, respectively. Haploidy occurs in a number of so-called **haplo-diploid** organisms, particularly in the Hymenoptera, in which females are normal diploids, while males are haploid, developing from unfertilised eggs. The evolutionary significance of these alternative life cycles is discussed in Chapter 12.

Polyploids can be divided into two classes: **autopolyploids** with all chromosomes derived from one species, and **allopolyploids** in which chromosome doubling follows interspecific hybridisation, so that the chromosome sets are derived from more than one species. When related species have chromosome numbers which are multiples of the same basic number, this may be indicative that polyploidy has played a role in the evolution of the group. Polyploidy is generally rare in animals, occurring primarily in groups such as the annelid worms, fish, amphibians and reptiles. For example, fish of the carp family (Cyprinidae) have chromosome numbers of either around 50, or around 100. By contrast, polyploidy is common among plants, where somewhere in the region of a third of known species are polyploids.

The most common cause of polyploidisation is the failure of separation in anaphase I during gametogenesis, although failure of disjunction during mitosis in somatic cells may also lead to tetraploid shoots which produce diploid gametes. When a diploid gamete fuses with a haploid gamete the resultant zygote will be triploid and sterile because during subsequent meiosis sets of three homologues become associated. Disjunction then leads to chromosomally unbalanced gametes with two of some chromosomes and one of others. Should two diploid gametes of the same species fuse to produce an autotetraploid, homologues group in sets of four on the meiotic spindle. Because regulation of the disjunction of these quadrivalents is not precise, unbalanced gametes with one, two or three copies of different chromosomes will be produced, and again the increase in ploidy will be associated with reduced fertility.

Because of these problems during meiosis, autopolyploids are uncommon, even in plants. However, some wild and many cultivated autopolyploids are known. A well-documented example of a wild autopolyploid is the marsh bedstraw (*Galium palustre*). This plant has three chromosomal races. Around

Oxford two types of population occur. Populations of diploids with 24 chromosomes grow in marshy habitats which dry out during the summer. Populations of octoploid plants with 96 chromosomes grow on permanently wet sites. The third race, a tetraploid, is known from Devon, where it grows in situations that only occasionally dry out. It seems that, in this case, the increase in ploidy is correlated with an ability to colonise wetter habitats, implying that the possession of extra copies of one or more chromosomes confers increased water tolerance.

Among crop species, autopolyploids have been extensively produced by treatment of diploids with colchicine ($C_{22}H_{25}O_6N$). Colchicine, an extract of the autumn crocus or meadow saffron (*Colchicum autumnale*), accelerates the rate of production of chromosome doubled mutations by inhibiting spindle production during **mitosis**, so that chromosome pairs fail to separate at anaphase. Plants of an even ploidy (e.g. tetraploid, hexaploid, octoploid) often have a reasonable level of fertility, while those with an odd number of sets of chromosomes (e.g. triploid, pentaploid) are usually sterile. For example, diploid, triploid and tetraploid strains of sugar beet are known. The triploid is the most vigorous and produces the highest commercial yield. However, the triploid is sterile, and so is routinely produced by crossing the diploid with the tetraploid, both of which are fertile.

While autopolyploidy is rather uncommon, the same cannot be said of allopolyploidy. In allopolyploids, by definition, chromosome doubling occurs in interspecific hybrids. Such hybrids are likely to be sterile due to non-homology of their chromosomes. For such organisms to persist they must be able to survive and reproduce mitotically. Many plants achieve this through one or other of the many forms of vegetative reproduction. So new individuals which can exist independently of their 'parent' may result from the production of bulbs, corms, stolons, rhizomes, leaf plantlets, tillers and so on. The important feature of all of these is that the new organism is derived as a result of mitosis, and, genetically, it is an exact replica of its parent (with the exception of rare mitotic mutations). Interspecific hybrid plants may thus persist for considerable periods. During this period of mitotic persistence of a reproductively sterile hybrid, the production of a shoot (in essence a cell line) with double the chromosome complement of the hybrid may restore the production of viable gametes by meiosis because each chromosome now has a homologous partner. Should this shoot produce both male and female reproductive parts and produce compatible gametes, the shoot will have the potential to give rise to progeny that are genetically distinct and isolated from the sterile hybrid, and from both of its parent species. This is easy to understand. If the sterile hybrid were the result of a cross between two diploid species, the new species would be an allotetraploid. Cross-pollination between either parental species and the allotetraploid will give rise to triploid individuals. These triploids will contain two sets of chromosomes from one of the original parental species, and one set from the other. As a consequence of problems of chromosome pairing at meiosis,

such triploids are likely to be sterile. So, the result of this process of interspecific hybridisation, followed by chromosome doubling, will be a new,
reproductively isolated species (see Baker, 1959).

Many examples of allopolyploidy are known among cultivated plants. For
instance, wild and cultivated wheats have chromosome numbers of 14, 28, 42
or 56, those with higher chromosome numbers being the product of allopoly-
ploidisation, as shown in Box 2.2.

Box 2.2 The evolution of modern hexaploid wheat

The probable evolution of modern bread wheat (*Triticum aestivum*) has
been reconstructed by analysis of closely related wild grasses. It appears
that the evolution of this important crop has involved three diploids,
each having seven pairs of chromosomes, two interspecific hybridisa-
tions, and two chromosome doublings.

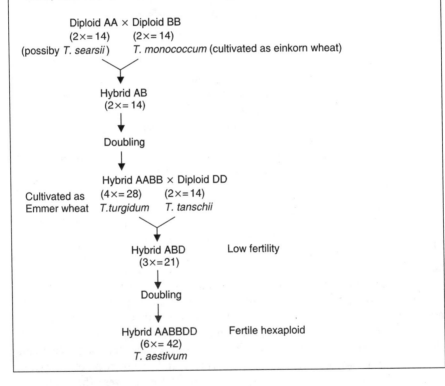

Allopolyploidy has also been observed in the wild. The British species of
cordgrass, *Spartina maritima*, has a diploid number of 56 chromosomes. In
1878, a hybrid between this species and an American species, *Spartina alterniflora* (with a diploid number of 70), was recorded in Southampton Water. This

hybrid, named *Spartina townsendii*, had a diploid number of 63 and was sterile, and, in the rare places that it still occurs, still is sterile. Subsequent chromosome doubling has produced a new fertile plant with 126 chromosomes. This polyploid is reproductively isolated from both its parent species, and so may be considered a distinct species (see Chapter 10). It is also very vigorous and in many localities has out-competed both *S. maritima* and *S. alterniflora*, notably on the south coast of England, where it is considered a pest species because it blocks harbours and coastal channels.

The role of heritable variation in evolution

The final source of genetic variation which contributes to an individual's genotype is that created in sexual organisms by recombination, both through the random assortment of individual chromosomal pairs and as a result of **chiasmata** formation and crossing-over. In effect, mutation acts to modify the basic components of the chromosome complement, while sex is a mechanism which allows genetic shuffling. It may be argued that mutation is the more fundamental source of variation. The question arises, which is the more important process in terms of evolution. Upon which of these types of variation does selection act? Do adaptations evolve as a result of selection acting upon novel mutations that by chance arise concurrently with environmental change? Or conversely, does selection act primarily upon varieties that are generated each generation through the random assortment of genes and recombination during sexual reproduction?

The case of industrial melanism in the peppered moth (*Biston betularia*) provides one of the best-known examples where a mutation produced a novel, advantageous phenotype which then increased in frequency through the action of natural selection (see Majerus (1989) for review). Similar examples include the evolution of drug or pesticide resistance (e.g. Georghiou and Taylor, 1976). Many other adaptations appear to have evolved as a result of selection acting upon variants created by the reassortment of pre-existing genes. Again, examples can be drawn from the evolution of melanism in the Lepidoptera. For instance, the darkest melanic varieties of the common marbled carpet (*Dysstroma truncata*) are produced by a set of three complementary genes with additive effects. It is probable, from the examination of old Lepidoptera collections, that the dark alleles of these genes all existed prior to widespread industrialisation. However, only over the last century and a half has pollution in industrial areas increased the fitness of those individuals carrying dark alleles for all three loci. It is these dark forms that now predominate in most industrial areas, with the less melanic forms being present in rural areas of southern Britain. Predominantly non-melanic populations are now confined to the Highlands and Islands of Scotland (Kettlewell, 1973).

Perhaps the most powerful examples of how much latent variability exists are provided by the history of domestication. Here, widely differing phenotypes have often been created over time periods which were too short for novel mutations to have played a significant role. Instead, the diverse breeds of dogs, cats, cows, sheep and the like we see today have all, by and large, been created by selective polarisation of existing natural variability. Consider the case of the dog. Dogs are wolves which were first domesticated about 13 000 years ago. The breeds we see today vary tremendously in size, structure and behaviour, causing many to believe that domestic dogs have mixed ancestry. However, molecular genetic analysis has confirmed that our pet pooches are, in essence, pet wolves, showing that their whole gamut of diversity has been obtained simply through directed selection of the products of genetic recombination.

Summary

This chapter considers the ultimate source of heritable variation, mutations. A diverse array of types of mutations, including deletions, duplications, substitutions, translocations, transpositions, inversions, and changes in chromosome number are described. The mechanisms underlying these differing types of mutation are examined, as are their rates and evolutionary implications. Finally, the relative roles of mutation and recombination in generating novel variants that selection may act upon is discussed.

Suggested reading

BERNARDI, G. (1993). The vertebrate genome: isochores and evolution. *Molecular Biology and Evolution*, **10**, 186–204.
A review of the evidence for and theories behind isochores.
EYRE-WALKER, A. (1993). Recombination and mammalian genome evolution. *Proceedings of the Royal Society of London, Series B*, **252**, 237–243.
A good illustration of how to make progress in the theory of evolution by combining speculation with data culled from the literature and genome databases.
HANCOCK, J. M. and DOVER, G. A. (1990). 'Compensatory slippage' in the evolution of ribosomal RNA genes. *Nucleic Acids Research*, **18**, 5949–5954.
Part of a continuing story about how evolution shapes a particular DNA sequence – a warning for those involved in phylogenetic reconstruction.
JEFFREYS, A. J., MACLEOD, A., TAMAKI, K., NEIL, D. L. and MONCKTON, D. G. (1991). Minisatellite repeat coding as a digital approach to DNA typing. *Nature*, **354**, 204–209.
A seminal paper on how to get the most information out of the genome's most variable sequences.
JEFFREYS, A. J., TAMAKI, K., MACLEOD, A., MONCKTON, D. G., NEIL, D. L. and ARMOUR, J. A. L., (1994). Complex gene conversion events in germline mutation at human minisatellites. *Nature Genetics*, **6**, 136–145.
Molecular biological elegance personified. A brilliant demonstration of the potency of molecular techniques for dissecting mutational processes.

RAND, D. M. (1994). Thermal habit, metabolic rate and the evolution of mitochondrial DNA. *Trends in Ecology and Evolution*, **9**, 125–131.
A good survey of some of the least appreciated aspects of molecular evolution: how the mitochondrial molecular clock can go wrong.

RUBINSZTEIN, D. C., AMOS, W., LEGGO, J., *et al.* (1994). Mutational bias provides a model for the evolution of Huntington's disease and predicts a general increase in disease prevalence. *Nature Genetics*, **7**, 525–530.
Illustrates how population genetics, theory and evolution can be combined to understand an important evolutionary process.

Chapter 3

Genetic changes in populations

Why do some die and some live?
                               (Alfred Russell Wallace, 1858)

Two mechanisms of evolution

At the end of life comes death. Over time, each generation of organisms is replaced by a new generation, composed of their descendants. At the same time, the genes of the old generation are replaced by their copies, carried by the new generation. For any particular gene, if all individuals carry identical copies, this process of replacement has no effect. However, many genes and DNA sequences are **polymorphic**; that is to say they exist in two or more forms. In such cases, each of the different forms, or **alleles**, may change in their relative frequencies between generations. This chapter examines the various systems which cause change, affect the speed of change, and determine its consequences.

A child is born carrying a mixture of its parents' genes. Aged seven, she is with her family on an aeroplane which crashes, killing everyone on board. This is a tragic, but random event, which is difficult to blame on the girl's genotype. No matter what genes she carried she would have died along with her unfortunate co-travellers. In a second case, a similar girl dies from a disease. Possibly the disease is a genetic disease; possibly the child is more susceptible to this disease than are other children. Here, the girl's genotype, her genes, may have had a role to play in her death.

These two scenarios of child death illustrate the two major forces affecting which genes are represented in the next generation. In the first, due to a simple random sampling process, a gene's alleles may increase or decrease in frequency. Since these changes in allele frequency are directionally random (an increase in allele frequency is as likely as a decrease), the process has been termed **random genetic drift**. In the second case, there is an element of predictability, in that some genotypes are more likely to survive than others. This is the evolutionary force identified by Charles Darwin, and is called **natural selection**. Together these two processes work to mould the genetic constitution of future generations and, ultimately, species, often acting within the same individual. For example, when a child dies from a genetic disease, all the child's genes which do not affect the outcome of the disease will suffer a chance death.

Genetic drift

Chance, and hence random genetic drift, plays an important role in the fate of all genetic characters, whatever their apparent importance to an organism. Every gene faces a lottery for passage into the next generation: it must be on the right segregating chromosome, get into the right sperm or egg, and be in an individual which survives any of a wide diversity of possible deaths to reach adulthood and which then finds a fertile mate to start the process again. By contrast, natural selection affects only a proportion of characters. Even then, it acts merely as a modulator, biasing drift in favour of either propagation or demise. Thus, although random genetic drift is usually associated either with genetic characters which do not affect fitness (selectively neutral characters) or with circumstances where death does not depend on genotype, such as in the air crash described above, in fact it affects all genes. Because of this, drift is arguably the more fundamental process, and will be dealt with first.

The process of random genetic drift acting on a polymorphic gene has often been likened to the process of a blindfolded person drawing coloured balls out of a bag. For a DNA sequence having two alleles, imagine a large bag containing equal numbers of red and blue balls. Drawing ten balls, we might find seven blue and three red. This may not be the most likely combination, but it is certainly a plausible one. Now let these balls found a new large bag population, containing the same proportions as the sample, 70 per cent blue and 30 per cent red. Repeating this procedure time and time again, each time drawing ten balls to found a new large population, it is easy to imagine the frequency of red and blue balls wandering up and down until one colour is fixed, i.e. all ten founders are of the same colour.

Random genetic drift works in exactly the same way as the drawing of coloured balls. Some genes make no contribution to the next generation because, by chance, their carriers fail to survive to maturity, or their carriers fail to mate, or they are carried in gametes that are unsuccessful in the fertilisation game. At each genetic locus, alleles which happen to be carried more frequently by successful gametes will increase in number, whilst those that are poorly represented will suffer a decline. As a consequence of this random sampling of genes each generation, alleles will drift up and down in frequency until, eventually, one or other becomes fixed. A gene, previously polymorphic for two alleles, would then be **monomorphic**. Theoretically, if all other things are equal, no matter how variable the initial population, one by one variants will be lost, and, given enough time, every individual in the population will become identical genetically.

Despite billions of years of evolution, this clearly has not happened. If genetic drift acts continuously to remove genetic variability, and genetic variation still exists, there must be another mechanism regenerating it. That process is mutation (see Chapter 2). A good analogy of the balance between genetic drift and mutation is to think of a large bath in which the plug does not fit too well and into which a tap is dripping. Drift is now the leaky plug, allowing genetic variability to leak out of the population, and mutation is the dripping tap, constantly replenishing. Since the rate of leakage increases with increasing water pressure as the bath fills, eventually an equilibrium will be achieved, in which the level of water in the bath reflects a balance between the rate of dripping and the rate of leakage.

The dynamics of genetic drift are relatively simple, being determined by three main parameters: the size of the population concerned and its underlying substructure (which together dictate the rate of change in the frequency of existing alleles), and the mutation rate (which feeds new variability into the system). This simplicity, particularly with respect to population size and mutation rate, is reflected in the degree to which mathematicians have derived predictive equations. Many of the most important contributions have been spawned by the long-standing debate over the relative importance of drift and natural selection in evolution (see Chapter 4). In particular, Motoo Kimura's (1968) **neutral mutation theory of molecular evolution (the neutral theory)** has received considerable attention. Only the most basic of aspects are included here, since our aim is a descriptive approach. However, the reference list at the end of this chapter includes texts which give a fuller treatment, including more of the mathematical nuts and bolts.

Factors which affect genetic drift

Population size

Population size is arguably the most important factor influencing how random genetic drift operates. Intuitively, genetic change will occur fastest in small populations. In the extreme case, if, instead of ten balls, we draw only a single ball from our bag, whatever colour that ball is will be fixed immediately. Even if we draw two balls, there is a 50 per cent chance that they will be either both blue or both red. Towards the other extreme, if the founding population of balls is 1000 or more, any change in frequency between the sample and the original bag will almost certainly be tiny. Thus, the rate of change of allele frequencies due to drift is inversely proportional to population size. The disproportionate influence of small population sizes on neutral drift gives rise to two important concepts: the **founder effect** and **genetic bottlenecks**.

The founder effect describes the situation where a large population grows up from a small number of founding individuals. Often, this process is synonymous with colonisation. There are two main ways in which the process of founding results in genetic change. First, some of the genetic variants present in the main population, particularly the rarer ones, may not be represented in the new population. Second, since founding is essentially a process of sampling, the frequencies of alleles will often differ between the founders and the parent population. To illustrate, consider an allele which is present in 1 in 50 individuals in the parent population, but which just so happens to be amongst 10 individuals which colonise a new habitat. This allele will have increased in frequency from 1 per cent to 5 per cent (assuming the organism is diploid), a fivefold increase in frequency. It is easy to appreciate that the strength of both these effects becomes stronger as the number of founding individuals decreases.

The term genetic bottleneck is descriptive, referring to any occasion where a population declines in size and then re-expands. The genetic consequences of a bottleneck are identical to those associated with a founder effect. Indeed, the simplest way to view a bottleneck is as a population refounding itself from a small number of individuals. Bottlenecks are usually associated with environmental disasters such as oil spills, periods of drought or disease. For example, British kingfishers suffered a dramatic crash in numbers in 1963, when the unusually severe winter froze most of their normal feeding sites. As with the founder effect, the crucial parameter is the minimum size the population experiences.

Spectacular fluctuations in population size are found particularly among invertebrates. For example, in 1976, following a warm summer and mild winter, the British seven-spot ladybird (*Coccinella septempunctata*) population exploded dramatically. It has been estimated that on a single day in July of that year, something in the region of 23 654 400 000 dead seven-spots (about four times the total current world human population) were washed up on the south and east coasts of Britain (Majerus, 1994). This is equivalent to a ton of ladybirds per mile of coast. This population explosion was followed by a catastrophic crash due to starvation, and populations did not revert to more normal levels until the early 1980s.

In contrast to these chance boom and bust events, some populations show more predictable, cyclical changes in population size. In such cases, bottlenecks are an integral feature of their genetic structure. For example, North American voles of the genus *Microtus* show regular population cycles of abundance and have formed the focus of a now classic study. Similarly, primitive Soay sheep on the remote island of Hirta in the St Kilda archipelago have no natural predators or grazing competitors and appear to go through chaotic cycles (Grenfell *et al.*, 1992). In such cases, the repeated exposure to bottlenecks is expected to reduce the standing level of neutral variability.

The fundamental mathematical equations which describe how genetic drift operates were derived assuming an **ideal population**, a homogeneous population in which mating is completely random and all individuals contribute equally to the next generation. However, real populations are not like this. In order to alleviate this problem, population geneticists came up with the concept of **effective population size**, usually denoted as Ne and defined as the size a functionally equivalent ideal population would be. In practice, Ne tends to be smaller, sometimes much smaller, than the true or census population size. Conversion between these two measures requires at least three important factors to be taken into account: the breeding system, size fluctuations and population structure.

To begin with, not all members of a population are equal. Some never get to breed and some may be very successful. For example, in a polygynous breeding system, a relatively small number of males father most of the next generation. In order to calculate the effective population size it is therefore necessary to count only those individuals which breed successfully and then to allow for any disparity between the sexes.

Second, any measure of population size must take into account the inevitable fluctuations which affect all populations. It can be shown that the effective size of a population over time is best approximated by the harmonic mean (see Fig. 3.1) calculated over all generations considered. Since the harmonic mean is dominated by its smallest terms, the effective size of a population lies closer to its minimum than to its maximum size. The take home message is that, with respect to random genetic drift, most populations that fluctuate in size behave much more like their low points than their peaks.

Third, perhaps the most complex factor is that relating to population substructure. It may be reasonable to treat a population such as that of sea birds nesting on a single cliff as a single homogeneous entity in which there is almost complete genetic mixing. Imagine, however, a population which is spread evenly over an area which is far greater than any one individual would ever travel in its lifetime. Perhaps snails on a prairie. In such a case, although there are no geographically defined boundaries, it is clear that genetic changes in one area will take generations to spread to other areas. Then the entire population behaves like a large number of overlapping, smaller populations, each one defined by a radius about its centre. Here, a functional Ne can be calculated based on the density of individuals, the degree of dispersal in each generation and any differences in behaviour shown by the two sexes.

Fig. 3.1 Effective population size in a fluctuating population. The effective size of a population which varies in size is given by its harmonic mean, which is the inverse of the mean reciprocal effective population size. Thus, for a given period of n generations:

$$N_e = \frac{1}{\left(\frac{1}{N_1} + \frac{1}{N_2} + \frac{1}{N_3} + \ldots \frac{1}{N_i} + \ldots \frac{1}{N_n}\right)}$$

where N_i is the effective population size in generation i.

Migration

In many population genetic models, populations are treated as discrete entities with unspecified but rigid boundaries. However, real populations are rarely like this. In most populations individuals are distributed patchily with variations in population density being associated with habitat heterogeneity. Some habitats are favourable, others are unfavourable. Individuals may then move, or migrate, from one favourable patch to another.

Migration can be viewed as partial mixing. To this extent, it acts as a sort of genetic glue, preventing or slowing the divergence of a subgroup from those others with which it exchanges members. Migration is a strong glue. Remarkably little exchange greatly restricts the extent to which two populations diverge by genetic drift alone. Interestingly, the genetic effect of migration is independent of population size. All that matters is the number of reproductive individuals that exchange each generation.

The reason that the only important parameter is the number of migrants is not intuitively obvious, but can be appreciated by comparing large and small populations. In a small population, drift operates fast, potentially leading to rapid divergence. However, each individual exchanged will make a relatively large contribution, thereby providing a strong force preventing divergence. In a large population, a small number of migrants will have little effect, but then the process of divergence by neutral drift will be very slow anyway. The two effects – tendency to diverge, and each migrant's ability to prevent divergence – are both tied to population size and balance each other precisely.

As a rule of thumb, as few as a single migrant per generation will slow population divergence. Considerable divergence may still occur, but this will be measurably less than if the populations were fully isolated, i.e. if there were no **gene flow**. If 10–20 migrants are exchanged, subpopulations may still exhibit appreciable allele frequency differences, at least for some alleles, but the populations will appear generally similar. At higher rates of migration the degree of genetic differentiation will be minimal.

Mutation

The term mutation embraces all possible forms of changes which affect a genome. As described in Chapter 2, mutations range in size from gross events, in which whole chromosomes or sets of chromosomes are gained or lost, or parts of chromosomes are gained, lost, inverted, duplicated or translocated, all the way down to the substitution of one nucleotide base for another. Sources and causes of mutation are, as yet, poorly understood. Some events are endogenous, originating from the cell's own activities. These include errors induced during DNA replication, chromosomal cross-overs and DNA repair, the integration and excision of mobile elements, slippage, gene conversion, RNA mediated events and many more. Exogenous sources include exposure to

mutagens, such as radioactivity, ultraviolet rays and various chemicals including those in cigarette smoke.

Mutation rates vary over many orders of magnitude. For example, the probability that a given nucleotide base will differ between parent and offspring is
around one in one billion. At the other end of the spectrum, minisatellites, on
which the technique of DNA fingerprinting is based, can mutate extremely
rapidly. In the most extreme example yet uncovered, as many as one sperm
in six carries a novel allele (Vergnaud *et al.*, 1991). For the most part, however,
we still have a lot to learn about the way in which changes come about, how
fast they occur, and their likely consequences.

Not all DNA sequences are affected by the same mutational mechanisms to
the same extent. Many are much more prone to mutations from one source
than from any other. For example, most mutations which affect microsatellites
are the gain and loss of repeat units, a process which is thought to occur as a
result of DNA strand slippage during replication (see Chapter 2). By contrast,
in minisatellites, which are structurally similar, mutations appear to be cata-
lysed by a specific protein which binds to a short, conserved sequence found in
each repeat unit (Collick *et al.*, 1991). Mitochondrial DNA is different again.
In most higher animals this molecule evolves mainly through the accumulation
of single base changes, but such changes occur up to ten times as frequently in
the mitochondrial genome as in equivalent regions of the nuclear genome. In
order to get a feeling for the rate at which any given sequence is likely to evolve,
therefore, it is vital to understand the mutational forces affecting it.

Random genetic drift: levels of variability and rate of change

Having considered the basic components which allow drift to operate, it is now
time to examine how these combine to determine evolutionarily important
parameters. There are three interesting questions to ask: what proportion of
new mutations are fixed, what is the mean rate of change in allele frequency,
and how much neutral genetic variability does the population maintain?

What proportion of new mutations become fixed?

It is both surprising and useful to know that the probability of fixation of a
particular mutation is not affected by population size, but instead depends
solely on its rate of creation, the mutation rate. The number of mutations
which become fixed in a population at each generation is the same as the
probability that any such mutation occurs in any individual per generation.
There are two ways to convince oneself of this.

First, population size has two effects which operate in opposite directions.
With respect to mutations, a large population carries more individuals and,
therefore, more chances that any given mutation will arise. Once a specific

mutation has arisen, however, there is a much lower probability that it will ever get to fixation. Since these two effects are opposing and directly proportional to population size, population size drops out of the equations, leaving only mutation rate.

The other way to look at this is by considering the process from the viewpoint of an individual gene in a single chromosomal lineage. If population size did have an effect, then this would require genes in big populations to accumulate mutations at a different rate compared with equivalent genes in smaller populations. Since there appears to be no clear mechanism by which a gene could 'know' the size of the population in which it exists, one has to conclude that mutations accumulate at a rate which is independent of population size.

How quickly do allele frequencies change?

The question of how fast allele frequencies change under the influence of drift has already been broached. Using the analogy of pulling coloured balls out a bag we showed that change occurs more rapidly in smaller populations. In fact, the rate of change is inversely proportional to population size. To get a feel for this, it is interesting to consider how fast genetic variability is lost following a sharp fall in population size, a situation of great relevance to conservation biology.

It can be shown that the fraction of heterozygosity which is lost each generation following a bottleneck is $1/2N_e$, assuming that the regeneration of variability by mutation is negligible. In other words, a population with $N_e = 10$ would lose one-twentieth of its variability each generation. Consider a fictitious species with a generation length of two years. Two centuries ago this unfortunate beast met with disaster. Oriental gurus suddenly realised that the only sure way to cure a sprained ankle was to apply a poultice containing extract of its left tear duct. Consequently, the species was hunted down to an effective population size of 1000, from which it never recovered. In this case we can estimate that over 95 per cent of its initial variability would still be present today. If the same animal had been hunted even further, down to an N_e of only 100, then it would still retain 60 per cent of its initial heterozygosity. These simple calculations illustrate just how long and extreme a bottleneck needs to be in order to have a significant effect. In practical terms, most species which are thought of as being endangered probably retain similar levels of neutral variability to those which existed prior to any human interference.

How much neutral genetic variability do populations maintain?

Levels of neutral genetic variability are positively correlated with population size. The reason for this again makes sense. In a small population, mutations arise infrequently, but each has a relatively high probability of becoming fixed and the time taken to go to fixation is relatively short. In a large population, there is no shortage of new mutations, but each one has a very slim chance of fixation and will take a long time to achieve this. The inevitable consequence is that large populations carry many more transition states, mutations which are drifting either to fixation or to elimination. In other words, genes in a large population tend to be more polymorphic than equivalent genes in smaller populations.

Measuring mutation rate and effective population size

An important conclusion from the preceding passage is that allelic diversity depends on both mutation rate and population size. Consequently, if the mutation rate of a sequence is known, this can be combined with an empirical measurement of allelic diversity to derive an estimate of effective population size. Given how important a parameter population size is, and how difficult it is to measure using direct methods, such a calculation is extremely useful. Estimates for mutation rates are even more difficult to obtain. Until recently, they were either non-existent, or pure guesses or, at best, imprecise.

The measurement of mutation rate is akin to searching for a needle in a haystack. Most DNA sequences will mutate only once every 10^4 meioses, and often less frequently than this. Our inability to generate pedigrees of such large size have prohibited quantification of this parameter. However, technological progress has recently opened up a way to measure mutation rates directly, without having to follow inheritance through innumerable generations. A single ejaculate contains millions upon millions of sperm, each one of which is the product of meiosis. Consequently, even rare mutations are likely to be represented, albeit by as few as a single copy. As described in Chapter 2, this possibility has been realised by the pioneering work of Alec Jeffreys and colleagues (Jeffreys *et al.*, 1994). With ingenuity, recently developed machines capable of transferring single sperm to individual test-tubes may allow male mutation rates to be measured in more and more DNA sequences.

From allele frequency to genotype frequency: the Hardy–Weinberg law

One of the simplest, yet most important, concepts in biology was published simultaneously by the biologists Hardy and Weinberg in 1908. Now referred to as the **Hardy–Weinberg law**, it can be summarised as follows. After one single generation of mating, the frequencies of each possible genotype will be given by the product of the frequencies of its component alleles (twice the product for heterozygotes) (Box 3.1). The frequencies of alleles and genotypes will be stable

Box 3.1: Proof of the Hardy–Weinberg law

In its simplest form the H–W law considers a single locus with just two alleles A^1 and A^2 in a very large population of sexually reproducing diploid organisms that mate randomly. The frequencies of A^1 and A^2 are p and q respectively (where $p + q = 1$). Let the frequencies of the genotypes A^1A^1, A^1A^2 and A^2A^2 be X, Y and Z respectively (where $X + Y + Z = 1$). Then, because the homozygote A^1A^1 carries two A^1 alleles, and the heterozygotes one A^1 allele, the frequency of A^1 (p) in the population will be

$$\frac{2X + Y}{2(X + Y + Z)}$$

Similarly, the frequency of the A^2 allele (q) will be

$$\frac{2Z + Y}{2(X + Y + Z)}$$

From these values, and on the basis of Mendelian segregation, the various possible matings can be tabulated (Table a), and the frequency of the three genotypes in the next generation can be obtained (Table b).

Table a:

females	Males X (AA)	Y (Aa)	Z (aa)
X (AA)	X^2	XY	XZ
Y (Aa)	XY	Y^2	YZ
Z (aa)	XZ	YZ	Z^2

Table b:

Possible matings	Frequency of mating	Proportion of offspring from each type of mating AA	Aa	aa
$AA \times AA$	X^2	X^2		
$AA \times Aa$	2XY	XY	XY	
$AA \times aa$	2XZ		2XY	
$Aa \times Aa$	Y^2	$\frac{1}{4}Y^2$	$\frac{1}{2}Y^2$	$\frac{1}{4}Y^2$
$Aa \times aa$	2YZ		YZ	YZ
$aa \times aa$	Z^2			Z^2
Total	1	$(X + \frac{1}{2}Y)^2$	$2(X + \frac{1}{2}Y)(Z + \frac{1}{2}Y)$	$(Z + \frac{1}{2}Y)^2$
	i.e	$p2$:	$2pq$:	q^2

continued

From the above, it can be seen that the ratio of the genotypes A^1A^1, A^1A^2 and A^2A^2 in the first generation will be $p^2 : 2pq : q^2$, irrespective of the frequencies of the genotypes in the parental population.

The second generation can then be generated using these new frequencies as shown in Table c.

Table c:

Possible matings	Frequency of mating	Proportion of offspring from each type of mating		
		AA	Aa	aa
$AA \times AA$	p^4	p^4		
$AA \times Aa$	$4p^2q$	$2p^2q$	$2p^2q$	
$AA \times aa$	$2p^2q^2$		$2p^2q^2$	
$Aa \times Aa$	$4p^2q^2$	p^2q^2	$2p^2q^2$	p^2q^2
$Aa \times aa$	$4pq^3$		$2pq^3$	$2pq^3$
$aa \times aa$	q^4			q^4
Total	1	$p^2(p^2 + 2pq + q^2)$	$2pq(p^2 + 2pq + q^2)$	$q^2(p^2 + 2pq + q^2)$
	i.e	p^2 :	$2pq$:	q^2

From this second generation it is apparent that the frequencies of both the alleles and the genotypes remain the same. According to the Hardy–Weinberg law, these frequencies will be maintained in all subsequent generations, unless they are perturbed by selection, migration or mutation.

thereafter. The law assumes that the genes involved are found in an infinite population of sexually reproducing and randomly mating diploid organisms not affected by selection, migration or mutation.

The importance of this law is twofold. First, it provides the mathematical ground rules for predicting expected genotype frequencies based on the H–W set of assumptions. Second, it acts as a form of universal null hypothesis against which data from real populations may be tested. Any deviations from expectations then indicate that the assumptions have been contravened; for example, that selection is acting or that mating is not random. It is important to note here that the finding of concordance between observed genotype frequencies and H–W expected frequencies does not necessarily mean that all the H–W assumptions are met. As we shall see, stable genotype frequencies can be maintained in a population by a balance of selective factors (see Chapter 4).

Causes of deviation from Hardy–Weinberg expectation

A natural extension of the H–W law is the concept of **Hardy–Weinberg equilibrium**, the state in which genotype frequencies in a population match the expectations of the H–W law. Two common scenarios arise where population samples show deviations from H–W equilibrium: through inbreeding and through **physical linkage**.

Inbreeding

As a general rule, inbreeding increases the number of homozygotes in the population. This is easy to envisage by considering the extreme case of self-fertilisation. Consider a plant that is heterozygous for a selectively neutral gene carrying two alleles, A and a. The F1 progeny will be in the Mendelian ratio of $0.25AA : 0.5Aa : 0.25aa$, or half homozygotes and half heterozygotes. In a selfing organism, homozygotes can only breed true and hence can be thought of as being fixed. Thus, there will be a ratchet-like loss of heterozygosity. This is true regardless of the initial allele frequencies and applies equally to diallelic and multiallelic loci. In all cases, the proportion of heterozygotes declines by a half in each generation with full selfing.

The principle that inbreeding causes a loss of heterozygosity can be extended to cover less extreme cases. In dioecious organisms, selfing is not possible. The closest matings in genetic terms are either between full siblings or between parent and offspring. As full sibs and parents and offspring only share half their genetic material, the decrease in heterozygosity resulting from these types of mating is half that seen in the case of self-fertilisation, i.e. a quarter per generation. Breeding between individuals that are less closely related and share a lower proportion of their alleles as a result of common descent still leads to a decline in heterozygosity. In progeny from half-sibs, which share one-quarter of their genes, the decline in heterozygosity is one-eighth. For first cousins, sharing an eighth of their genes by descent, it is one-sixteenth, and so on.

Taken to the limit, the term inbreeding does not have to involve matings between known relatives, but covers virtually every conceivable deviation from random mating. This is best seen by comparing real populations with the qualifications stated by Hardy and Weinberg. Real populations cannot be infinite, and within them any given individual is only ever likely to encounter a small subset of all possible partners. Most real populations also have some form of substructure or subdivision. Just as with matings between known relatives, inbreeding due to non-random mating also results in an increase in overall homozygosity. This effect is most frequently encountered when it results from population subdivision, when it is referred to as the **Wahlund effect**.

The Wahlund effect is best illustrated by an example. Imagine a large population, with allele A at a frequency of 0.5 (ignore other alleles), which is split into two equal smaller populations, one where the frequency of A is 0.1, the other where A is at 0.9. After one generation of random mating the frequency of AA homozygotes will now be the square of the allele frequencies, i.e. 0.01 and 0.81. The mean frequency of AA homozygotes is thus $(0.01 + 0.81)/2 =$

0.41, which is considerably higher than 0.25 (i.e. 0.5^2), the frequency found in the original population.

Thus the mere act of subdivision can increase the number of homozygotes, even when there is no change in mean allele frequency. The simple fact that mating occurs within groups rather than between groups means that partners are on average more related than they would be if the two subpopulations were combined. Turned around, this means that any excess of homozygotes found in a sample of individuals can be interpreted in terms of a mixture of individuals from two different populations. Of course, other explanations are possible, e.g. selection might be acting against heterozygotes or there may be assortative mating preferences.

Given that an excess of homozygotes indicates non-random mating, and, further, that this is most likely to be the result of population substructure, some form of formal quantification is clearly desirable. Appropriate formulations were first derived by Sewall Wright in the 1920s and 1930s (Wright, 1932, 1935, 1948), and involve three measures, collectively referred to as the **F statistics** and distinguishing each of the three possible levels of inbreeding (see Wright, 1951). F_{IS} is known as the **inbreeding coefficient** and reflects the reduction in heterozygosity of an individual organism relative to the subpopulation in which it exists. This is closest to the image that the term inbreeding (mating between relatives) conjures. F_{ST} is the **fixation index** and reflects the reduction in heterozygosity due to population subdivision. Finally, F_{IT} is the overall inbreeding coefficient of an individual, and effectively combines the first two measures.

The mathematics involved in the calculation of F statistics is straightforward, since it relies simply on comparing observed and expected heterozygosities (see Box 3.2). Together, the F statistics often prove informative about the way in which variability is partitioned within a sample or samples. They apply best to polymorphic, perfectly neutral markers. Unfortunately, each of the three most commonly used genetic markers suffers at least one major drawback. Minisatellites tend to be just too polymorphic. Consequently, homozygotes are rare whatever the population structure and the method loses power. Protein isozymes may be affected by natural selection, and so are often not perfectly neutral. In addition, isozymes tend to be rather limited in the levels of variability they exhibit, and hence many loci must be screened to get a reliable estimate. Finally, microsatellite markers (see Chapter 9), which seem set to become the most-used currently known genetic markers in population biology, often harbour hidden 'null' alleles and are subject to frequent back-mutations. Both these aspects need to be considered.

In summary, Wright's F statistics provide a convenient and accessible way of quantifying population substructure using allele frequency data. However, their use with current genetic markers should be tempered with some caution, since none of the major classes is ideal. On the more positive side, continuing theoretical work is yielding modified measures which appear increasingly capable of allowing for specific anomalies.

Box 3.2: Calculation of F statistics

Consider some genotype data for one locus, collected for individuals sampled from two or more putative subpopulations. From such a data set, three quantities can be derived:

1. H_i = the observed heterozygosity of an individual, calculated as the mean frequency of heterozygotes averaged over all subpopulations.
2. H_s = the expected heterozygosity of an individual in a subpopulation, calculated separately for each subpopulation and then averaged.
3. H_t = the expected heterozygosity of an individual in the whole population, calculated assuming all samples are drawn from a single, homogeneous, randomly mating population.

In all cases, expected heterozygosities are calculated as one minus the expected homozygosity, i.e.

$$1 - \sum_n p_i^{\,2}$$

where p_i is the frequency of the ith allele of n.

Now, the three F statistics are calculated as follows:

$$F_{IS} = \frac{H_s - H_i}{H_s}$$

$$F_{ST} = \frac{H_t - H_s}{H_t}$$

$$F_{IT} = \frac{H_t - H_i}{H_t}$$

Where F_{IS} is the reduction in heterozygosity of an individual relative to the subpopulation it is in due to non-random mating (mating with relatives), F_{ST} is the reduction in heterozygosity due to population subdivision and F_{IT} is the total reduction in heterozygosity of an individual due to the effects of non-random mating and population subdivision combined. Together, these three measures are referred to as hierarchical F statistics.

Directional change in allele frequency: the force of natural selection

When a particular variant allele of a gene alters the lifetime reproductive success of the individual which carries it, that allele will be subjected to natural selection. Those alleles which increase the average number of descendants produced will tend to increase in frequency relative to their homologue. The obvious converse is that alleles which cause fewer than average progeny to be produced will decline in frequency.

Natural selection will operate if three fundamental criteria are met. First, there must be phenotypic variation for the particular character in question. Second, some of this phenotypic variation must be due to genetic variation, and hence be heritable. Third, individuals of one phenotype must produce, on average, either more or fewer descendants than those of another. In other words, there must be a correlation between fitness and phenotype.

Natural selection and random genetic drift will inevitably operate concurrently. Which of these processes has the greater influence in a specific instance depends crucially on two factors: the fitness differences between phenotypes, and the population size. Taking extreme cases, a number of genetic diseases cause certain infant mortality. Such mutations, when expressed, will leave behind them no progeny, and therefore will be eliminated as soon as they arise. Drift cannot act. Similarly, a new mutation which confers resistance to a common debilitating disease will be highly advantageous to its carriers, and will spread rapidly to fixation. Again, in the face of such a strong effect, drift is unlikely to influence the process significantly.

At the other extreme, many (probably the majority of) mutations confer only a slight advantage or disadvantage on the individuals that bear them. Such mutations behave rather like an aphid trying to fly on a blustery day. Although there is a preferred direction of travel (selection is directional), this is often overwhelmed by the vagaries of drift. Only on a perfectly calm day is the aphid's flight controlled and directional. In evolutionary terms, the calm day is equivalent to a population that is large enough for the effect of drift to be negligible. Of course, just as the aphid is a weak flier and takes a long time to progress from A to B, so too do mutations that cause a minimal change in fitness take proportionately longer to progress to fixation.

Whatever the magnitude of the selective advantage, beneficial mutations are at their most vulnerable in the first few generations after they arise. During this period they will be represented by few copies and, although tending to increase in frequency, can easily be eliminated through chance events. For example, a mosquito which inherits a new mutation conferring insecticide resistance would be extremely successful but could easily be lost if the pool in which its progeny dwell is discovered and they are devoured by a predator. However, once the adults have appeared, matured and dispersed, much of the danger is past. All the evolutionary 'eggs' are no longer in one basket. Once these initial stages have been negotiated, drift will play an ever decreasing role as the mutation increases in frequency towards fixation. The larger the selective advantage, the more rapidly fixation will be achieved and the safer the mutation will be from stochastic loss near its origin.

Selection in diploid organisms

It is easy to envisage the effect of selection in haploid organisms because alleles may be viewed as simply having some selective effect or being selectively neutral. However, most higher organisms are diploid, i.e. they carry two copies of most of their genes. For neutral alleles this has little consequence greater than doubling the number of copies present in the population. With regard to selection, however, the consequences are important. The critical factor is the fitness of heterozygotes.

Let us consider a gene with two alleles, A^1 and A^2. There are three possibilities. Most commonly, the heterozygote, A^1A^2, will have fitness within the bounds of the fitness of the homozygotes, neither better than either, nor worse than either. If the heterozygote is phenotypically identical to one of the homozygotes (i.e. one of the alleles is **dominant**), then the fitness of the heterozygote will equal that of the homozygote it resembles. This is probably the most frequent scenario. The second possibility is that the heterozygote has higher fitness than that of either homozygote, a condition termed heterozygote advantage. Lastly, the heterozygote may have lower fitness than either homozygote. This situation, called heterozygote disadvantage, is rare.

The subtleties of natural selection can be gained from consideration of the first scenario, where heterozygote fitness is neither more nor less than either homozygote. We can consider the fate of a beneficial new mutant allele A^2 in two extreme cases. First, the mutant allele may be phenotypically dominant, and the heterozygote have the same high fitness as the (as yet unformed) A^2A^2 individual. This mutation will be subject to selection immediately, because the newly formed heterozygote has higher fitness than the rest of the population. Chance permitting, the allele will immediately begin to climb towards high frequency. However, the allele approaches fixation slowly. The outgoing **recessive** allele finds ever greater protection in the heterozygous form as it declines in frequency. Selection upon the old allele does not act effectively at high frequencies, because the deleterious phenotype it creates is rarely expressed.

The other extreme is where the beneficial mutant allele is phenotypically recessive, and the heterozygote both appears like, and has fitness equal to, the rest of the population in which it first occurs. Mutations like this behave very differently, because they spend the vast majority of their early existence coupled to a dominant homologue which prevents expression of the beneficial trait. This follows directly from the H–W equations, which state that the frequency of any given homozygote is equal to the square of the frequency of its allele. Only when the alleles reach sufficiently high frequency such that the homozygote A^2A^2 appears will the beneficial effects of the A^2 allele be observed. This threshold frequency is remarkably high. Drift plays an important role in the early evolution of this allele, despite the fact that it is beneficial. However, if drift does produce an increase in frequencies such that homozygotes are formed, the allele will increase rapidly to fixation.

The same analysis can be performed for novel deleterious alleles. Deleterious alleles which are dominant and therefore expressed in heterozygotes will be eliminated fairly rapidly, the speed depending upon the magnitude of the disadvantage. Deleterious alleles which are recessive are not rapidly eliminated. They are rarely present in the deleterious homozygous state, and are thus rarely exposed to selection. Genetic drift can operate and they can reach significant frequencies in the population. This analysis is confirmed by the observation that most of the common genetic diseases of humans are recessive mutations. Further, the mortality from these diseases reflects the fact that they can reach high levels in the population. Genetic diseases affecting around one in 10 000 babies are not uncommon, and reflect the fact that one in every 50 or so carry the disease gene (are heterozygotes). Some, such as cystic fibrosis, affect up to one in every 2500, with one in every 25 individuals carrying the deleterious allele.

The fitness of the heterozygote thus has a profound effect on the rate and course of evolution. Drift plays an important role in the early evolution of recessive mutations. They may increase to high frequency despite being deleterious, and conversely may disappear despite being advantageous. Dominant mutations behave more deterministically, being rapidly lost when deleterious, and rapidly spreading when beneficial.

Maintenance of allelic diversity

Although natural selection is usually thought of as either fixing favoured alleles or removing detrimental ones, there are also circumstances in which natural selection can help to maintain diversity. One such situation is that of the peppered moth, *B. betularia*. In highly polluted regions the melanic form of this moth has a selective advantage, and appears at high frequency. However, gene flow from neighbouring rural regions maintains a constant input of the less-favoured typical form. Selection and migration combine to maintain polymorphism (see Brakefield (1987) and Majerus (1989) for reviews).

This case of polymorphism relies on gene flow between areas in which selection acts in opposite directions. In other instances, selection can conserve genetic diversity on its own. For this to come about, each allele must be favoured in a way which prevents either elimination or fixation. Two phenomena will be considered here: when selection favours heterozygotes, and when selection favours rare phenotypes. Both of these types of situation can maintain stable allele frequencies, giving a **stable** or **balanced polymorphism**.

Heterozygote advantage (overdominance) When the heterozygote form is favoured over both of the two possible homozygotes, the population inevitably remains polymorphic. Neither allele will go to fixation because both occur in the most favoured genotype, the heterozygote. The result is an equilibrium state in which two alleles are maintained in the population at frequencies which depend on the relative disadvantages of the two homozygotes compared to the heterozygote.

The phenomenon of heterozygote advantage has been postulated as being responsible for maintaining genetic variability in natural populations (e.g. Ford, 1940). However, rather few good examples exist, the classic case being human sickle cell anaemia. Here, individuals who are homozygous for the normal oxygen-carrying haemoglobin gene are prone to malaria. Homozygotes for the sickle allele produce mutant haemoglobin molecules which aggregate into rods and distort the shape of the red blood cells into crescents. Until recently, these individuals would invariably die from anaemia before reaching reproductive maturity. Their sickle-shaped cells stick in the capillaries and have poor oxygen-carrying capacity. Heterozygotes suffer mild anaemia, but their unusual haemoglobin molecules confer considerable resistance to malaria. As a result, in regions with a high incidence of malaria, heterozygotes are the most fit form and form the basis of a stable polymorphism.

Cases of heterozygote advantage may also occur if selection pressures vary over the life history of the animal. Josephine Pemberton and colleagues (1991) have examined the Isle of Rum population of red deer for a number of different protein isozyme polymorphisms. One particular enzyme, mannose phosphate isomerase, was found to be encoded by two alleles, called slow (*s*) and fast (*f*) on the basis of their electrophoretic migration rate. By typing calves which died in the first year of life, they found that *ff* homozygotes were much more likely to die than either *fs* heterozygotes or *ss* homozygotes. So, why has the *f* allele not been eliminated? The answer lies in later life. Females who carry an *f* allele reproduce earlier and are more fecund than *ss* homozygotes. There is heterozygote advantage, *fs* being fittest across the life history as a whole.

This scenario illustrates a case of temporally staggered heterozygote advantage. Early in life one homozygote is selected against, then later the other is selected against. Temporal variation in the environment may also confer an advantage on heterozygotes. If the environment changes with time, as it does say as the seasons change, different alleles may be favoured at different times of year. Again heterozygotes may be at an advantage over homozygotes because they carry two different alleles. A chromosomal polymorphism in *Drosophila pseudoobscura* appears to be maintained in this way (Box 3.3).

Similar balancing forces potentially exist in the spatial dimension. In the case of the peppered moth considered above, a polymorphism was maintained more or less by gene flow between habitat types characterised by opposing selective forces (either for or against the melanic form). However, if the environment contains heterogeneity on a finer scale, forcing individuals to move frequently between patches, polymorphism could be maintained in another way, by

---

Box 3.3: Temporal climatic variation and the maintenance of chromosomal polymorphism in *Drosophila pseudoobscura*

Many species of *Drosophila* are known to be polymorphic for chromosomal inversions. These have been well researched as they are easy to see by cytological examination of the **polytene** nuclei of larval salivary glands. The Standard (ST), Arrowhead (AR) and Chiricahua (CH) chromosome 3 inversions of *D. pseudoobscura* occur together as a polymorphism in southern California. The frequencies of these chromosome arrangements vary in a regular manner throughout the year (Dobzhansky, 1951, 1956). In an extensive and elegant series of laboratory and field studies, Theodosius Dobzhansky and his colleagues were able to show that a variety of environmental factors which vary with season, including temperature, food type and amount of intraspecific competition, affect the fitnesses of these chromosomal arrangements (Dobzhansky, 1951, 1961; Dobzhansky and Spassky, 1954; Levene *et al.*, 1954). This accounts for the seasonal variation in the frequencies of ST, AR and CH. However, the reason that the polymorphism is maintained is because the various inversion heterozygotes generally have higher fitness than their homozygotes (Dobzhansky and Spassky, 1954). Indeed, Dobzhansky (1947) was able to demonstrate that in experimental cultures from any single location, the heterozygotes consistently occurred at frequencies higher than their Hardy–Weinberg expectations.

---

heterozygote advantage. Wherever particular alleles confer increased fitness in particular patches, heterozygotes, by carrying two different alleles, may be at an advantage.

There is considerable debate as to the number of polymorphisms which are actually maintained by heterozygote advantage. On the one hand, despite decades of research, there are still only a handful of good examples. Such rarity appears to make sense biochemically because for many genes it is difficult to imagine plausible mechanisms which would cause heterozygotes to be favoured over both homozygotes. On the other hand, theory developed by Fisher (1930) concerning the evolution of dominance suggests that heterozygote advantage should arise when a gene has more than one function. Briefly, he proposed that if one of the two alleles was beneficial with respect to one function, but the other allele was beneficial with respect to the other, selection will act in favour of modifier genes which emphasise the positive aspects of each allele. This would lead to a state where possession of both alleles was advantageous (see Chapter 12). The heterozygote might well be fitter than either homozygote because it enjoys the positive aspects of both alleles.

Is the view that heterozygote advantage is common tenable given the dearth of examples? The view is probably still a fair (if unproven) one, because the paucity of cases may be attributable to the difficulty inherent in detecting it. It is perhaps no coincidence that sickle cell anaemia involves two important human diseases. In what other organism would we have had enough information at our fingertips to link resistance to a parasite, anaemia and a blood protein? To make this link requires combining genotype data gained from genealogy with detailed information about fertility and causes of death. An ability to cover the entire life history may be particularly important, as shown by the study of red deer on Rum. To find heterozygote advantage, a large-scale project needs to be set up with genetic analysis running alongside detailed demographic study.

Frequency-dependent selection   In cases of heterozygote advantage it is supposed that in a given environment, one phenotype (the heterozygote) is fittest. However, many situations are known in which fitnesses are not constant, but instead correlate with some environmental variable such as the frequency of the phenotype itself, population density or the abundance of a second species. Such selection is referred to as **frequency-dependent selection**.

When the fitness of an allele is positively correlated with its frequency then the likely outcome is that it will spread with accelerating rapidity until it becomes fixed. However, when the fitness of an allele is negatively correlated with its frequency, the result is a stable polymorphism. Any deviation from the point of equilibrium will face selection acting in the opposite direction: the allele which suffered a decline will increase in fitness whereas the allele which increased in frequency will become less fit (Box 3.4). Depending on the nature of the selective forces involved, allele frequencies will either stabilise more or less at the point of equilibrium or tend to oscillate about it. This balance is best shown by an example.

Consider a predator, say a bird, that eats snails. Many birds are known to hunt for food by forming a search image: they learn from the food items they find and become better at finding similar items, or actively search for such items to the exclusion of others. The consequence of this simple behaviour pattern is frequency-dependent selection. If our snail-eating bird lives in an area where yellow snails predominate, it will form a search image for yellow snails. Now imagine that a mutation occurs, giving rise to a pink form which is identical in crypsis and all other aspects to the yellow form. When a bird comes to hunt, the first snail it encounters is almost certain to be yellow, promoting a search image for yellow snails and making it less likely that a pink form will be found and eaten. In time, this selective advantage will allow the pink form to rise in frequency until it is as common as, or commoner than, the yellow form. As the frequency of the pink form continues to rise, yellow snails will become increasingly difficult to find and there will come a time when the search image will change to pink. Thus, by simply eating disproportionately more of the commoner morph, the bird's feeding habits will actively maintain a stable colour polymorphism. Indeed, further colour morph mutants, should they arise, will also be assimilated into the polymorphism. The simple balance between being common and eaten, or rare and ignored, will always favour new forms

Box 3.4: Negative frequency-dependent selection

Assume a polymorphic species has just two forms (dark and light) controlled by a single biallelic gene. The alleles are A^1 and A^2 with the former fully dominant to the latter. Let there be a stable equilibrium frequency of 0.5 for each of the alleles (frequency of $A^1 = p = 0.5$, frequency of $A^2 = q = 0.5$).

If selection is negatively frequency-dependent, then in a simple case the fitnesses of the phenotypes and their genotypes may be:

Phenotype	dark	dark	light
Genotype	A^1A^1	A^1A^2	A^2A^2
Frequency	$p^2 = 0.25$	$2pq = 0.5$	$q^2 = 0.25$
Fitness	1	1	$1.5 - q$

It is then easy to see that the polymorphism will be maintained at the equilibrium frequencies by considering what happens when q changes. If the light form becomes more common, q increases and so the fitness of the A^2 homozygote declines, with an ensuing drop in its frequency which is thereby brought back towards the equilibrium frequency. The reverse happens if the dark form increases in frequency. Then q becomes less than 0.5, with the result that the fitness of the A^2A^2 genotype becomes greater than that of either of the other genotypes, and the light form increases at the expense of the dark form. It is only at the equilibrium frequency, in this case when $q = 0.5$, that the fitness of all the genotypes is the same. The evolutionary explanation of the maintenance of a $1:1$ sex ratio in most dioecious organisms follows exactly this rationale (see Chapter 4).

(except perhaps in the extreme case when a highly polymorphic species would exert a selective pressure on the bird to abandon search images altdgether).

The behaviour of this system around equilibrium depends critically on how tenaciously the bird clings to its search image. If the bird is perceptive and switches rapidly, the equilibrium frequencies of the morphs will be quite stable. However, if the bird clings to its image long after it has ceased to be easy to find, there will be overshoot, and the system will tend to oscillate continuously about the equilibrium.

Although difficult to prove in the field, wherever search images play an important role in hunting, the opportunity must exist for balanced polymorphisms to be maintained. A similar rationale applies to cases of polymorphic **Batesian mimicry** (see p. 78). Batesian mimics avoid predation because although otherwise unprotected, they resemble closely a species, the model, which has some strong defensive trait such as a sting, a bite or toxicity. Given perfect mimicry, the success of the strategy depends critically on the frequencies of model and mimic. When mimics are rare, predators are much more likely to find trouble at their first encounter, and therefore to develop avoidance. As mimics increase in frequency, there is an ever-increasing risk that a predator will first encounter the mimic. In the ensuing confusion, both mimic and model will suffer. A novel form which mimics a different model species will be favoured at this time because it will be rare compared to its model. Eventually a new equilibrium will be reached involving two mimetic forms and two models. As yet, the importance of this evolutionary chain of events is unclear. On the one hand, there are a number of examples of butterflies which are both Batesian mimics and polymorphic. On the other hand, the pathway appears to require a somewhat improbable single mutation to create a decent mimic (see Chapter 4).

Frequency-dependent selection does not have to depend on interactions between organisms. Polymorphism may also be maintained by interactions between members of the same species. For example, females of some species (e.g. the scarlet tiger-moth, see p. 84) prefer to mate with males of a different phenotype to their own. The consequence of this is that the large number of females of the common morph will prefer to mate with males of the rarer phenotype, and in so doing will increase its fitness. Conversely, males of the common morph will be preferred by only a small number of females, those of the rarer phenotype. Many types of self-incompatibility systems in plants operate in the same sort of way, pollen bearing a particular incompatibility allele being effective only on plants carrying a different allele.

These are just three types of situation which lead to negatively frequency-dependent selection. There are many others. However, these three demonstrate the essentials of the way fitness may vary with allele frequency, and how selection may maintain variation.

Selection at more than one locus

So far, we have discussed only situations where selection is acting at a single locus at a time. Nature is rarely so kind as to leave us with such a simple situation. Selection can act on many loci simultaneously. The results of selection at many loci are not always easily envisaged. Whereas with one locus and two alleles, there were three possible types of fitness relation, there are clearly more for two loci with two alleles, and many more possibilities when more than two loci are involved.

We can simulate selection at two unlinked loci each with two alleles easily on a computer. The basic process is simple and is shown in detail in Fig. 3.2. There are nine possible genotypes. First we set initial frequencies f for each of the nine genotypes. These are equivalent to the frequencies of zygotes. Following this, we go through the process of maturation and reproduction. Each of the nine genotypes has a fitness, w, associated with it. The relative entry of each

(i) Frequency of different zygote genotypes
(proportion of population).

	AA	Aa	aa
BB	f1	f2	f3
Bb	f4	f5	f6
bb	f7	f8	f9

Relative fitness of genotypes

	AA	Aa	aa
BB	w1	w2	w3
Bb	w4	w5	w6
bb	w7	w8	w9

(ii) Frequency of different mature adult genotypes
(proportion of population).

	AA	Aa	aa
BB	f1.w1/N	f2.w2/N	f3.w3/N
Bb	f4.w4/N	f5.w5/N	f6.w6/N
bb	f7.w7/N	f8.w8/N	f9.w9/N

N is the normalising coefficient, which transforms relative frequencies into proportions. It is the sum of frequencies, i.e. N = f1.w1 + f2.w2 + f3.w3...

(iii) Frequency of different gamete genotypes
(assuming no linkage).

$$f(AB) = \frac{f1.w1 + 0.5.f2.w2 + 0.5.f4.w4 + 0.25.f5.w5}{N}$$

$$f(Ab) = \frac{0.5.f4.w4 + 0.25.f5.w5 + f7.w7 + 0.5.f8.w8}{N}$$

$$f(aB) = \frac{0.5.f2.w2 + f3.w3 + 0.25.f5.w5 + 0.5.f6.w6}{N}$$

$$f(ab) = \frac{0.25.f5.w5 + 0.5.f6.w6 + 0.5\ f8.w8 + f9.w9}{N}$$

(iv) Frequency of new zygotes
(assuming random union).

	AA	Aa	aa
BB	$f(AB)^2$	$2.f(AB).f(aB)$	$f(aB)^2$
Bb	$2.f(AB).f(Ab)$	$2.f(AB).f(ab) +$ $2.f(Ab).f(aB)$	$2.f(aB).f(ab)$
bb	$f(Ab)^2$	$2.f(Ab).f(ab)$	$f(ab)^2$

Fig. 3.2 Simulation of selection acting on two genes, A and B, each with two alleles.

genotype into the production of successful gametes is then a simple transformation of initial frequency by relative fitness. We can then calculate the frequency of gametes which go on to form the next generation by applications of Mendel's first and second laws. Alleles at a locus partition equally between gametes, and genes at different loci segregate independently. So, for instance, an individual that is $AABb$ forms two types of gamete, AB and Ab in equal proportion. If we then presume that gametes fuse randomly, we can calculate the new frequency of each genotype. We have gone through one round of selection. With a computer, it is simple to iterate the process to examine selection over longer periods of time.

If enough of these simulations are run, it becomes clear that they fall into two categories. In the first case, neither the speed nor the direction of selection is altered from what we would have expected from consideration of each locus separately. This happens if the genotype at one locus has no effect on the fitness effects of the other. So, for instance, in the example below, individuals of genotypes of bb, Bb and BB have relative fitnesses w_{bb}, w_{Bb}, w_{BB} against any given background at the other locus and vice versa. The fitness of the individual is simply calculated by multiplying the fitnesses of each locus considered alone:

	AA	Aa	aa
BB	$w_{AA}.w_{BB}$	$w_{Aa}.w_{BB}$	$w_{aa}.w_{BB}$
Bb	$w_{AA}.w_{Bb}$	$w_{Aa}.w_{Bb}$	$w_{aa}.w_{Bb}$
bb	$w_{AA}.w_{bb}$	$w_{Aa}.w_{bb}$	$w_{aa}.w_{bb}$

This will clearly be quite common. For instance, if gene A is concerned with respiration and gene B with colour pattern and crypsis, there is no reason that fitness should be anything other than that obtained by compounding the effects of each. The fitness effect of the genotype at one locus is not affected by the genotype at the other.

However, in other simulations, either the speed or end-point of evolution is altered from that expected under simple consideration of the loci separately. Fitness here is an interactive property of the two loci. Either the relative fitness of genotypes BB, Bb and bb differ between backgrounds at the A locus, or the relative fitness of genotypes AA, Aa and aa differ between backgrounds at the B locus (or both). We say there is an epistatic fitness interaction between the loci. To illustrate this principle, consider two genes which affect moth colour pattern. Two novel mutants A and B arise, both dominant. Possession of each allele separately is advantageous, but possession of both together makes the moth virtually invisible to all birds. The relative fitnesses of the genotypes might look like this:

	AA	Aa	aa
BB	9	9	3
Bb	9	9	3
bb	2	2	1

Here, although the end point of selection is the same, the positive fitness interaction means the two alleles go to fixation more rapidly than they would were they apart. Were the fitness of individuals bearing both advantageous alleles less than the product of each separately (say four, not nine), then fixation for both would be achieved less rapidly. This might occur if they both produce very good crypsis on their own, and addition of the other elements only provides a minor further advantage.

However, different loci do not necessarily augment each other's fitness effects. Carry our cryptic moth analogy further. There are two possible mutants A and B for colour pattern. Each one on its own is very advantageous, creating a beautifully cryptic moth. Together, however, they are a disaster. The moth sticks out, and birds detect and pick them off easily. The fitness matrix takes this form:

	AA	Aa	aa
BB	0.7	0.7	3
Bb	0.7	0.7	3
bb	2	2	1

Here, it is uncertain which form will predominate. It depends upon historical chance. If the mutant B occurs before A, selection will produce a population which is entirely $aaBB$. If the mutant A occurs before B, then selection will produce a population which is entirely $AAbb$. This is despite the fact that mean fitness is higher at the other equilibrium. Sewall Wright (1932) produced the analogy of an **adaptive landscape** to describe this situation. It may be seen that the fitness landscape has two peaks, fixation of a and B, and fixation of A and b, with a valley of lower mean fitness when both mutant alleles are present (for then the forms bearing both A and B exist). These peaks are of different height, but the population cannot cross between them by selection. Selection drives the population uphill to the peak, even if this peak is not the one of highest mean fitness. It is notable that **fitness peaks** also exist when selection is occurring at a single locus and there is heterozygote disadvantage.

Consider the situation where individuals of genotype aa have fitness 1, but a novel mutant allele A arises which, when homozygous (AA), is highly advantageous (say fitness 2). However, if the heterozygote Aa is less fit (say fitness 0.8), then the novel allele cannot invade. The population is stuck at a local optimum.

Interaction between loci for fitness can create peaks on the adaptive landscape, and it is clearly true that where there are peaks, populations may get stuck on lower peaks, away from the global optimum. Do populations permanently get stuck up peaks which are not of highest mean fitness? Several arguments suggest not. First, the environment may change in a locality, altering the landscape, and setting off change. Even if the selective regime reverts back to its original state, the population has shifted peaks. If it makes contact with an unchanged population, then it is possible that the whole mass may move to the higher peak (equally, it is possible they move to the lower one!). Sewall Wright

(1977) in his shifting balance theory suggested that when selection was weak, random genetic drift in highly subdivided populations could move allele frequencies downhill enough in a subpopulation that it would allow movement between peaks. Drift associated with founder effects could also be important in this context. These randomising influences allow a shift from local peak to global peak within local populations, and with time, within the population as a whole.

Selection where the phenotype is the product of genes at more than one locus

Many phenotypic traits are the product of genes at more than one locus. The same principles apply to the action of selection in these cases as apply when two loci produce different aspects of phenotype but the loci interact with respect to fitness. For instance, if we are considering selection upon height, we can consider a case where there are two genes, each of which can have a dominant allele which adds 10 cm to height. If the genes do not interact in producing the phenotype (an individual with both tall alleles is 20 cm taller than otherwise), and if fitness relates linearly to height, then the consequences of selection are straightforward. However, two circumstances may produce fitness interaction. The first of these is if the genes are additive in their phenotypic effect, but this does not translate into linearly changing fitness. If being 20 cm taller is disadvantageous, but being 10 cm taller is beneficial, then the result of selection is unpredictable. There are two adaptive peaks: fixation for either tall allele. The population will end up at one or other of the peaks. Which equilibrium is reached depends largely on chance. The second situation which may produce a fitness interaction is when the two genes are not additive in their phenotypic effect, but instead interact to produce the phenotype. A common case will be when mutants at two loci together produce a novel form, but when apart have no phenotypic effect. Here, even if the novel form is advantageous, selection will not drive a population to this peak on its own. Drift is required to take the frequencies of the mutant alleles to a sufficient level that individuals bearing both mutations are formed.

Allele frequency and genotype frequency in the two loci case

Allele frequency in the single locus case could be transformed into genotype frequency using the Hardy–Weinberg law. A **panmictic population** will remain in Hardy–Weinberg equilibrium unless acted upon by selection, or if mating is non-random. In the case of a single locus under selection with random mating, the population departs from equilibrium each generation if the heterozygote is of anything but intermediate fitness. However, random mating restores genotype frequencies at the locus to equilibrium. We can create similar rules for the two locus case (Box 3.5). We refer to non-random association between loci as **linkage disequilibrium**.

When examining associations between loci, it is important to realise that random mating does not absolutely restore the population to equilibrium each generation. Rather, recombination and random assortment push genotype frequencies toward equilibrium. The amount that mating breaks down associations will depend crucially on whether the loci in question are on the same chromosome (are physically linked), and if they *are* on the same chromosome, how closely they are linked. This is because linkage increases the probability of a particular combination being co-inherited, and tightly linked loci have a greater probability of being co-inherited than ones which are loosely linked, because recombination occurs less frequently.

This effect can be seen most easily by considering a haploid population which, for some reason, is in complete linkage disequilibrium between two loci A and B. Each locus has two alleles (A and a, B and b) at frequency 0.5 each. However, type AB constitutes 50% of the population, and type ab 50%. Types Ab and aB do not occur. Now, what happens after a generation of random mating? One half of all the diploid progeny will result from AB fusing

---

Box 3.5: Linkage disequilibrium

Linkage disequilibrium is most easily envisaged by examining the frequency of different gamete genotypes. Consider two loci, A and B, each with two alleles, A and a, B and b. The frequencies of the these alleles are p and q, r and s respectively, where $p + q = 1$, $r + s = 1$. If there is no association between loci, then we expect the following frequencies of gametes:

AB	Ab	aB	ab
$p.r$	$p.s$	$q.r$	$q.s$

With no linkage disequilibrium, the product of the frequencies of gamete types $AB.ab$ is the same as the product of the gamete types $Ab.aB (p.q.r.s)$. The coefficient of the disequilibrium, D, is a measure of the departure from this situation. It is the product of the frequencies of AB and ab minus the product of the frequencies of Ab and aB, i.e. if we have gamete frequencies:

AB	Ab	aB	ab
w	x	y	z

where $w + x + y + z = 1$, then $D = w.z - x.y$.

with *ab*, and will be heterozygous at both loci. If the loci are tightly linked, then the chromosomes *AB* and *ab* segregate separately during meiosis, and the disequilibrium is maintained. If they are unlinked, then the loci in heterozygous individuals assort randomly, creating each of the four genotypes. The association decays by 50% through mating (remember, 50% of the individuals were homozygous for the two genotypes and therefore had no segregation). The change in linkage disequilibrium in each generation is given by the formula $D_{new} = D_{old}*(1 - r)$, where *r* is the amount of recombination, and genes on different chromosomes have $r = 0.5$, tightly linked genes have *r* around 0, and loosely linked genes have *r* between these two values.

Disequilibrium is thus expected to decline over time through recombination (for linked loci) and random assortment (for unlinked ones). But what generates disequilibrium in the first place? The most obvious force creating associations is epistatic fitness interactions. Just as deviation from Hardy–Weinberg equilibrium occurs when the heterozygote is not of intermediate fitness, disequilibrium is generated each generation if there are epistatic fitness interactions between loci. If, in a haploid, *AB* or *ab* are the two fittest forms, then the parental population will contain an excess of these above random expectation. The magnitude of the excess will relate to the size of the selective advantage and the degree of interaction between loci.

Epistatic interactions which generate associations between loci are most frequently seen when the loci in question are tightly linked, because here recombination breaks down the associations formed very slowly, and **epistasis** can thus build up substantial disequilibrium. One of the best characterised examples is the mimetic butterfly, *Papilio memnon*. This butterfly is palatable, and is polymorphic, mimicking more than one species. There are many aspects to the mimicry. The mimic is the same as its model in colour, pattern and morphology. Many of the different genes involved in the mimicry are closely linked, and despite the fact that there are many polymorphic genes, pure mimetic forms predominate. The genes involved in mimicry are in strong disequilibrium. Although other forms are known, they occur at very low frequency in the wild. This is because they are at a severe selective disadvantage, because they half mimic one species and half another. A high degree of disequilibrium is produced because of the strength of the fitness interaction and the tightness of the linkage between the genes.

In parallel with the one locus case, another factor that generates disequilibrium between loci is non-random mating. If females of genotype *AA* are more likely to mate with males of type *BB*, then the *AABB* type will occur more frequently than consideration of allele frequencies would suggest. This type of association has recently been shown in the seaweed fly, *Coelopa frigida*. Female seaweed flies prefer males bearing genes in a particular chromosomal inversion. Among progeny of females that exercised this preference, Gilburn and colleagues (1993) found an association between the female mating preference and the preferred chromosomal inversion. A more common source of non-random mating is self-fertilisation in plants. This is likely to generate strong disequilibrium between loci.

Disequilibrium may also arise through chance events. Simplest of all is mutation. A mutation which arises *de novo* in a population can only be associated with one allele at any other locus. It must initially be in disequilibrium. Over time, the association will deteriorate, but tightly linked loci are likely to remain in disequilibrium for an appreciable period of time.

Chance events associated with genetic drift may also create disequilibrium. As we saw earlier in this chapter, when populations become subdivided they will tend to become differentiated due to founder effects and drift, regardless of what is happening through natural selection. Consequently, alleles become over- or under-represented in a subpopulation relative to the population as a whole, and the population as a whole will be in disequilibrium. This situation is well illustrated by human populations. A mixed sample of Japanese and Swedes would suggest that genes controlling hair colour, skin colour, eye shape, stature and many other traits were all linked. Clearly these traits do not all lie on the same chromosome, but particular sets of traits do co-occur in individuals because the human population is not panmictic, and genetic differences exist between the various racial groups. Were the two populations to become entirely mictic, it is to be expected that the disequilibrium would break down rapidly for all but the most tightly linked genes.

Hitch-hiking

In the consideration of the dynamics of neutral alleles earlier in this chapter, it was tacitly assumed that every locus could be considered separately with respect to drift and selection. This is true if the loci are randomly associated. However, if two genes are in disequilibrium, then the fate of these genes becomes linked. Consider the case where alleles at one locus, *A*, are neutral with respect to selection, but genotypes at another locus, *B*, have different fitnesses. Allele frequency at the neutral locus will be affected by selection acting at the other locus. The allele at locus *A* which is most commonly associated with the favourable genotype at the other locus will increase in frequency. We say that **hitch-hiking** occurs. An allele which itself gives no selective advantage increases in frequency, being carried by an allele under selection at another locus.

The process of hitch-hiking can be seen most easily for a new beneficial mutation (say long legs) which arises next to the gene for toe hairiness. The toe-hairiness locus has two alleles, hairy and smooth. The long-leg mutation occurs on a chromosome bearing the smooth toe allele. Smooth and long-leg will be in complete linkage disequilibrium. Long-legged individuals have higher fitness, so the long-leg mutation increases in frequency and as it does so, so does the smooth-toe allele. However, at each generation, there is the possibility of meiotic recombination, breaking down the association. The strength of the hitch-hiking effect decreases with time. The rate at which it does so depends upon the tightness of the linkage between the loci.

The effect of hitch-hiking increases with increasing selective advantage of the novel allele, and with the period of time over which disequilibrium is maintained. The effect will be greatest for loci which are tightly linked, and in circumstances where non-random mating continually regenerates the association.

Pleiotropic effects of genes

When selecting upon one characteristic of an organism, it is common to observe that other aspects change also. For instance, when Dmitry Belyaev (1979) applied artificial selection for tameness to a captive silver fox population, he found that after 18 generations they had reduced aggression and increased the tameness of his foxes. However, the foxes had changed in other ways. They wagged their tails, barked, and the females had abnormal ovulation patterns. How could this occur? One possibility is that the genes for these characteristics were in linkage disequilibrium, as described above, so selection acting on one changed them all. However, this is unlikely to be the case. What is more likely is that selection for tameness translated into selection upon the titre of hormones in the blood, and this had multiple effects on the phenotypes produced. These aspects of phenotype respond to selection at the same time because they are all the product of the same gene(s).

In cases like this, we say that the genes have **pleiotropic** effects. Whenever two aspects of phenotype are genetically correlated, there are two possible explanations. First, they may be encoded by separate loci which are in disequilibrium. Alternatively, they may be encoded at one locus with multiple effects.

An appreciation of pleiotropy is important for an evolutionary biologist. Morphological and behavioural traits can change not only because of selection acting directly upon them, but also because of correlation with selectively advantageous features. We should not see each difference between species and each change in a species as adaptive without first enquiring as to whether change could be merely a result of correlation with another feature of phenotype under selection.

Further to this, pleiotropic effects can alter the response of a population to selection. When traits are genetically correlated, they do not respond individually to selection, however desirable that might be for the organism. Rather, selection on a character will only occur if change brings benefit to the organism as a whole.

Summary

In this chapter we have attempted to review the principal processes of evolutionary change. Mutations provide the raw materials on which a combination of selection and drift can then act. Random drift is the force which determines the fate of most mutations, the frequencies of which will move slowly up or down, eventually becoming fixed or lost. Selection acts as a positive vetting agent, barring harmful changes but producing the deterministic spread of some beneficial ones. The action and direction of selection depend upon the fitness of the different genotypes at a locus. Where genes at different loci interact to produce fitness, then the speed or direction of selection can be complex.

Suggested reading

ARNOLD, S. J. and WADE, M. J. (1984). On the measurement of natural and sexual selection theory. *Evolution*, **38**, 709–779.

A clear and concise treatment.

ENDLER, J. A. (1986). *Natural selection in the wild*. Princeton University Press.

An excellent critical review of evidence of natural selection in natural populations.

HARTL, D. L. (1988). *A primer of population genetics*. Sinauer, Sunderland, Massachusetts.

A lucid, yet gentle introduction to the basic maths behind population genetics.

MAYNARD SMITH, J. (1989). *Evolutionary genetics*. Oxford University Press.

Covers most population genetic concepts in a clear and reasonably accessible manner.

PRICE, T. and LANGEN, T. (1992). Evolution of correlated characters. *Trends in Ecology and Evolution*, **7**, 307–310.

Detailed and precise.

SPIESS, E. B. (1989). *Genes in populations*, 2nd edn. John Wiley, New York.

Gives the details of the nuts and bolts of theoretical population genetics.

Chapter 4

Selectionism, mutationism and neutralism

> The students of adaptation forget that even on the strictest application of the theory of Selection it is unnecessary to suppose that every part of an animal has, and every thing which it does, is useful and for its good. We, animals, live not only by virtue of, but also in spite of what we are.
>
> (Bateson, 1894)

Introduction

In this statement, Bateson provides a reveille of one of the most enduring and, at times, heated, biological controversies of the twentieth century. This controversy concerns the relative importance of the two principal processes of evolution: selection and random genetic drift. At the genetic level, the essence of this controversy may be couched as a question. Which evolutionary changes are the result of selection? The corollary is that those evolutionary events not driven by selection must have occurred through genetic drift. This controversy has had three phases: the classical and the balanced views of the genome favoured by **mutationists** and **selectionists** respectively; the selection versus random genetic drift of the 1930s, 1940s and 1950s; and the neutral mutation challenge to selection of the 1960s and 1970s. This chapter discusses how these conflicting views arose and became modified in the light of empirical evidence.

The historical perspective is worth considering both from the point of view of the evolutionary issues covered, and also for the insights it provides into the way theoretical science advances. For the latter reason, the issues will be considered chronologically, noting not only how the debate advanced, but pinpointing the theoretical advances and some of the experimental results that impinged significantly upon the debates. The controversies have been very stimulating and fruitful, with great wealths of empirical evidence being accumulated to try to resolve them. Here it is only possible to describe some of the many studies that have impinged on these debates. Others may be consulted by reference to the suggested reading at the end of the chapter.

The Mendelian challenge to Darwinism

Few students of evolution now realise that the rediscovery of Mendel's work was viewed initially as a major challenge to Darwinism. Over the last half century, the prominence of the **Neo-Darwinian synthesis**, with its meld of Mendelian genetics and Darwinian selection, has eclipsed the fact that, from the turn of the century, evolutionary biologists divided broadly into two schools, the Darwinian selectionists and the mutationists. The selectionist school took the view that many mutations produced only very slight effects on the phenotype, and so gave rise to heritable, continuous variation. It was the action of selection upon this variation that was the primary mechanism of evolutionary change. Since the process of evolution here occurs through many almost imperceptible changes, this school is often referred to as the **gradualists**. Mutationists took an opposing stance. They asserted that small continuous variations were simply background noise and had little part to play in evolution. Rather adaptations (and species) arose as the product of single mutations with large phenotypic effects. Indeed, as late as 1927, Punnett (1927) asserted that continuous variation was non-heritable fluctuation which had no part to play in evolution. At the core of this controversy lay the nature of the creative forces of biological life. Which process – mutation or selection – is the prime mover, leading to the production of new adaptations or species?

The division between these two schools of thought led to opposing views of the genome. The way in which early geneticists envisaged the genome was that it comprised genes, most of which were monomorphic. Mutations that arose were either deleterious and so were selected out of the population, or they were beneficial and spread rapidly to fixation (Punnett, 1915). Rare heterozygote genotypes were viewed as examples of transition state loci in which new mutations were replacing old ones. This became known as the mutationist, or **classical view of the genome**. An alternative model, the selectionist, or **balanced view of the genome**, entailed a high proportion of loci being polymorphic, the various alleles being maintained in a population by a balance of opposing advantages and disadvantages.

Polymorphic Batesian mimicry

In 1908, E. B. Poulton (1908) introduced an example that was to remain at the centre of this evolutionary debate for more than two decades. It was used as evidence to support both sides. The example was one that Darwin had considered one of his own best examples of evolution by natural selection: the mimetic resemblance of a defenceless insect to a toxic, distasteful, or otherwise well-protected insect. This type of resemblance is termed Batesian mimicry, after the Victorian scientist H. W. Bates, who described many such similarities (Bates, 1862). Poulton (1908) questioned how the chance and random process of mutation could, in effect, make the wings of one butterfly look like those of another.

The finding, by Punnett and others, that the different forms of polymorphic Batesian mimics were apparently controlled by different alleles of single genes, provided strong support for the mutationist's viewpoint. In a number of species of swallow-tail butterflies (e.g. *Papilio polytes*, *Papilio memnon* and *Papilio dardanus*), completely different mimetic forms, occurring in the same population, would freely interbreed, and yet produce no intermediate forms (Leigh, 1904; Poulton, 1906, 1908; Jacobson, 1909; Baur, 1911; Lamborn, 1912; Leigh, 1912; Carpenter, 1913; Fryer, 1913; Punnett, 1915; Swynnerton, 1919). Despite unquestionably large phenotypic differences, these forms apparently differed from each other in respect of just a single mutation.

Punnett (1915) concluded that these strikingly accurate mimetic forms must have come into existence in a single step through mutation. Punnett did not deny a role for natural selection. He saw it as the force which increased the frequency of the most fit form at the expense of other forms. However, he did not see natural selection as playing any part in the formation of the mimetic resemblance. The existence of more than one mimetic form in some populations (e.g. *Papilio polytes*, in Ceylon (Sri Lanka)) was difficult for Punnett to explain, for the expectation was that one of the forms would be at least marginally fitter than all others, and so would spread to fixation. He was forced to conclude that predators had developed the ability to distinguish between the mimics and their models, and that the different forms had become selectively neutral.

Punnett's difficulties with the existence of apparently stable polymorphisms in *P. polytes* and other species did not diminish the challenge that the single gene control over mimetic resemblance made to the selectionist lobby. For gradualists, the two empirical observations, that a perfect Batesian mimetic resemblance often involved several quite different features (wing colours, body colours, distribution of pigments, shape of wings, flight pattern, resting behaviour and oviposition behaviour) and that all the differences between forms were controlled by a pair of alleles of a single gene, must have appeared irreconcilable. However, in 1927, Sir Ronald Fisher (1927) managed to reconcile the problem.

Fisher (1918) had already shown that continuous variation could be the product of the cumulative effect of mutations of many Mendelian genes, which individually had a small phenotypic effect. To explain the evolution of Batesian mimicry by a series of small evolutionary steps he proposed that the phenotypic expression of genes could be modified by the action of other genes. A mutation that arose which gave a non-mimetic species an imperfect resemblance to a potential model would spread because this resemblance would fool some predators. Furthermore, while the mimic was rare, compared to the model, the predators that were duped would be unlikely to evolve the ability to learn to distinguish mimic and model for the selective pressure to do so would be so slight. The imperfect mimetic form would thus become established in the population. The expression of this new imperfect mimetic form would then vary due to the action of **modifier genes**. Selection would favour those modifier genes that produced an increase in the accuracy of the mimicry (Fisher, 1927).

The system Fisher proposed would have genetic constraints. For example, new modifier genes that improved the mimicry would spread only if they were expressed when the initial mutation was expressed. This restriction is necessary for it must be assumed that the phenotypes of other forms of the species, whether mimetic or non-mimetic, have evolved through natural selection. Consequently, it is unlikely that a modifier which improved the mimicry of the novel form would also have a beneficial effect on the expression of other forms. It is far more likely that any influence that such modifiers had on the expression of these older forms would be maladaptive, and, as these older forms would be commoner than the new mimic, selection would act against modifiers whose action was not limited to those in which the new allele was expressed.

Two mechanisms were proposed which could lead to this limitation of expression. First, the initial mutation could act as a **switch gene**, turning on the expression of a series of specific modifiers. Just as *Sry*, the mammalian sex-determining gene, makes individuals of otherwise similar genetic background develop very differently (either into males or into females), so a switch gene could alter the phenotype of otherwise genetically similar *Papilio* individuals. Alternatively, the genes controlling the various components of the mimicry could become tightly linked on the same chromosome (so comprising a super-gene), such that recombination between the loci involved would occur only very rarely, and the genes would usually be inherited together.

Empirical evidence for both of these mechanisms has been obtained independently in different species of swallow-tail butterfly. In *P. memnon*, genetic analysis and the finding of rare recombinants has shown that the most influential gene controlling alternative forms is a super-gene consisting of at least six loci (Clarke *et al.*, 1968; Clarke and Sheppard, 1971). In *P. dardanus*, the great range of mimetic female forms, up to a dozen different mimics in some populations, are controlled by a multiple allelic series at the *H* locus (Clarke and Sheppard, 1959a,b, 1960a,b, 1962; Clarke, 1963). Each allele of the *H* locus switches on a different set of specific modifiers, each set being fixed in a particular population. Good mimicry is thus dependent on the original mutation in combination with the selected genetic background of modifiers.

On the assumption that a switch mechanism operates in *P. dardanus*, Ford (1937, 1953) predicted that different modifiers would have been selected to fixation in different parts of Africa. He deduced further that perfect mimetic resemblances could be broken if mimics from different regions were crossed. These predictions were tested and fulfilled as shown in Box 4.1.

Polymorphic Batesian mimicry has become a paradigmatic example of Darwinian selection for three reasons. First, it played a major role in the first mutationist–selectionist controversy. Second, for evolutionary biologists, the

Box 4.1: Evidence that modifiers affect the expression of the main mimicry gene (*H* locus) in *Papilio dardanus*

Clarke and Sheppard verified Fisher's switch gene modifier system for the evolution of mimetic polymorphism by crossing forms of *Papilio dardanus* from different populations (Clarke and Sheppard, 1959a,b, 1960a,b, 1962; Sheppard and Cook, 1962; Clarke, 1963).

One series of tests involved the forms *hippocoon*, *hippocoonides* and *cenea*. The forms *hippocoon* and *hippocoonides* mimic two geographic races of the butterfly *Amauris niavius*, f. *niavius* and f. *dominicanus*, respectively. These differ in respect of a number of traits, including the size of white marks on the forewings and hindwings, the width of the dark border, the extent of white spots on this border, and the presence or absence of dark rays in the white basal area. These two forms are not found together, *hippocoon* being West African, while *hippocoonides* occurs in East and South Africa. The *cenea* form mimics *Amauris echeria* and occurs sympatrically with f. *hippocoonides*, but not with f. *hippocoon*.

By crossing South African *hippocoonides* females with males from a West African *hippocoon* stock, Clarke and Sheppard were able to demonstrate that these forms are allelically the same with respect to the main mimicry *H* locus of *P. dardanus*, being homozygous for the allele *h*, and that the differences between them are due to a small number of modifiers which segregated independently in the F2. They then crossed South African *cenea* with each of these forms. When crossed to *hippocoonides*, the F1 indicated *cenea* to be controlled by a unifactorial dominant (H^C), the F2 producing the expected 3:1 segregation ratio of *cenea* : *hippocoonides* with no intermediates. Conversely, when crossed to West African *hippocoon*, the F1 were variable and intermediate although more similar to *cenea* than to *hippocoon*. Furthermore, the F2 showed a great range of variation with no clear-cut segregation. These crosses are summarised below.

The crosses between these three forms of *P. dardanus* lead to the obvious deduction that both those modifiers that cause the H^C allele to be dominant to the *h* allele and the modifiers that fine tune the resemblance of the various forms of this species to its models are variable between populations. The *cenea* allele is dominant to the *hippocoonides/hippocoon* allele, but only if the correct dominance modifiers are present in the population (see p. 285). In West Africa they are not.

continued

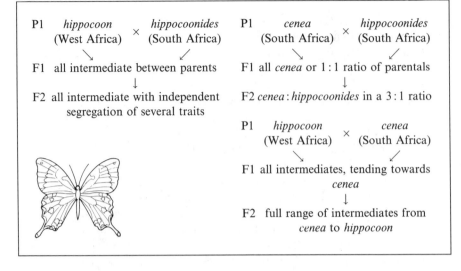

P1   *hippocoon*   ×   *hippocoonides*
     (West Africa)   (South Africa)

F1   all intermediate between parents

F2   all intermediate with independent
     segregation of several traits

P1   *cenea*   ×   *hippocoonides*
     (South Africa)   (South Africa)

F1   all *cenea* or 1 : 1 ratio of parentals

F2   *cenea* : *hippocoonides* in a 3 : 1 ratio

P1   *hippocoon*   ×   *cenea*
     (West Africa)   (South Africa)

F1   all intermediates, tending towards
     cenea

F2   full range of intermediates from
     cenea to *hippocoon*

relationship between mimics and their models provided an almost unique view of the optimum outcome of an evolutionary process. The optimum phenotype for a mimic is to resemble the model precisely. Third, conspicuous polymorphisms, such as those seen in some Batesian mimetic butterflies, were easily accessible, and yet were difficult to explain from an evolutionary standpoint.

The maintenance of conspicuous polymorphisms

Conspicuous polymorphisms played a central role in the second phase of the selection versus random genetic drift arguments of the 1930s, 1940s and 1950s. These arguments are often referred to as the Fisher–Wright controversy after the two main protagonists, Sir Ronald Fisher and Professor Sewall Wright.

It is notable that Punnett (1915) resorted to a neutralist explanation to resolve his difficulties in understanding the existence of different mimetic forms being maintained in the same population. Such recourse was characteristic of the general view of polymorphism in the early part of this century, and indeed before. Conspicuous morphological polymorphisms were a quandary for the early Neo-Darwinists, because it was difficult to envisage how two or more forms could be maintained in a population unless they were selectively neutral. Indeed, when Darwin defined the concept of natural selection in *The Origin of Species*, he specifically disassociated the concept from natural polymorphisms:

This preservation of favourable variations and the rejection of injurious variations I call natural selection. Variations neither useful nor injurious would not be affected by natural selection, and would be left a fluctuating element, as perhaps we see in the species we call polymorphic.

Arguments over the type of evolutionary mechanisms that might be responsible for the existence of the obvious polymorphisms have continued from Darwin's time to the present day. Up until the early 1940s, most scientists who studied evolution believed conspicuous polymorphisms had to be selectively neutral. Variations in the frequencies of forms over time were therefore attributed to random genetic drift (see Chapter 3).

Both Fisher and Ford disagreed strongly with this interpretation of polymorphism. Fisher (1922) had already shown mathematically that two or more alleles of a species could be maintained in a population by selection. At equilibrium, the selective forces for and against each allele are balanced precisely so that each type of allele has the same overall fitness. Fisher proposed two basic types of mechanism that would maintain polymorphism. First, heterozygote advantage, whereby a fitness advantage of a heterozygote over its respective homozygotes leads to both the alleles carried by the heterozygote being maintained in the population (see Chapter 3). Second, negative frequency-dependent selection, in which the fitness of a form is inversely correlated to its frequency. In other words, if the frequency of a form declines, its fitness increases (see p. 66).

It was generally assumed by early evolutionary biologists that selection is a very weak force because evolution appeared to them to be so slow. However, Ford took a different view. Having considered Fisher's mathematical analysis of polymorphism, Ford speculated that change might be slow, not because selective forces are weak, but because they are strong and balanced. Polymorphisms maintained by balancing selection would change only very slowly as the overall relative fitnesses of the forms changed in response to alterations in the balance of selective advantages and disadvantages. Ford then came to a remarkable and surprising deduction. He saw that the most temporally stable and geographically widespread polymorphisms would be those that would be the subject of the strongest balancing selection. In such cases, the frequencies of genotypes and phenotypes would be maintained at stable equilibrium points, so that they could deviate only slightly from these frequencies before the balancing selection brought them back. The way in which the most widespread type of stable polymorphism, the occurrence of two distinct sexes, is usually maintained illustrates the power of balancing selection beautifully.

Selection maintains the sex ratio

In most species in which sex is determined genetically, populations contain approximately equal numbers of males and females. Why should this be so? One answer is that in most species sex is determined by the segregation of X and Y chromosomes during meiosis in the **heterogametic sex**, and that this will generate a 1:1 ratio of X- and Y-bearing gametes. This is true as far as it goes.

However, mechanisms for producing other sex ratios do exist and, in some cases, appear to be stable (see pp. 158 and 172).

An explanation based on the stability of the common 1:1 ratio would be more convincing and again we turn to Fisher (1930) for an explanation. Assume that the sex ratio of the next generation is determined by genes acting in the parental generation. This might result from genes acting in the hetero-gametic sex to alter the ratio of male- and female-determining gametes produced, or from genes acting in the **homogametic sex** to alter the success of the two types of gamete in fertilisation. Consider what happens if a species' sex ratio is displaced from parity, either through some chance, sex-biased mortality, or through the action of a sex ratio modifier gene. Say, there are now fewer males than females. In forming the next generation, each zygote will have one mother and one father. It follows that the average contribution made by each male is now greater than the average made by each female. In other words, males are now fitter than females and will be selected for. Furthermore, the bigger the skew in the sex ratio, the larger will be the selective differential between the sexes. Only when the sex ratio returns to parity will the restoring force disappear. In this way, we can see that the evolutionary stable equilibrium point is to have equal numbers of males and females, for it is only at this ratio that the reproductive value of a son is precisely equivalent to that of a daughter.

It was Fisher and Ford's adherence to the view that selection maintained polymorphism that placed them in opposition to those who considered polymorphisms to be non-adaptive, and particularly to Sewall Wright from Chicago, one of the chief advocates of random genetic drift. As this debate fired up, Fisher and Ford began to collect evidence to resolve the controversy. Their work on the scarlet tiger-moth (*Callimorpha (Panaxia) dominula*) (Fisher and Ford, 1947), and later, Ford's work on the meadow brown butterfly (*Maniola jurtina*) (see Ford (1964) for a review), Cain and Sheppard's work on the banded land snails *Cepaea nemoralis* and *C. hortensis* (Cain and Sheppard, 1950, 1952, 1954; Sheppard, 1952a) and Clarke and Sheppard's work on *Papilio dardanus* and other polymorphic Batesian mimics (see p. 80), were conducted specifically with this aim in mind. These studies now form the classic platform for many of our views on the evolution and maintenance of conspicuous **genetic polymorphisms** (see Ford (1975) for review).

Polymorphism in the scarlet tiger-moth

The scarlet-tiger is a colonial, day-flying moth that, in Britain, inhabits marsh-land and hedgerow habitats. In one British colony, at Cothill in Berkshire, it is naturally polymorphic. Three forms occur, all controlled by two alleles of a single locus with no dominance. The normal form, *dominula*, and the rarest form, *bimacula*, are homozygotes while the heterozygote, *medionigra*, is phenotypically intermediate between these. (It should be noted that in the literature, the names given to the two alleles are usually *dominula* and *medionigra*, despite the *medionigra* form being the heterozygote.) By examining old museum collections, Ford established that the maximum frequency of the *medionigra* allele prior to 1929 was in the region of 0.024. In both 1936 and 1938, Ford visited the colony and found several specimens of the *medionigra* form, concluding that this form had increased in frequency (Ford, 1964).

Fisher and Ford (1947) surveyed this colony from 1939 to 1946, in the initial two years simply by direct observation, and from 1941, using multiple mark–release–recapture techniques. The data they collected allowed them to estimate the frequency of the two alleles, the average survival rate, and the population size (allowing for recruitment into the adult population at the beginning of the flight season and the natural decline in adult population towards the end of it). The estimates obtained for these parameters (Box 4.2) were the first on which the alternative explanations of polymorphism, selection or drift could be tested.

Box 4.2a: the colour pattern polymorphism in the scarlet tiger-moth at Cothill

The study of the colour pattern polymorphism in the Cothill colony of the scarlet tiger-moth, from 1939 until 1946, provided data on both the changes in the frequencies of the alleles, and the population size, over time. These data allowed the first assessment, based on empirical evidence, of the relative roles of selection and drift on a conspicuous polymorphism. Details of captures of the three forms of the scarlet tiger-moth at Cothill, Berkshire, over this period are given in the table below. The population size estimates from multiple mark–release–recapture data, and frequency of the *medionigra* allele are also given.

Fisher and Ford (1947) argued that the changes in allelic frequencies were too great to be accounted for by drift, given the estimates of population size obtained.

Year	Captures			Total	Population size	Frequency of *medionigra* allele
	dominula	*medionigra*	*bimacula*			
1939	184	37	2	223	?*	0.092
1940	92	24	1	117	?*	0.111
1941	400	59	2	461	2000–2500·	0.068
1942	183	22	0	205	1000	0.056
1943	239	30	0	269	1000	0.056
1944	452	43	1	496	5000–6000	0.045
1945	326	44	2	372	4000	0.065
1946	905	78	3	986	6000–8000	0.043

(From Fisher and Ford, 1947).
*Although no population estimates were made for these years, Ford notes the moths to have been flying in good numbers, and Fisher and Ford assumed the populations to have been 1000.

Box 4.2b: Experimental support for the hypothesis that balancing selection maintains polymorphism in the scarlet tiger-moth

Despite Fisher and Ford's argument that changes in the frequencies of the forms recorded between 1939 and 1946 were too great to be accounted for by drift, and thus must be the result of selection, they did not investigate the selective factors involved. Furthermore, their interpretation was controversial, being challenged in particular by Wright (1948). To resolve the controversy, Philip Sheppard manipulated the frequencies of the alleles controlling the polymorphism in two colonies. In one he raised the frequency of the rarer allele, *medionigra*, to an abnormally high level. In the other, he introduced this allele at a very low level. His rationale was that if the polymorphism was maintained by selection, the frequencies of *medionigra* in these two colonies would converge.

In 1951, at Hinksey, near Oxford, Sheppard set up an artificial colony of the scarlet tiger-moth in a habitat that appeared suitable for the moth, but from which the moth had not previously been reported. The colony was founded by releasing 4000 eggs from crosses between the homozygote *dominula* and the heterozygote *medionigra*. The initial frequency of the so-called *medionigra* allele in this population was thus 0.25. In 1952, 30 larvae were collected, these producing 21 *dominula* and nine *medionigra* adults, giving a frequency for the *medionigra* allele of 0.15. By 1960, the colony was well established, with adults flying in good numbers both in 1960 and 1961. The frequency of the *medionigra* allele, based on substantial adult samples, had declined further from the initial release level to 0.0625 in 1960 and 0.0731 in 1961 (see Table a).

Table a: Phenotype frequencies, and the frequency of the *medionigra* allele from the artificial colony at Hinksey

Year	dominula	medionigra	bimacula	%m allele
1951	2000 eggs	2000 eggs	–	25.00
1952	21 larvae	9 larvae		15.00
1960	269	32	3	6.25
1961	217	35	1	7.31

In 1954, Sheppard introduced the *medionigra* allele into an established natural colony of the scarlet tiger-moth at Sheepstead Hurst, over a mile from the Cothill colony. Here the *medionigra* and *bimacula* forms do not occur, except, presumably, as rare mutants (of 11 102 individuals caught between 1949 and 1954, all were of the *dominula* form). To introduce the *medionigra* allele into this colony, Sheppard scattered eggs from 50

continued

crosses, between f. *dominula* males and f. *medionigra* females, in suitable situations. The following year, two of a sample of 875 moths were of the *medionigra* form. Samples of adults caught at this colony on a number of occasions over the next 15 years (Table b) show that the frequency of the *medionigra* allele increased to around 0.02.

Table b: Phenotype frequencies, and the frequency of the *medionigra* allele from the artificial colony at Sheepstead Hurst

Year	dominula	medionigra	bimacula	%m allele
1949–53	all	–	–	0
1954	introduction of *medionigra* allele			
1955	873	2	0	0.11
1960	398	9	0	1.11
1961	405	9	0	1.09
1964	739	13	0	0.86
1969	509	25	0	2.34

In both of these colonies the frequency of the *medionigra* gene has moved towards the level at which it is maintained in the Cothill colony,  the convergence in the Hinksey colony being from a frequency in excess of the Cothill equilibrium frequency, and, in the Sheepstead Hurst colony, from a frequency below equilibrium.

Fisher (1930) had already shown that it is possible to calculate the extent to which, for a given population size, random genetic drift could alter allele frequencies between generations, in a population of given size. Using this method, and an assumed breeding population of 1000 adults each year from 1939 to 1946, Fisher and Ford (1947) demonstrated that the changes in the frequency of the *medionigra* allele were too large to be attributable to genetic drift alone. In their conclusion they wrote:

Thus our analysis, the first in which the relative parts played by random survival and selection in a wild population can be tested, does not support the view that chance fluctuations in gene ratios, such as may occur in very small isolated populations, can be of any significance in evolution.

There followed criticisms of this work and the interpretation of the data obtained (Wright, 1948), and defensive counter criticisms (Fisher and Ford, 1950), such that the so-called Fisher–Wright controversy became one of the most acrimonious and infamous biological debates in the twentieth century.

Subsequent work on the scarlet tiger-moth has vindicated the Fisher–Ford view, with two opposing selective factors being identified. On the one hand, experiments were designed which allowed the relative fitnesses of the different genotypes to be assessed. Young larvae of known genotype were marked radioactively (by feeding larvae on comfrey containing S^{35}) and released, later assaying samples of final instar larvae and adults, using a Geiger counter. The conclusion reached was that the *dominula* form showed a strong survival advantage over the *medionigra* form (Sheppard and Cook, 1962). On the other hand, mating choice experiments showed that the forms mate disassortatively. Individuals prefer to choose mating partners of any genotype other than their own (Sheppard, 1952b). Such a mate preference system produces selection that is inherently negatively frequency-dependent, conferring an advantage on rarer genotypes. In the Cothill colony, this selection is strong enough to maintain the *medionigra* and *bimacula* forms, despite their innate survival disadvantage.

Further support for the balancing selection explanation came from Sheppard's work using artificial and manipulated colonies of the scarlet tiger-moth, as detailed in Box 4.2b.

Hind-wing spot numbers in the meadow brown butterfly

Fisher and Ford's 1947 paper on *Panaxia dominula* reported only the first of a series of studies relating to the selection–drift argument. In the late 1940s, Ford and colleagues began an extensive study of the meadow brown butterfly (*Maniola jurtina*) on the Isles of Scilly off the south-west coast of England. The underside of the hind wings of meadow brown butterflies has a pattern of between zero and six small black spots. Data on the hind-wing spot number distributions of populations on a number of the Isles of Scilly were collected. These distributions, particularly in females, have been found to vary between populations throughout Britain and Europe. Figure 4.1 shows the female hind-wing spot number distributions for several of the islands. Three of the islands (St Mary's, Tresco and St Martin) are relatively large, having areas in excess of 275 hectares. Five other islands are much smaller with areas of less than 16 hectares. It is apparent, from the normal female spot number distributions, that those of the three large islands are rather similar, while the distributions for the small islands (e.g. Tean and White Island) show considerable variation.

Proponents of non-adaptive polymorphism claimed that the meadow brown data supported their viewpoint (Dobzhansky and Pavlovsky, 1957; Waddington, 1957). The population sizes of meadow browns on the large islands are much greater than on the small islands. Therefore, it seemed reasonable to propose that the diversity of spot number distributions on the smaller islands was the outcome of random genetic drift or founder effects. This would be the case particularly if, as Waddington (1957) argued, the populations on the small islands, from time to time, experienced population bottlenecks (see Chapter 3), or actually became extinct and were refounded by a limited number of butterflies. Then, genetic drift could significantly affect the genetic constitution of the local races derived from them. On the larger islands, Waddington argued that the minimum population size would be large enough to buffer the

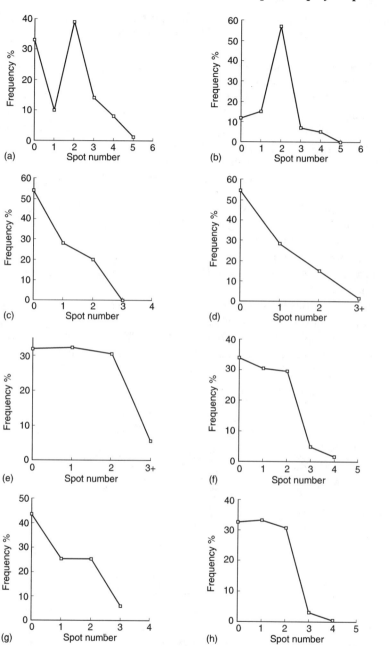

Fig. 4.1 Female spot number distributions in the meadow brown butterfly, *Maniola jurtina*, on the Isles of Scilly (after Ford, 1964). (a) Tean, pre-1951. (b) Tean, western area, 1954. (c) White Island, whole island, pre-1958. (d) White Island, northern population, 1958. (e) White Island, southern population, 1958. (f) St Martin's, 1955 + 1956. (g) St Martin's, 1957. (h) St Martin's, 1958.

gene pool against such effects. Consequently, the similarity of spot number distributions between the larger islands was simply a reflection of the original Scillonian race.

Work on the effect of founding population size in cultures of *Drosophila pseudoobscura* lent further support to the non-adaptive thesis. In 1957, Dobzhansky and Pavlovsky set up 20 cultures of F2 hybrids between Texan and Californian flies which differed in respect of a chromosome 3 inversion (Dobzhansky and Pavlovsky, 1957). Half of the cultures were founded by 20 flies, the others by 4000 flies. After 18 months the frequency of the Texan chromosome type had declined to between 20 and 35 per cent in the cultures derived from large founding populations, and to between 16 and 47 per cent in those derived from a small number of founders (Fig. 4.2). This result was consistent with drift theory, where the genetic variance in a population derived from a small number of individuals is greater than when a similar population is derived from many.

Ford (1964) later demonstrated that the meadow brown data could also be explained in terms of natural selection. He proposed an explanation based on strong selection, which, he argued, was consistent with both the variation in the spot number distributions between the small islands, and the uniformity of spot number distributions on the large islands. Ford (1964) cites many pieces of evidence to support his contention. Here it will be sufficient to give just three examples to illustrate the central theme of his selectionist explanation. These are described in detail in Box 4.3.

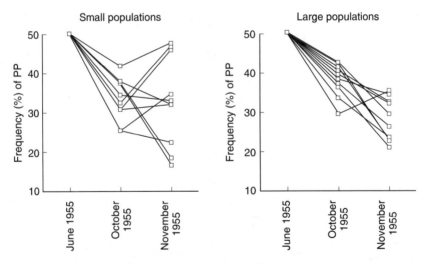

Fig. 4.2. The effect of population size on changes in frequency of chromosomal inversion polymorphism in *Drosophila pseudoobscura*. The percentage frequencies of Pike's Peak inversion chromosomes in 20 replicate experimental populations. Stocks were initially heterozygous for the PP and the Arrowhead inversions. In the ten 'small populations' each generation was founded by 20 flies. In the ten 'large populations' each generation was founded by 4000 flies. (After Dobzhansky and Pavlovsky, 1957.)

Box 4.3: Examples of evidence used by Ford to support his contention that selection, rather than drift, is the primary determinant of hind-wing spot numbers in the meadow brown butterfly (*Maniola jurtina*) on the Isles of Scilly

The number of small spots on the underside hind-wing of the meadow brown butterfly is both temporally and spatially variable. On the Isles of Scilly, the female spot number distributions vary between small islands but tend to be similar between large islands. These findings were used to support the view that the variation was the result of drift. Ford disagreed with this view, presenting a wealth of data to support an alternative explanation based on selection. Three of his data sets are presented here as examples.

1. On Tean, one of the smaller of the Isles of Scilly, cattle were grazed regularly until 1950. The habitat of the island comprised three areas of long grass and scrub, divided by belts of short, heavily grazed grass. These belts were unsuitable habitats for the meadow brown, and it has been shown that a band of as little as 100 metres of such terrain can act as an effective barrier to migration between populations. The population on Tean was therefore divided into three sub-populations, each to a degree isolated from the other two. Prior to 1951, the female spot number distributions for these three populations were all bimodal with most females having two spots, and many of the rest no spots (Fig. 4.1a).

    The removal of the cattle in 1950 led to a dramatic change in the vegetation on the island. By 1953, lack of grazing had made the whole island suitable for the butterfly. By 1954, increases of gorse, bramble and bracken in the central long grass and scrub area made this belt unsuitable, the butterflies being split into two populations at either end of the island. The south-eastern population retained its bimodal spot number distribution in 1954. However, at the western end, the population had become two-spotted unimodal (Fig. 4.1b), and was significantly different from samples taken in the same region in the previous year. To produce a change of the magnitude seen, the selection against zero-spotted individuals has been estimated to have been 64 per cent. Subsequently, this situation stabilised and persisted.

    Ford (1964) proposed that the elimination of zero-spotted females was much too great to be accounted for by genetic drift. Rather he suggested a selectionist explanation, arguing that the population at the south-western end of the island retained its bimodal distribution

continued

because the habitat in this area became similar to the three original areas inhabited by meadow browns prior to 1953, that is to say long grass and scrub. The western population, however, inhabited an area that was primarily long-grass with little scrub except at the extreme western tip of the island. This western habitat was appreciably less suitable for zero-spotted females.

2. White Island consists of two ridges of high ground running at right angles to one another and connected by a low isthmus. The two areas of high ground have different microclimates due to their different aspects. Prior to 1958, the population of meadow browns on this island was continuous, with most females having no spots (Fig. 4.1c). During the winter of 1957/58, gales drove the sea over the isthmus, destroying the vegetation and effectively isolating the meadow brown populations on the two ridges. The two populations diverged in respect of their spot number distributions, the northern population remaining zero-spot unimodal (Fig. 4.1d), and the southern population becoming 'flat-topped' with equal numbers of zero-, one- and two-spotted females (Fig. 4.1e) in 1958. The populations retained these distributions for a decade, but, by 1970, the vegetation across the isthmus had regenerated sufficiently to support meadow browns, so the population had again become continuous. The spot number distribution over the whole island then reverted to zero-spotted unimodal. The magnitude of the changes observed is difficult to explain except in terms of a selection-driven adaptive response to alterations in the habitats of the populations.

3. On the large islands, Tresco, St Martin's and St Mary's, changes in habitat may also lead to rapid changes in spotting. Prior to 1957, the female spot number distributions on these three islands were flat-topped (Fig. 4.1f). During the first half of 1957, the Isles of Scilly suffered a severe drought, which affected adversely the grass crop on which overwintering meadow brown larvae feed. In the summer flight period of 1957, females on both Tresco and St Martin's had zero-spotted unimodal distributions (Fig. 4.1g). By 1959, when the vegetation had recovered, the distributions on these islands reverted to the flat-topped distributions characteristic of the larger islands (Fig. 4.1h). Strong selection pressures are necessary to produce the observed changes. For example, on Tresco, to produce the unimodal zero-spotted distribution of 1957 from the 1956 flat-topped distribution would require the selective elimination of 60 per cent of higher spotted females.

Ford cites two instances in which division of a continuous population on a small island led to divergence between the new subpopulations. His explanation of this divergence was that the division had allowed each subpopulation to become adapted to its local conditions. This adaptation was reflected in the female spot number distributions. The third set of data cited, showing extreme spot number changes in a single generation on two of the large islands, completely undermines the intermittent drift and founder explanations of the original data, for these explanations are specifically based upon the permanence of the original spot number distributions on the large islands.

The essential difference between large and small islands in respect of the selectionist explanation is that the small islands comprise just one or two different habitats that are suitable for the meadow brown, while the large islands comprise a mosaic of a larger number of suitable habitats. On the small islands selection leads to the butterflies becoming adapted to locally specific conditions. In contrast, on the large islands, gene flow between regions limits the potential for adaptation to specific local conditions and the net result is a compromise, a single average condition.

Studies on other species showing conspicuous morphological variation have also tended to vindicate the selectionist viewpoint. Of particular note are the banded land snails, *Cepaea nemoralis* and *C. hortensis*. Variation in colours and banding patterns in these species were, for a long time, held up as examples of polymorphisms that were obviously non-adaptive. As Mayr (1942) wrote:

There is, however, considerable indirect evidence that most of the characters that are involved in polymorphism are completely neutral, as far as survival value is concerned. There is for example, no reason to believe that the presence or absence of a band on a snail shell would be a noticeable selective advantage or disadvantage.

And later:

Even more convincing proof for the selective neutrality of the alternating characters is evidenced by the constancy of the proportions of the different variants in one population. The most striking case is that of the snails *Cepaea nemoralis* and *C. hortensis*, in which Diver (1929) found that the proportions of the various forms from Pleistocene deposits agree closely with those in colonies living today.

Of course, Ford would interpret such long-term temporal constancy in the frequencies of variants as evidence of strong selection, the various forms being held at specific equilibrium frequencies by a balance of selective advantages and disadvantages. Subsequently, a number of selective factors have been shown to affect the frequency of both colour and banding variants in both species. These include the level of predation of forms in different habitats, both the crypsis of forms and their abundance having an influence (see for example, Cain and Sheppard, 1954), the climate in terms of temperature, sunshine levels and humidity (Cain and Currey, 1968; Jones, 1973, 1982; Cowie and Jones, 1985; Jones and Hunter, 1993), and the viability of different forms (Sedlmair, 1956).

The resurgence of neutralism

By the mid-1950s, largely on the back of evidence obtained by the Oxford group of ecological geneticists, led by Ford, the selectionists' explanation of conspicuous polymorphisms held sway. However, two sets of findings, one from a series of theoretical studies, the other a series of empirical observations, prompted a rebirth of the selectionist–neutralist argument.

Genetic loads

Central to the first of these is the realisation that natural selection involves death. Darwinian selection, in its starkest form, is not 'the survival of the fittest'; it is the death of the less or least fit. The proportion of the population that die each generation as a result of the action of selection on a genetic system is termed the **genetic load**. The concept of genetic load was introduced independently by J. B. S. Haldane (1937) and later by H. J. Muller (1950) in a discussion of the effect of deleterious mutations, resulting from radioactive fallout, on the human population.

In 1957, Haldane published a series of theoretical explorations of the number of selective deaths inherent in various types of evolutionary system on the assumption that selection was the primary evolutionary mechanism (Haldane, 1957). He considered three systems which would impose load on a population: the eradication of novel deleterious mutations (**mutational load**); the substitution of an old allele by a novel and more fit allele (**substitutional load**); and the loss of homozygotes in polymorphisms maintained by heterozygote advantage, whereby two or more alleles are maintained in a population because heterozygote genotypes are fitter than homozygotes (**segregational load**).

Haldane showed that, for a particular locus, the mutational load was simply twice the mutation rate for that locus. His calculation demonstrated that this load was independent of both the fitness of the new mutation (he assumed that the fitness of the new mutant would be less than one, i.e. the mutation is at least mildly deleterious, as is usually the case), and of the mating system. That the mutational load is independent of these factors is easily understood. Suppose a dominant lethal mutation occurs in a gamete: the zygote formed by that gamete and another dies, the result being the loss of two haploid genomes. But, if the mutation has a milder affect, reducing the fitness of its bearer by a lesser amount, the load of selective deaths is ultimately the same because a deleterious mutation will still be selected out of the population eventually.

In respect of substitutional load, Haldane pointed out that it takes a specific number of selective deaths to replace one allele by another. To take a well-known example, in 1849, a novel melanic form of the peppered moth (*Biston betularia*) was recorded in Manchester (Edelston, 1864). This form was at a selective advantage over the original white and black form of the moth in this highly polluted industrial city, and began to increase in frequency, replacing the non-melanic. Haldane demonstrated that the number of selective deaths necessary for complete replacement is dependent upon the initial frequency of the novel advantageous form and the size of the population through which it is spreading, but is independent of the relative fitnesses of the old and new forms. An allele which confers a relatively small advantage on its bearer will cause relatively few selective deaths each generation, but since it will take longer to reach fixation, selective deaths will occur for more generations. On the strength of his calculations, Haldane argued that this substitutional load imposed a severe restraint on the number of new alleles that can be fixed in a population per generation, and hence on the rate of evolution by natural selection.

The third type of load, segregational load, was a particular problem for advocates of the view that the most common way that polymorphisms are maintained is through heterozygote advantage. Each polymorphism of this type requires selective deaths at each generation, because unfit homozygotes will segregate out continually and be selected against. Many polymorphisms will require many deaths. For each locus the load depends on the relative fitnesses of heterozygotes and homozygotes. Unless the disadvantage to homo-zygotes, compared with heterozygotes, is extremely small, a severe constraint is imposed upon the number of polymorphic loci in a population that may be maintained by heterozygote advantage.

These results lead to one of three possible conclusions. First, the genome consists of many polymorphic loci, but the alternative alleles for the majority of these loci are selectively neutral to one another and are neither maintained at equilibrium frequencies, nor changing in frequency in a directed way, by selec-tion. Second, the genome is mainly monomorphic. The few genes that are polymorphic are the result of new deleterious mutations, or the spread of novel beneficial mutations, or the maintenance, by balancing selection, of stable polymorphisms. Third, the assumptions upon which the estimations of genetic loads are based may be untrue.

For the most part, the third option has been ignored, and the concept of a genetic load has been used to argue that balanced polymorphisms are rare. If they were common, the required number of selective deaths would be too great. Thus, whereas some polymorphism can be maintained by selection, over and above this it must be due to predominantly neutral alleles drifting either towards extinction or fixation.

Calculations about the magnitude of the genetic load suffer from a poor understanding about how loci interact in respect of fitness. For example, is it exactly twice as bad to carry two slightly deleterious mutations as it is to carry one? For the most part, we can only guess and apply simple models, assuming that the selective effects are either additive or multiplicative. However, the reality may be different. One argument is that heterozygotes are generally fitter than homozygotes. If this is so, one could imagine than the next generation is founded mainly by those individuals which show greatest heterozygosity. The flip side of this is that those animals which die will show greater than average homozygosity, and the population will therefore lose many deleterious alleles in one go. In this way, cleansing is made more efficient, and a larger load can be carried for the same cost. Of course, the difficulty of heterozygotes segregating out homozygotes remains.

In the context of genetic load arguments, it is important to remember that all aspects of the genome and its function are evolved states. Early models tended to concentrate on working out the magnitude of genetic loads given certain simple assumptions. However, the way in which any load manifests itself must itself be subject to selection. It follows that any modifiers which improve the efficacy of genomic cleansing are likely to be selected for and optimised. Such systems might involve changes to the breeding system such that heterozygosity is maximised, or changes to the genes themselves, such that new mutations are less detrimental. At a higher level, the whole cellular machinery could evolve raised fault tolerance. Fault tolerance is the ability to continue functioning with any given number of defects. By analogy with engineering systems, increased fault tolerance could improve greatly the efficiency of cleansing, because the greater the number of faults that can be tolerated, the larger will be the number of deleterious mutations that will be shed when the threshold is exceeded.

Are such models which mediate the effects of genetic loads realistic? Is it not reasonable to assume that the fitnesses of many genes – this one influencing fertility, that one conferring resistance or susceptibility to a pesticide – are, in fact, independent? Or is it more realistic to assume that say the most fit half of a population survive, while the other half die. Peter O'Donald argued that both types of model were equally unrealistic, and the true position would surely be somewhere between these two extremes (O'Donald, 1969). As yet, the situation is not resolved. However, molecular genetic analyses on long-living and early-dying individuals of a population could be used to assess whether animals that have low survival are more or less homozygous than average.

Data from gel electrophoresis

By the mid-1960s, **gel electrophoresis** had begun to yield substantial amounts of data that allowed variation in the immediate products of genes, proteins, to be studied. Gel electrophoresis separates protein molecules by size and charge, and consequently allows small changes in their amino acid sequences to be detected. Early results, such as those of Lewontin and Hubby (1966) on *Drosophila pseudoobscura*, revealed that there is much more variation in populations than anyone had supposed previously.

Many subsequent electrophoretic surveys have confirmed this finding. The proportion of genes that are polymorphic varies between taxonomic groupings, from about 15 per cent in mammals to over 40 per cent in *Drosophila*. The critical point is that in almost all species studied, there are a large number of polymorphic loci. The proposition that the genome consists almost exclusively of monomorphic loci must surely be rejected.

Given the existence of this huge wealth of genetic variation, it seems inconceivable that it is all maintained by selection, even allowing for factors that may mediate the effects of genetic loads. Yet, in the vast majority of studies that have quantified the relative fitnesses of different genetic forms of a species, differences in fitness were found.

The neutral mutation theory of molecular evolution

Mitoo Kimura (1968) added to the selection–drift debate by proposing that the majority of amino acid substitutions in proteins and nucleotide substitutions in DNA are selectively neutral. Consequently, the frequencies of these molecular variants will increase or decrease in a population through random genetic drift.

Kimura was very specific about the types of mutations he was considering. He divorced the theory from both those mutants that have a detrimental effect on their bearers and from the rarer category comprising those that confer a fitness benefit. His theory accepts, therefore, that the evolution of structural, physiological, developmental, behavioural and ecological characters is, and has been, controlled largely by natural selection. Kimura argued that many mutations are unlikely to affect their bearer's phenotype, and so will have little effect on fitness. For example, some nucleotide substitutions will create a synonymous codon, due to the so-called redundancy of the genetic code (that some nucleotide sequence changes do not produce a change in an amino acid). Alternatively, some changes in the sequence of amino acids in an enzyme or other protein may not affect the function of that molecule. In addition, a considerable proportion of DNA is apparently functionless, and it seems reasonable that nucleotide substitutions in these regions will be selectively neutral. Kimura's theory then proposes that most evolution at the molecular level involves the random substitution of neutral alleles, one by another. Although beneficial mutations will occur and spread under the influence of selection, compared to the production and gradual random replacement of neutral mutations, such mutations will arise so rarely that they will have very little effect on the overall rate of nucleotide or amino acid substitutions.

The use of the term 'neutral' in naming this hypothesis, while perfectly accurate in a literal sense, has nevertheless been somewhat unfortunate. Evolutionary biologists are used to considering mutations as being deleterious, neutral or beneficial, with neutral mutations being associated with intermediate fitness. However, as Kimura made clear, the neutral mutations that his theory is concerned with are 'good genes' that are adaptively equivalent to one another because they do not significantly modify the morphology, metabolism or behaviour of their carriers.

It is also important to understand how similar in fitness two alternative alleles have to be for them to be effectively neutral. It does not have to be the case that they have to be mathematically identical in fitness, simply that any differences in fitness between the two are so slight that random genetic drift causes greater changes in frequency than does selection. Given that change due to drift is inversely related to effective population size, it is clear that at small population size, alleles which produce individuals that differ quite significantly in fitness may still be effectively neutral, whereas at large population sizes, even small differences in fitness effects are significant. Mathematically, two alleles are effectively neutral whenever the difference in fitness between them, when multiplied by four times the effective population size, is appreciably less than one.

In respect of the neutral theory, we need to ask four questions. What is the rate of substitution of selectively neutral alleles? What predictions does the theory make that can be tested by observation or experiment? How has, and how should, the rationale of the neutral theory of molecular evolution influence the way we study evolution? And what portion of genetic variation existing in populations is neutral?

The rate of substitution of neutral alleles has been discussed in Chapter 3. The neutral mutation theory of molecular evolution makes a number of testable predictions about the rate of substitution in different DNA sequences. For example, sites that are constrained by function will evolve more slowly than unconstrained sites. In a typical gene, this means that we expect the rate of substitution will be greater in introns than in exons. Within exons, the rate should be greatest for third codon position nucleotides (which show most **degeneracy**) and least for those in first and second codon positions (which show least degeneracy). Both these predictions have been confirmed experimentally (e.g. for codon position, see Kimura (1983); for introns versus exons see Bodmer and Ashburner (1984)).

The neutral theory also predicts that evolution through the substitution of neutral alleles will occur faster in proteins in which many amino acid changes do not alter the basic function of the molecule, than in those in which the amino acid sequence is functionally constrained. Consideration of the importance of specificity of amino acid sequences in different proteins should then allow predictions about their relative rates of amino acid substitutions. For example, Kimura (1983) compared the evolution of histone H4 with fibrinopeptides. The former, he argued, would be highly constrained because it binds to DNA and is essential to chromosome structure, while the latter would be less constrained, as their only function appears to be in binding to fibrinogen during blood clotting, after which they are cleaved and discarded. The three orders of magnitude difference in rates of amino acid substitutions in these two molecules (0.008×10^{-9} for histone H4, and 8.3×10^{-9} for fibrinopeptides) reflect these different levels of animo acid sequence conservation.

Such evidence for the neutral theory is open to the criticism based, at least in part, on its subjective assumptions – the relative importance of specific animo acid sequences to the proper functioning of a protein. However, some evidence, particularly the observed acceleration of nucleotide substitution rates in pseudogenes (non-transcribed genes) compared with their functional counterparts, is difficult to explain through selection. Thus, the α-3 globin gene in mice, a pseudogene derived by duplication from the α-1 globin gene, and its conspecific functional gene ancestor are more different than are the α-1 globin genes of mouse and rabbit. This would be expected on the basis of the neutral theory because it is highly likely that the amino acid sequence of the α-1 globin genes was functionally constrained, while the non-functional α-3 globin pseudogene is, by definition, unconstrained.

Finally, as the possibility of a neutral mutation becoming fixed in a population is random, and (according to the neutral theory) the number of neutral mutations far exceeds the number of selectively positive or negative mutations, the neutral theory predicts that the rate of substitution of mutations is constant. This prediction gives rise to the concept of the **molecular clock** of evolution.

The molecular clock, if it can be shown to exist, has great implications for the study of evolutionary biology (see Chapter 11). The phylogenetic relatedness of species would be reflected by the degree of genetic differentiation between them. Furthermore, if the clock can be calibrated accurately, the date at which divergence events took place in the past could be estimated.

There has been much debate about how generally the molecular clock may be applied. Problems emanate from the variation in substitution rates between proteins, from the difficulty in demonstrating which of different categories of nucleotide or amino acid substitutions are effectively neutral, from differences in the mutation rate, from whether the rate at which the clock ticks is constant with respect to time or generations, and from how the clock should best be calibrated. These will be discussed in Chapter 11, together with a more detailed discussion of the assumptions of the clock, of how phylogenies may be constructed and of the tests of the clock and its accuracy. However, it is pertinent to note here that, despite initial criticism from the selectionist lobby, the majority of evolutionary scientists now accept the neutral mutation theory of molecular evolution, with its associated molecular clock. Controversy still exists in respect of the relative importance of neutral evolution and adaptive evolution, but Kimura's neutral theory and Darwin's selection theory are not mutually exclusive, for they seek to explain different types of evolutionary change.

We have shown that since Darwin and Wallace proposed the theory of natural selection to account for biological evolution and the origin of species, a number of scientific discoveries have challenged Darwinian views. By using selected examples in a roughly chronological order, we have told the story of these controversies. While we are aware that many other pieces of both theoretical and experimental work had bearing on these issues, our aim has been to relate the basic principles involved. In the main, natural selection theory has either resisted its challengers, or new elements have been added to our interpretation and understanding of evolution.

Summary

In this chapter we examine some of the various challenges that have been made to selection theory this century. The initial challenge was brought by the mutationists, who raised the question of whether selection acted mainly upon variation produced by major mutational events or the small differences seen in continuously varying traits. Polymorphic Batesian mimicry, particularly in butterflies, was central to the arguments for and against each view. This led to a more general examination of the roles played by selection and random genetic drift in the evolution and maintenance of genetic polymorphisms. Empirical evidence collected over three decades suggested that, at least in respect of morphological polymorphisms, the role of drift was secondary to that of selection. Theoretical considerations of the load of selective deaths that a population could support if all or most genetic variation were maintained by selection suggested that there must be a fairly low upper limit to the amount of genetic variation that could be sustained in a population by selection. The discovery, through gel electrophoresis, of very high levels of genetic variation, led to the postulation of the neutral theory of molecular evolution. Although initially seen as a challenge to the Neo-Darwinian synthesis, it is now generally recognised that the neutral theory and selection theory are not mutually exclusive, and most evolutionary biologists recognise that both have played significant, if somewhat different, roles in evolution.

Suggested reading

ENDLER, J. A. (1986). *Natural selection in the wild.* Princeton University Press, Princeton. Contains a wealth of rigorously considered empirical evidence of the role of natural selection.

FORD, E. B. (1964). *Ecological genetics*, 1st edn. Methuen, London. Interesting and easy-to-read account of the development of ecological genetics. Relies fairly heavily on material from the Oxford school and the interpretation is occasionally subjective.

PROVINE, W. B. (1985). The R. A. Fisher–Sewall Wright controversy. *Oxford Surveys in Evolutionary Biology*, **2**, 197–219. An excellent objective account of this controversy.

SARKAR, S. (ed.) (1992). *The founders of evolutionary genetics.* Kluwer Academic, Dordrecht. A series of perspectives of Fisher, Haldane, Muller and Wright, and the effect that they, individually and through their interactions with each other and others, had on the development of evolutionary genetics.

SHEPPARD, P. M. (1975). *Natural selection and heredity*, 4th edn. Hutchinson, London. Interesting and easy-to-read accounts of the development of ecological genetics.

TURNER, R. G. (1985). Fisher's evolutionary faith and the challenge of mimicry. *Oxford Surveys in Evolutionary Biology*, **2**, 159–196. A fascinating essay on Fisher and his contribution to evolutionary genetics.

Chapter 5

The evolution of altruistic and co-operative behaviour

Natural selection is often described as 'the survival of the fittest'. This conjures up images of individuals continuously struggling with one another to survive and reproduce: a 'dog-eat-dog' world, where selection has favoured individuals who look after themselves and their descendants. Yet there are many examples of apparently selfless behaviours. In the vampire bat, individuals which have been successful in foraging often share their blood meal with others who have failed, a meal which can rescue the recipient from death by starvation (Wilkinson, 1984). In pea aphid colonies, individuals parasitised by the wasp *Aphidius ervi* immediately drop off the food plant, thereby forfeiting all their future (albeit now limited) reproductive success. Their suicide decreases the probability of the invader attacking the aphid's parental colony, either by killing the parasite within, or removing it to a safe distance (McAllister and Roitberg, 1987). Young but reproductively mature Florida scrub jays often delay their reproduction for a season or two, and instead help their parents to raise young (Woolfenden and Fitzpatrick, 1984).

We say such behaviours are **altruistic**, for the actions appear to benefit the recipients at a cost to the fitness of the donor. Fitness here is measured in terms of the number of direct descendants. An altruistic behaviour is thus one which, on average, produces a decrease in the number of grandchildren produced by the individual. Such behaviours can take many forms, as illustrated above. The most extreme cases of altruism occur among the eusocial animals, such as the ants and the termites. Here, some of the individuals do not reproduce at all, but expend all their effort in increasing the reproduction of others. Such behaviour is difficult to comprehend in terms of selection acting to maximise individual reproductive output, because individuals showing these behaviours clearly leave fewer direct descendants.

The same quandary is produced by some examples of **co-operative behaviour**. Co-operative behaviours are those that result in an increase in the fitness of each of the parties involved. For example, a hyena may obtain more food if she teams up with another individual to hunt. In teaming up, she may also increase the rate of food gathering of the individual she joins, because two hyenas are more than twice as efficient at hunting as individual foragers. This is because a wildebeest (or other prey) trying to escape from one attacker often falls into the path of the other and is caught. The evolution of co-operative behaviour in this instance can be explained in terms of selfish individuals gaining benefit from their interaction.

However, co-operation is not always so easy to comprehend. In some cases, although co-operation has benefits, defecting from a co-operation has an added advantage: a benefit gained from the party interacted with, but no favour returned. Consider the hunting scenario above, but add a risk to the hunter, such as that of injury in subduing the fleeing wildebeest. Now, if one of a group of hunters cheats, stands off a little to avoid the dangers of the hunt but still shares equally in the spoils, this behaviour will cause an increase in its success. If a cheating behaviour of this kind is controlled genetically, the selfish behaviour will be expected to spread at the expense of co-operation; leading to a breakdown of the mutualism. This quandary is often termed the prisoner's dilemma (Box 5.1). When cheating is possible, we must probe deeply to understand why co-operation is maintained.

The problems of the evolution of altruism and co-operation are functionally similar. To understand altruism we must explain the evolution of a selfless act, one which decreases the lifetime reproductive success of the individual. To understand co-operation in the face of an advantage from cheating, we must also explain why selfishness has not evolved. This chapter is concerned with how altruism evolves, and why co-operative interactions exist in the face of potential gains from cheating behaviour.

The importance of kinship

'A mother's job is to care for her child', run the words of an old lady in the Simon and Garfunkel song 'Bookends'. While making no comment about the social issues surrounding the role of fathers and mothers in parental care, consider a novel mutant gene which increases parental care afforded to individual progeny. If the progeny of those with this gene have a greater chance of survival than those not cared for by their parents, and the effects of this effort on the future reproduction of the parent is small, then the mutation will spread. This is because the number of progeny raised by the mutant individual is increased, and these progeny have a high chance of bearing the mutant gene. Parental care spreads because it increases individual fitness.

The same argument may apply equally to aiding the survival of a sibling. In a strictly monogamous diploid organism, a sibling is equivalent, in terms of relatedness, to an offspring. We can imagine, therefore, that the argument presented for natural selection favouring mutations which increase the survival of progeny should also apply to mutations which increase the survival of siblings, and indeed other relatives. As J. B. S. Haldane commented, he would jump in the river and give up his life if this would save two brothers from drowning (for they will have identical copies of around 50 per cent of his

Box 5.1: The prisoner's dilemma

The problem of the evolution of co-operation in the face of cheating can best be illustrated by a game called the prisoner's dilemma. The game runs thus. Two people at a time are placed in a room and are each given two pieces of card, one bearing the letter M (standing for mutualist), and one bearing the letter C (standing for cheat). They are told to simultaneously raise one of them, and are told:

1. that if they both raise M, both will be given £3;
2. that if one raises M and the other raises C, then the individual raising C (the cheat) will be given £4, and the individual raising M (the co-operator) will receive no reward;
3. that if they both raise the letter C, then they will both be given £1.

It is clear that in a population of cheats where everyone raises the letter C, a mutualist is a poor man: M individuals rarely encounter each other, and so are continually giving money to cheats. In evolutionary terms, mutualistic individuals cannot invade a population of cheats. However, in a population of mutualists, where everyone raises M, a cheating individual becomes rich. They receive a large pay-off for their selfish behaviour on virtually every encounter. In evolutionary terms, cheats can invade a population of mutualists.

The expectation, therefore, is that when cheating pays, cheating should spread, even though a population of co-operating individuals would enjoy greater wealth. This scenario is known as the tragedy of the commons. Where a resource is shared, a co-operative society runs more efficiently than an all-for-all society, but is invaded by cheats if cheats gain a reward from their behaviour. Human utilisation of fish and whale stocks perhaps illustrates this point all too well. There is much discussion within the fishing and whaling industries about conservation of stocks and maximising sustainable yield to safeguard both long-term jobs for seamen, and supplies of food to the consumer. However, the modern histories of these industries are a tale of over-exploitation for short-term gain. The optimal strategy for humans as a whole is undermined by individual interests.

The prisoner's dilemma thus exemplifies a problem evolutionary biologists must solve, that of why co-operation is observed in the face of a pay-off to cheating.

genes), or four half-brothers (for they will have identical copies of around 25 per cent of his genes), or eight cousins (for they will have identical copies of around 12.5 per cent of his genes) (Haldane, 1955).

This extension to the idea of individual selection was termed **kin selection** by John Maynard Smith (1965). Strictly, natural selection should not be expected to maximise individual fitness, measured in terms of the effect of the individual on the number of direct descendants. Rather, natural selection is expected to maximise the effect of the individual on the survival and reproductive success of all descendants, direct or otherwise, weighted by the likelihood that those descendants share the same genes. This, of course, is determined by the closeness of the descendants' kinships to the individual. Such a view of fitness, originating from the work of Bill Hamilton (1964), is termed **inclusive fitness**. Hamilton discussed the conditions under which kin selection would promote altruistic behaviour towards related individuals. Commonly termed Hamilton's rule, altruistic behaviour is expected to evolve if

$$b.r > c$$

where: b is the benefit the recipients of the altruistic behaviour receive, measured in terms of the effect of the action on their individual fitness; r is the average relatedness of the recipients of the action to the altruist; and c is the cost to the altruist, measured in terms of the decrease in individual fitness suffered.

Hamilton's rule emphasises the importance of relatedness in dictating when altruism will evolve. The definition of r above is phrased somewhat unusually as 'the average relatedness of the receiver of the act to the donor'. This is because acts of altruism are often directed to individuals of a particular type; for instance, to individuals with which an individual grew up. These will in general be siblings. However, where a female mates with more than one male, offspring may be either full siblings (share both a common mother and a common father), or half-siblings (share just a common mother). Assuming no ability to discriminate between siblings and half-siblings exists, the relatedness coefficient that will dictate whether altruistic behaviour spreads must incorporate this uncertainty by adopting the average relatedness to recipients of the altruistic act.

In addition to relatedness, Hamilton's rule includes terms for the benefit received by the recipient of the action and the cost of the action to the donor. Benefit to a recipient and costs to the donor are measured in terms of the effect of the action on the number of direct descendants that each individual leaves. The level of benefit and cost will depend primarily on the reproductive potential of the participants.

In terms of the benefit received by the recipient, altruism is more likely to be directed towards individuals with a high future reproductive potential, rather than ones with a low future reproductive potential. For example, altruism towards children will be more strongly selected for than that which is directed towards parents. This is not because of any significant difference in the likelihood that they bear the altruism gene. Rather, it results from the fact that an altruism gene will spread only in proportion to the benefits it can bring. There is little point in helping grandparents because the potential increase in fitness is near to zero. By contrast, a gene which promotes helping grandchildren will tend to spread rapidly because grandchildren have their entire fitness potential still in front of them, yet to be realised.

The other side of the altruism coin also shows variation with age. The cost incurred by an old individual in giving help is less than the cost incurred by a young one. Being near the end of their lives, old individuals have little reproductive potential of their own to lose, whereas help by a young individual may imperil all their future reproduction. The same act of altruism may have a large cost to a young individual, but only a minor cost to an older one. Therefore, old individuals are more likely to offer help.

Kin selection often helps to explain the evolution of altruistic behaviour. It equally explains why co-operation between relatives is more commonly observed than co-operation between unrelated individuals. The pay-off to co-operation and cheating behaviour should not be seen in terms of individual fitness, but in terms of inclusive fitness. In an encounter between two related individuals, cheating may increase personal reproduction (a cheat will produce more direct descendants), but will do this at the expense of the reproduction of relatives. Two brothers facing the prisoner's dilemma are more likely to co-operate than two strangers. In the terms of the prisoner's dilemma, a loss of income to one's brother is, in part, a loss incurred by oneself. Cheating on a sibling will only evolve when the personal gain of cheating is greater than half of the loss incurred by the cheated upon sibling. In short, co-operation is more likely to occur between relatives because cheating a relative both incurs a cost and brings less reward.

Empirical studies provide evidence of the importance of kin selection

Hamilton's rule makes intuitive sense. If the beneficiary of an altruistic action is related only distantly to the altruist, the likelihood that it bears the same altruistic gene by descent will be small. It is altruistic acts that are directed towards bearers of the same gene that are most likely to spread, and thus altruistic acts will most commonly be directed towards close relatives. Studies of behaviour within particular species of mammal, such as that by Paul Sherman on Belding's ground squirrel (*Spermophilus beldingi*), and studies across species of mammal with different ecologies, provide strong support for the importance of relatedness in dictating when altruism evolves.

Belding's ground squirrel   Belding's ground squirrels live in montane meadows in the western USA. The females are semi-colonial, young females remaining in their natal area once weaned, and young males tending to move away. When a colony is threatened by the approach of a potential predator some of the squirrels stand up on their hind legs and give an alarm call, a high-pitched screech. Sherman investigated the pattern of calling and observed that females sounded the alarm much more frequently than males. Furthermore, females that had close relatives nearby called more often than solitary females or females that had recently migrated into a new area (Sherman, 1977). These observations fit precisely with expectations based on considerations of inclusive fitness. Alarm calls are made most often when the beneficiaries are likely to bear the same genes by descent. In a related species, *Spermophilus tereticaudus*, Dunford (1977) found that males changed their behaviour when they left home. While residing in their natal area they made alarm calls, but as soon as they dispersed to areas where no relatives resided, they stopped sounding the alarm. Again, this fits the prediction of inclusive fitness theory.

The altruistic and co-operative behaviour of Belding's ground squirrel extends beyond alarm calls. Wandering male and female ground squirrels can be cannibalistic, killing and eating young. Sherman (1981) found that only unrelated young were eaten, and that related females co-operated to defend burrows against the threat posed by unrelated intruders.

Comparisons of behaviour in different species of primate   In many species of mammal dispersal of individuals between groups is confined to members of one sex. Individuals of the other sex that remain with the natal group to reproduce are said to be philopatric. In species where males disperse and females remain with the natal group, females in a group are likely to be related to each other, and males are unlikely to be related to any individual in the group. Conversely, in the (fewer) species in which females disperse and males remain with the natal group, males are likely to be related to other males, but females are unlikely to be related to any of the other individuals in the group apart from their own immature progeny.

In species where only males disperse, females in a group are likely to be related to each other and would be expected to show co-operative behaviours. Observations confirm this prediction. In grey langurs and yellow baboons, for instance, females co-operate and male–male competition is intense (Hrdy, 1977; Altmann *et al.*, 1988). In social systems where females disperse the reverse is true. Here, males are more tolerant of each other, and may be seen to co-operate. Female–female co-operation is less in evidence. For instance, in the chimpanzee, males seldom fight in the way that male mammals usually do, and equally unusually, females often behave antagonistically to one another (Wrangham, 1986). The pattern also holds for species where all individuals disperse. Here, few individuals of either sex are related, and the expectation based on consideration of inclusive fitness is for competition between males to be intense, and co-operation between females to be rare. This is observed in mountain gorillas (Stewart and Harcourt, 1987).

Can kin selection explain the evolution of eusociality?

Altruistic behaviour towards relatives may be favoured by natural selection. However, can kin selection theory explain the extreme cases of altruism, where individuals forfeit all personal reproduction to help raise siblings. Organisms in which this occurs are called **eusocial**. Eusociality has evolved several times in the bees and wasps, once in the ants, once in termites, and several times in aphids. One mammal, the naked mole rat, is also known to show eusocial behaviour.

Clues about how eusociality has evolved come from the study of semi-social systems, where fertile individuals frequently help relatives. One such study was undertaken by Glen Woolfenden and John Fitzpatrick on the Florida scrub jay (see Woolfenden and Fitzpatrick (1984) for a review). In this species, young fledglings do not always disperse on reaching reproductive maturity. Rather, they help feed and care for their younger siblings. The reason for the evolution of this behaviour lies in the fact that the number of suitable nest sites is limited. Competition between males for nest sites is therefore intense, and males that have just reached reproductive maturity are unlikely to obtain a nest site in competition with older, more experienced males. If a male has little chance of securing a nest at which to raise his own family, then kin selection favours the next best behaviour, help towards the raising of siblings.

Consider now an ecological change which makes nest sites even rarer. The probability of securing a nest site may now remain low throughout his life. At no stage would leaving the parental home increase inclusive fitness. Complete loss of reproduction may result, and the organism becomes eusocial.

This route to eusociality is driven by kin selection in which the advantages gained by raising relatives come to outweigh the benefits of reproducing oneself. Eusociality will be favoured when the ecology of a species is such that generations overlap, opportunities for personal reproduction are limited, and help to parents is effective in increasing their reproductive success. It is perhaps this last factor which explains the large number of incidences of eusociality amongst insects. A female insect can be almost unbounded in her ability to produce eggs. All that is required is nutrients to provision them. Within a termite colony, the queen can produce eggs at a rate of one every five to ten seconds. There is a near-linear relationship between the amount of food delivered to her and the number of eggs she produces. Help can have a very large effect on parental reproductive success in such insects, and this probably best explains the frequency with which eusociality has evolved in this group.

The above 'stay at home to help mother' scenario represents one route (called the subsocial route) by which eusociality can evolve. Other routes to eusociality have been proposed. In one of these the starting point is (usually) sister individuals founding a nest co-operatively. Following this, individual selection favours females which become reproductively dominant, for they produce more of their own daughters at the expense of nieces and nephews. With complete dominance, only one female reproduces, and her sisters help her reproductive efforts. If there is overlap of generations, then offspring have a choice. They may stay at home and help raise kin (for reproductive dominance will extend to them too), or leave in the hope of founding a nest on their own. If opportunities for personal reproduction by dispersing individuals are scarce, then the kin selective benefit of raising siblings may outweigh the possible benefits of personal reproduction, and favour individuals which stay at home. Eusociality has evolved because kin selection first promotes co-operative nesting by sisters, and then (if there is overlap of generations) promotes offspring which stay at home to help, rather than leave. This scenario beginning with the co-operative breeding of sisters is termed the parasocial route to eusociality.

Assessment of relatedness by animals

Kin selection appears to provide the key mechanism by which altruism and co-operation can evolve. However, people often question the general importance of kin selection, because they question the ability of animals to assess relatedness. There are two answers to this criticism. First, kin recognition is not necessarily required for altruism to evolve. Second, empirical studies show that many animals do have a capacity to recognise kin.

Animals may not have to be able to assess relatedness for altruistic behaviours to spread   In many cases, altruism may evolve in the absence of kin recognition. The evolution of helping in the Florida scrub jay does not require recognition of kin. Young jays simply have to aid offspring in the further broods produced by their parents. These are automatically kin; they do not have to be recognised as such. Similarly, in species such as the grey langur where only males disperse, the females remaining in the group will be related to a certain extent, and altruistic behaviours spread. Social organisation produces what might be termed a 'relatedness structure' which obviates the need for kin recognition. It is notable that such recognition of kin by situation, say a bird feeding chicks in the nest it built, can be abused. For example, the foster parents of the young cuckoo feed it because it hatched in their nest, despite the fact that it bears little similarity to a real chick of their own.

Animals can and do assess relatedness   Many examples of altruism in the field suggest that animals do assess and recognise kin. Most commonly, kin recognition is a trait learnt early in life with individuals recognising individuals they were reared with, and treating them as kin. This recognition based on early experience (imprinting) has been shown to occur widely. One example is that of Belding's ground squirrel, discussed earlier. Here, cross-fostering experiments were performed in which siblings were reared either all together by a different mother, or were reared in a brood with unrelated individuals. Once grown, the squirrels were observed to co-operate with litter mates as if they were kin, irrespective of their true genetic kinship, and co-operated more with these individuals than with true siblings from which they had been separated just after birth. However, Belding's ground squirrels have other mechanisms of kin recognition. Siblings that were reared apart did not co-operate as readily as siblings reared together, but were more likely to co-operate than unrelated individuals reared apart. This suggests that, although early experiences are most important in kin recognition, other factors also exist (Holmes and Sherman, 1982). Here some form of phenotype or genotype matching that allows genetic kin recognition is likely to be involved. Mice appear to have just such a system of genetic relatedness assessment involving the highly polymorphic major histocompatibility complex (MHC) (Box 5.2).

Box 5.2: The major histocompatibility complex: a genetic system of kin recognition in mice

All mammals have an immune system which accepts like but which reacts against non-like. Part of the mechanics for the recognition of self involves a cluster of over 20 linked genes (on chromosome 6 in humans) termed the major histocompatibility complex, or MHC. These genes are highly polymorphic, some having more than 50 alleles, and play an important role in mounting the immune response against pathogens and parasites. However, the MHC is perhaps best known for its other effect, that of producing the tissue type of an individual. MHC products are recognised by the producer's immune system as self, but the MHC products of other individuals are not. In transplant surgery, surgeons aim to produce a match between the tissue type of donor and recipient so as to decrease the probability of rejection of the transplanted organ. Even with anti-rejection drugs, replacing an organ with one of the wrong tissue type can be fatal.

As anyone waiting for a donor organ to become available knows, the number of tissue types in the population is high, and the probability of a good match with a random individual is low. This is largely because the MHC contains so many polymorphic genes. In cases like kidney transplants where there are two organs per individual, the best solution is the donation of an organ from a sibling. A sibling, due to their relatedness, is far more likely to have a closely matching tissue type than a random member of the population.

The MHC locus is thus perfectly suited for use as a cue in a kin recognition system. Extensive studies of the common house mouse (*Mus domesticus*) have shown that mice do use the MHC in kin recognition. Mice detect the MHC status of an individual from the odour of broken-down MHC products (cell surface glycoproteins) in their urine. This status is used as an indicator of relatedness. Female mice of similar MHC type are more likely to breed co-operatively, and mice of both sexes use the MHC in mate choice, to ensure outbreeding.

What mice compare MHC odours to when assessing their relatedness to another individual is not known, although there is some evidence that parental MHC odours, learnt early in life, may be used. However, other

continued

studies suggest an additional role of comparison of incoming odours to their own. What is certain is that a system of genetic identification exists, that this allows identification of the relatedness of individuals they have never previously encountered, and that MHC odours are used as the behavioural cue.

See Lenington (1994) for a review of the role of the MHC in kin recognition, and Potts and Wakeland (1993) for a review of the general factors shaping the evolution of the MHC in mammals.

A beautiful study which shows the ability of animals to apportion their effort effectively depending on relatedness is the investigation by two behavioural ecologists, Nick Davies and Ben Hatchwell, and two molecular ecologists, Terry Burke and Mike Bruford, into the parental care afforded to progeny by male dunnocks, *Prunella modularis* (Burke et al., 1989). Dunnocks form breeding partnerships of various types, one male with two females, one male with one female, one female with two males, or even involving two or three of each sex. This study looked at the interactions between members of breeding groups involving two males, one more or less dominant to the other, and a single female. The question asked was: under what circumstances do subordinate males help to care for chicks? Behavioural observations were made to assess the amount of effort males placed into helping to feed the young. Then DNA fingerprint analysis of both the adult birds and the nestlings was used to determine how many of the offspring in the nest were fathered by the subordinate male. A direct correlation between the number of nestlings fathered and the contribution made to rearing the young was found (Fig. 5.1).

Spatial structure and group selection

Kin selection provides one reason why altruism may evolve. The presence of relatives means that inclusive fitness is increased by altruistic acts. However, in certain kinds of spatially substructured populations, the evolution of altruism and co-operation may be favoured for other reasons. If groups are integral and have different survival and reproduction rates depending on the genetic constitution of the individuals within them, then groups may be said to have different fitness. **Group selection**, that is selection acting to increase the 'fitness' of a group of individuals independently of the fitness of each individual in the group, may occur. Fitness of a group here is a product of the survival of the group, and the tendency of the group to grow and produce new groups.

Group selection will frequently not alter our perspective on evolution, for what is selectively advantageous at the individual level will also be selectively advantageous at the level of the group. Any new aspect of design which

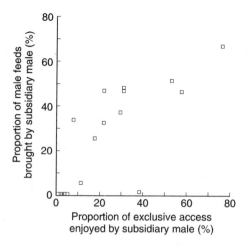

Fig. 5.1 Parental effort in the dunnock, *Prunella modularis*. Dunnock nests attended by more than one male were monitored, and the access of each male to the female monitored. DNA fingerprint studies subsequently confirmed a strong relationship between the amount of access of the male to the female(s) and his probability of paternity of the young in the nest. Effort made by each male in rearing progeny was also monitored. It was observed that the effort contributed by the subordinate male correlated strongly with his access to the female, and thus with his probability of paternity. Males feed progeny in relation to their probability of kinship with the progeny. (After Burke *et al.*, 1989.)

produces an increase in the fitness of one individual in a group without decreasing the fitness of others will serve to increase both individual and group fitness. Hence many adaptations appear to be 'for the good of the species'. However, when the trait in question increases the lifetime reproductive success of an individual (i.e. it increases individual fitness), but does this at a cost to another individual in the group (i.e. it decreases group fitness), the direction of group selection and individual selection differ. Similarly, opposition of forces will occur when a trait increases group fitness but decreases individual fitness.

Territoriality is a case in point. Obtaining and keeping possession of a territory may increase the lifetime reproductive success of the individual who gains access to more resources, but may reduce group reproductive success. Effort placed by individuals into guarding territories against incursions by other members of the group decreases the productivity of the group and therefore it decreases group fitness. It might be said that the fitness of the group in such cases was inversely related to the frequency of the territorial behaviour and hence to the frequency of the genes within the group that produce it.

When individual selection and group selection are opposed, the evolutionary trajectory may not be that expected from pure individual selection. In the 1960s, a school of thought, led by V. C. Wynne-Edwards, championed the view that group selection was a more potent force than individual selection

(Wynne-Edwards, 1962). Altruism and co-operation could be seen not as a product of kin selection, but as a product of group selection. Animals would not be selfish in their reproduction and endanger the survival of the group, but would evolve to restrict their reproduction so as not to over-exploit resources. Groups which over-exploited resources would suffer high extinction rates. Selfish individuals would be selected against, either as a result of the extinction of their group, or because the fitness of all members of a 'selfish' group was lower than that of individuals in groups comprising just altruistic or co-operative individuals. The selection against groups containing selfish individuals, according to the group selectionist school, provides an explanation for the continued existence of altruistic and co-operative behaviours. According to them, natural selection will act not for the good of the individual, but for the good of the species.

Modelling has shown that in most cases when group selection and individual selection differ in direction, Wynne-Edwards' analysis is wrong. The result of selection is that expected from analysis of individual fitness, not from analysis of group fitness. This is because the effects of a few selfish individuals on group productivity tend to be small compared to the effect of selfish individuals on the survival of altruists within the group. The loss of altruists from a group by selection at the individual level is unlikely to be matched by a group selective advantage (increased productivity of altruist-dominated groups). A plethora of empirical studies, mainly focusing on analyses of bird reproductive biology, bear out this analysis. Individual birds were not observed to limit their reproductive rate in the interests of the group, but rather produced the number of eggs which maximised individual fitness (Box 5.3).

When individual selection and group selection are in opposition, selection usually produces individuals with maximal individual fitness. However, as David Wilson (1975, 1980) noted, there do exist some circumstances when spatial structure plays a bigger role. This is when groups are small, integral (low migration into and out of groups, and thus low rate of entry of selfish individuals), and short-lived (low potential for selfish individuals to take over a group). Under these conditions, the results of selection may differ from those expected in a panmictic population. Wilson called such small short-lived groups of individuals **trait groups**.

Parasites often exist in trait groups. A host may initially be parasitised by only a few colonist individuals. The productivity of the host for a parasite may depend on host longevity, which may be inversely related to the number of parasites present. If this is true, then there will be selection at the group level for a low reproductive rate of the parasite within the host, as this will increase host life span and therefore increase the period the host is infectious for the parasite. However, more rapidly reproducing parasite colonists will be over-represented in the pool of infective particles in the group. Where more than one colonist is present, natural selection at the individual level may promote strains which have increased reproduction, because these will out-compete the more benign types by weight of numbers.

Box 5.3: Parental effort in birds: individual selection, not group selection

Under group selection, individuals are expected to evolve rates of reproduction which are limited to ensure groups do not increase in size such that they exhaust their resources. Under individual selection, provisioning of young is expected which maximises personal lifetime reproductive success, irrespective of the effect of any population growth on group survival. A long-running investigation led by David Lack into the reproductive biology of the great tit, *Parus major*, focused on whether the reproductive biology of this species is compatible with individual or group selection. The findings of Lack's group were that individual birds produce and provision clutches of a size compatible with maximisation of individual fitness, and that parental effort is not compatible with considerations of group selection (Perrins, 1965; Lack, 1966).

This study took place in Wytham Wood, near Oxford. The birds were provided with nest boxes, and all birds were ringed so they could be identified and their fate determined. It was found that most pairs laid between seven and ten eggs in their clutch. Lack then asked whether the number of eggs laid was more in line with maximisation of individual fitness or group selection. Field studies showed that there was a negative relationship between clutch size and the weight of chicks at fledging, and further, that early survival of the fledglings was strongly dependent upon weight. Chicks in larger clutches received less parental attention, were smaller on fledging, and had higher mortality in their first year. However, the effect of a loss of a little parental care on fledgling survival was not great. Fledglings from clutches of eight eggs had only slightly lower survivorship than individuals raised in clutches of three, but considerably higher survivorship than fledglings from clutches of twelve. From these data, the individually selected optimum could be calculated, and turned out to be around eight eggs per clutch. There is a good match in these data between the clutch size produced by great tit pairs and the expectation from individual selection.

Although there are valid criticisms of this study (the study measured reproductive success in one season only, and therefore does not strictly make observations on lifetime reproductive success), further studies on a variety of species have borne out the basic conclusions. There is a match between observed parental effort and parental effort expected from considerations of selection at the individual level. Individuals do not limit their reproduction for the good of the group, but behave as would be expected under individual selection.

The result of selection in this case may be more that expected from considerations of group fitness than individual fitness. Since hosts carrying fast-replicating parasite strains export fewer parasite individuals, an aggressive strain may, under certain conditions, either not invade, or not go to fixation. This is despite being more successful than less virulent strains within the host. The crucial balance is between the increased horizontal transmission of the slower-replicating strains in individuals founded only by individuals of the less virulent type, and the transmission advantage of the faster-replicating strains when in competition. More benign strains may be maintained in such a situation, despite their competitive inferiority within an infected host.

One possible example of this is the evolution of virulence in the Myxoma virus, the cause of myxomatosis in rabbits. In Australia, this infection spreads between hosts via a mosquito vector. When first introduced into Australia, the virus was nearly always lethal. Hosts killed by the virus were not infectious for it, as mosquitoes do not bite dead rabbits. Highly virulent forms were selected against, and the virulence of the virus decreased over time. Today in Australia, rabbits with myxomatosis do not die as quickly, and some individuals are never killed by the virus, being maintained as infectious individuals over long periods of time (Fig. 5.2). Group selection is likely to be an important factor in the decline of virulence, the number of rabbits infected from a single infected individual being higher when the virus strains inside the rabbit are less virulent (May and Anderson, 1983).

In summary, group selection is common. However, it is seldom an important force because it does not usually alter the end-point of evolution from that expected from individual selection. When individual and group selection are in the same direction, the result is that expected under individual selection. When individual and group selection are in opposition, unless groups are short-lived and established by very small numbers of founding individuals, then again the evolutionary outcome will be that expected from individual selection. Group selection only promotes co-operation and altruism in opposition to individual selection when rather restrictive conditions of population structure and gene flow are met.

Reciprocal altruism

Co-operation between unrelated individuals in panmictic populations is usually associated with a lack of scope for cheating. However, there are cases of co-operation where cheating is possible, but appears not to have evolved. A study by Fischer (1980) of the black hamlet fish, *Hypoplectrus nigricans*, provides a good example. This fish is a simultaneous **hermaphrodite,** and in a sexual interaction, each individual may take the role of either the male or the female. Because eggs are more costly than sperm, given the choice a selfish individual should choose to be permanently male. However, the observation is that shortly after an individual donates eggs, it receives a reciprocal gift of eggs from the

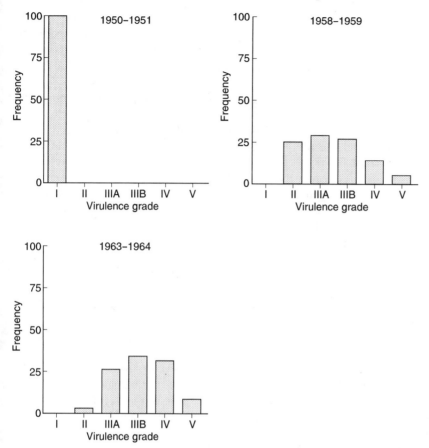

Fig. 5.2 Evolution of virulence of the myxoma virus of rabbits. Viruses, collected from wild rabbits in Australia at three different times, were injected into rabbits from a control population and the severity of the symptoms experienced by the rabbits scored on a scale from one to five (I = most severe). Samples of the virus collected shortly after its initial introduction to the wild rabbit population are the most virulent, always killing their host. Virulence decreased over time, as expected from considerations of optimal exploitation of the host, rather than optimal competitive ability. (Data from May and Anderson, 1983.)

partner. Usually the pair remain together, the last recipient of eggs offering eggs on the next occasion. The two fish are what Robert Trivers termed reciprocally altruistic (Trivers, 1971). Of course, the fish that receives the gift of eggs first could cheat, by not reciprocating. However, this is not common.

Why is it that these fish do not cheat? Reciprocal altruism is a case of the game of co-operation detailed at the start of this chapter, the prisoner's dilemma. To review, the prisoner's dilemma is a game which illustrates why mutualism is rare if cheating is possible (Box 5.1). Recapping, the participants have the cards M (mutualist) and C (cheat). In the context of the black hamlet fish, to raise M is to donate eggs after being given sperm; to raise C is to refuse to do this. The evolutionary problem is to ascertain why the population has not declined into pairs of prisoners all cheating for fear of being exploited.

The game described in the introduction was of prisoners which meet just once and play the game. In such a circumstance, the population will evolve such that all individuals are cheats. However, if individuals meet frequently, and can remember the results of previous encounters, then the result of the prisoner's dilemma is more difficult to predict. There is increased pay-off for mutualisms, for these may be established in the long term. This game, the iterated prisoner's dilemma, has been the subject of an extensive study by Robert Axelrod in association with Bill Hamilton (Axelrod and Hamilton 1981; Axelrod 1984). They found that, when a variety of strategies of playing this game were proffered, the one that gave the highest pay-off was a simple one called 'tit-for-tat'. This stratagem involved attempting mutualism initially, and then only attempting mutualism if the partner proffered mutualism in the previous encounter. The penalty for being duped was paid only once, but the benefits of mutualism could be reaped over a long period of time.

If the problem of co-operation is iterated, and individuals remember the results of previous encounters, the population does not necessarily descend to all cheats. A tit-for-tatter loses little in the long-run, because it will not be cheated all its life, but it may still make gains from individuals prepared to be mutualistic. Interestingly, black hamlet fish appear to play the 'tit-for-tat' stratagem, for if either party refuses to reciprocate, then the pair separates. The iterated prisoner's dilemma is an example where, in the long term, the individuals are not being truly altruistic. It is an unusual case of mutualism where cheating is possible, but does not pay as a long-term stratagem.

Iteration of a game may also promote co-operation because it allows the spread of law enforcement and punishment of defecting individuals (Clutton-Brock and Parker, 1995). When two individuals encounter each other only once, punishment of a cheat merely adds to the cost incurred by the duped individual. However, when individuals encounter each other repeatedly, punishment of a defector may be beneficial because, although it has a short-term cost, it reduces the probability of future cheating. This will commonly be the case, because once attacked, a defector will be less likely to cheat with the risk of further damage. For punishment to evolve, however, it must be cheap. Five-foot weaklings should not attempt to punish the local boxing champion. Punishment will evolve as a mechanism by which dominant individuals in a group ensure future co-operation from subordinates. As a human analogy, most people did not have the opportunity to refuse to pay a bill issued by the Chicago mob twice! The existence of dominance and punishment probably explains the maintenance of co-operation of individuals in many mammalian social systems.

Summary

The existence of altruism, acts which decrease individual fitness, and co-operation when defection would give an advantage to the individual, are counter to expectations based on a simple individual selectionist interpretation. Three classes of explanation are proffered here. First, and probably most importantly, selection acts not on individual fitness, but on inclusive fitness. Behaviours which aid the survival of relatives at a cost to individual fitness may spread under certain circumstances. In a similar manner, cheating on a relative may not be favoured. Where the spatial structure of the population is such that interactions tend to be with relatives, altruism and co-operation between individuals may evolve and be maintained. Second, the spatial structure of a population may promote co-operation under certain (somewhat restrictive) conditions because of group selection. Third, if the structure of a population is such that interactions between the pairs of individuals occur frequently, then cheating may be disadvantageous under individual selection, either because co-operation produces long-term dividends between equals, or because punishment by dominants on cheating subordinates ensures future co-operation.

Clearly, the factors which promote the evolution of co-operation and altruism are not mutually exclusive. One important step in the explanation of co-operative behaviour through kin selection is to measure the parameters of relatedness in groups. In the context of group selection, it is necessary to ascertain the number of individuals founding a group. The tools of molecular ecology have increased the accuracy of measurement of these parameters and begun to push study in this field beyond its previous limits (see Chapter 9).

Suggested Reading

CLUTTON-BROCK, T. H. and PARKER, G. A. (1995). Punishment in animal societies. *Nature*, **373**, 209–216.
   A review of the issues surrounding retaliation, with sections detailing how retaliation may discourage parasites and cheats and maintain co-operative behaviour.
GRAFEN, A. (1991). Modelling in behavioural ecology. In KREBS, J. R. and DAVIES, N. B. (eds) *Behavioural ecology, an evolutionary approach*, 3rd edn, pp. 5–31. Blackwell, Oxford.
   This essay contains an authoritative review of some of the complexities of the concept of inclusive fitness not discussed here.
KREBS, J. R. and DAVIES, N. B. (1993). *An introduction to behavioural ecology*. Blackwell, Oxford.
   Chapters 11, 12 and 13 give an excellent overview of the issues surrounding the evolution of co-operation altruism, and the empirical studies underlying them.
WILSON, D. S. (1983). The group selection controversy: history and current status. *Annual Review of Ecological Systematics*, **14**, 159–187.
   An appraisal of the importance of group selection.

Chapter 6

Sexual selection

Darwin (1859) gave a concise definition of **sexual selection:**

> This depends not on a struggle for existence, but on a struggle between males for possession of the females; the result is not death to the unsuccessful competitor, but few or no offspring.

The battle of the sexes

Male and female reproductive interactions – courtship, copulation and care of offspring – may be thought of as part of a single common cause uniting two inherently selfish partners. It is perhaps as a result of human preoccupation with romance and the pleasure of sex that the interactions between males and females preceding fertilisation are often viewed as the outcome of the common evolutionary interests of the two prospective parents. This may sometimes be true, but often it is not. There is little doubt that in most species both male and female reproductive behaviours have evolved to maximise reproductive output. However, the fact that the sexes act co-operatively and harmoniously to achieve this aim has been increasingly questioned during the last two decades. Views that courtship rituals serve to 'synchronise sexual arousal', 'form and solidify pair bonds', and 'act in species recognition', are being replaced by the views which emphasise the 'battle of the sexes', that is to say, the conflicts of interests between the sexes.

Individuals of both sexes are expected to behave in a way which maximises the number of their own genes passed to the next, and subsequent, generations. Some of these behaviours are co-operative, because, as both partners pass on their genes via the same progeny, the well-being of progeny is in the interests of both. For instance, in Bewick's swans, *Cygnus comumbianus bewickii*, a high degree of parental care is required to raise offspring, and this is given by both mother and father in concert. Their success in raising broods increases with the length of the pairing, and new pairings are less successful, even if they involve experienced individuals (Scott, 1988). However, Bewick's swans represent an unusually harmonious interaction between the sexes, and in many species antagonistic behaviours are common. Although the sexes have common interests in certain aspects of courtship and mating (for instance, selecting a partner of the correct species), the interests of males and females in choice of mates, levels of promiscuity, provisioning of zygotes, and caring for the offspring thereafter, often differ.

The fundamental cause of sexual conflict is the difference in investment in progeny by males and females. In most organisms, males produce small, energetically cheap sperm, whereas females produce large, resource-rich eggs (Parker *et al.*, 1972). In organisms where there is care beyond the resourcing of eggs, this is often provided by the female only. Given that an individual who is caring for progeny will not be interested in sex, species in which only females care for offspring contain, at any point in time, fewer females than males who are interested in copulation. The term **operational sex ratio** (OSR) was introduced by Steven Emlen and Lee Oring to encompass this point (Emlen and Oring, 1977). The OSR is perhaps best seen as the ratio of sexually active males to sexually active females in a population at any moment in time. It depends on the sex ratio of the adult population, the relative amounts of care males and females allocate to offspring, and the degree of reproductive synchrony amongst members of a given sex.

In species where males do little caring, the OSR may be strongly male-biased. If both sexes care equally for offspring, the OSR tends to be around 1. Species in which the male expends the greater effort in caring (often referred to as showing **sex role reversal**), usually have an OSR that is female-biased. This situation is seen in the lily-pad-walking American jacanas, *Jacana spinosa*. Here, the females defend large resource-based territories which may contain several nests both made and tended by males. Each female mates with all the males within her territory, then lays eggs in their nests. The reproductive rate of the female is thus higher than that of the male.

In calculating the OSR, therefore, it is important to examine the effect of care on the readiness of each sex to mate. In cases where both sexes care, this may require careful observation. For example, in the three-spined stickleback, *Gasterosteus aculeatus*, both male and female parents care, although in rather different ways. The female resources the eggs, which she lays in a nest made by the male, who thereafter cares for and protects them and the resulting fry. However, care by the male is energetically cheap compared to the resources that the female has expended in provisioning the eggs. Furthermore, he may tend broods from several females simultaneously, without significant additional effort. This disparity in expenditure given by each parent per offspring means that, despite both sexes caring for the offspring, the OSR is male-biased.

Sexual selection: a consequence of bias in the operational sex ratio

When the OSR is biased, one sex is 'in demand' for copulation and one sex is 'in excess'. Selection will favour behaviours in members of the sex in excess which increase their access to partners. This 'sexual selection' may take two forms. First, it will favour individuals which can, by competing with members of the same sex, monopolise access to sexual partners (**intrasexual selection**). In species where females care and males do not, this is termed **male competition**, or sometimes male–male competition. The equivalent in sex-role-reversed species is, of course, female competition. The second form of sexual selection is the selection acting on members of the common sex to attract partners to mate with them (**intersexual selection**). Where females care and males are the common sex, this form of sexual selection is termed **female choice**. Male choice may be found in sex-role-reversed species.

Sexual selection is most easily recognised in **polygynous** or **polyandrous** species, where some traits clearly serve only to increase the number of mates obtained. However, sexual selection also occurs in **monogamous** species. Here, behaviours will be selected which allow access to the partners of highest quality. In many monogamous birds, for instance, females which breed early in the season have the highest reproductive success, raising more progeny per brood, or having a higher probability of raising additional broods in a season. Clearly, males which pair with early breeding females will gain an advantage, and selection will favour traits which improve access to, and acceptance by, such females. In monogamous species where the parents expend roughly equal effort in care, the lack of bias in the OSR means that sexual selection may occur in both sexes. Both sexes will be selected to gain access to or choose high-quality mates. A recent study of the crested auklet, *Aethia cristatella*, by Ian Jones and Fiona Hunter, has provided good evidence for such 'mutual sexual selection' (Jones and Hunter, 1993).

Male competition and female choice are not mutually exclusive

Although male competition and female choice were outlined as two distinct mechanisms of sexual selection by Darwin (1859), they are not mutually exclusive. Both may operate in the same species and on the same trait. Traits evolved through male competition may allow some males to become victors over others. Yet females may still choose between victors. In elephant seals, for example, a female has little chance of physically resisting the advances of a non-chosen male because of his sheer bulk. However, she may exercise some degree of choice by protesting loudly when a lower ranked male attempts to force his attentions upon her to attract the attention of the dominant male, who will attack and drive away the interloper. In the house sparrow, *Passer domesticus*, the size of the black throat patch of males acts both as a signal of dominance status in aggressive interactions between rival males, and as an adornment that is attractive to females.

Not only are male competition and female choice not mutually exclusive within a species, but they may be instrumental in the evolution of each other. In species such as the Uganda kob, *Kobus kob thomasi*, for instance, males congregate on a site called a **lek**, and compete vigorously for territories in the centre of the group. The reason for this competitive behaviour is simple: females prefer to mate with males at the centre of the lek. Male competition is driven by the preferences shown by females (Balmford *et al.*, 1992).

Sexual selection by male competition

The evidence that reproductively mature males compete with one another for access to females is overwhelming. Battles for dominance between rival deer stags, male elephant seals or in the rowdy groups of male humpback whales (cover picture) could hardly be interpreted otherwise. The most obvious examples of traits which have evolved through male competition are those involving male weaponry (the single enlarged claw of fiddler crabs, the spurs on the legs of cockerels, the antlers of male stag beetles, the horns and reinforced foreheads of musk oxen, and possibly the narwhal's tusk) or the large size of males relative to females (particularly among birds and mammals). Evidence comes not just from observations on single species, but also from comparison between species. For instance, amongst species of primate, both the relative size of canine teeth and the relative body mass of males and females strongly correlate with the degree of polygyny (Clutton-Brock *et al.*, 1977; Harvey *et al.*, 1978).

In some instances, weaponry and bulk are used in contact fights between rival males, as is seen in rival rutting deer. More commonly, however, dominance is established by ritualised combats composed of threats, dances, songs, and other displays of prowess. Examples of noise displays include stags bellowing, bitterns booming and gorillas drumming their chests. Examples of animals in which males 'fight' using colour traits include butterflies, chameleons, octopuses and the flash coloration of primate eyelids. In many plants, sexual selection favours flowers which attract insects which act as pollinators. Here, the precise chroma and hue of flowers may have a role in male competition, particular colours being more attractive to pollinators, so that pollen from such flowers is the more likely to be transferred. Our own visual acuity and auditory capacity mean that we have recognised and identified many visual and auditory traits involved in male–male interactions. However, other types of stimuli which we find more difficult to perceive (for example, scent, ultraviolet and infrared light, tactile, electric and magnetic impulses) are also likely to be involved in some organisms. Indeed, any trait by which one individual of a species may perceive and assess another may become the subject of sexual selection by male competition.

The ritualisation of combats between males reduces the need for expenditure on defensive traits which would be necessitated by the evolution of weapons which could cause serious injury or kill during contact combat. It is notable that deer's antlers tend to interlock during combat rather than sliding past each other towards vulnerable flesh. One possible reason why the unicorn is mythical is that males equipped with a single spike horn would annihilate one another! It is rare that mortality is anything but an accident.

Red deer: a case study of male competition

The evolutionary consequences of male competition have been one of the main areas of study during a long programme of research on the red deer, *Cervus elaphus*, by Tim Clutton-Brock and others, on the island of Rum, off the west coast of Scotland. Red deer stags have antlers, which they use in fighting other stags. These antlers are not involved in foraging, and are probably not used much in defence against predators. Instead, it seems that this huge weight of bone and other tissue, regrown each year at a great cost in calcium alone, is present specifically for the purpose of fighting other stags during the rut and thus gaining access to females.

The size of red deer stags and their antlers reflects a trade-off between reproductive benefits and survival costs. The benefits of success in combat must be high to warrant such extravagant resource expenditure. Since red deer are polygynous, the 'expense' is apparently justified. Successful stags mate with not one, but with many females and, therefore, father many off-spring. In contrast, stags which are not successful in the rut usually father none. Consequently, there is tremendous variance in male reproductive success and an intense selective pressure for large male size. This is reflected in the differing pattern of development of males and females. Sons suckle their mother more intensely than daughters. Resources, particularly in the early years of a male's life, are channelled primarily into growth, whereas in young females, more is placed into fat reserves. There is clearly a trade-off here between sexual selection and natural selection (Clutton-Brock *et al.*, 1981) (Fig. 6.1). Indeed the lower fat reserves of males are reflected in high male mortality during harsh winters. Thus, the huge reproductive benefits of large size and antlers, which lead to high male rank in the mating game (sexual selective benefit) are balanced by a cost in terms of gambling with overwinter survival (natural selective cost).

The difference in the reproductive success of high-compared to low-rank males is so great that natural selection has favoured females which have the ability to influence the sex of their calves. Further study of the red deer showed that high-rank hinds (mature females) in good condition, carrying ample fat reserves, were more likely to produce sons than daughters. In contrast, those in poor condition tended to produce daughters (Clutton-Brock *et al.*, 1984). The reason for this is intuitively obvious. Because of the intensity of male competition, male reproductive success is strongly dependent upon size and condition. A large, good condition male will dominate others and gain a disproportionately high number of matings, whereas a small male usually gains few if any matings in his lifetime.

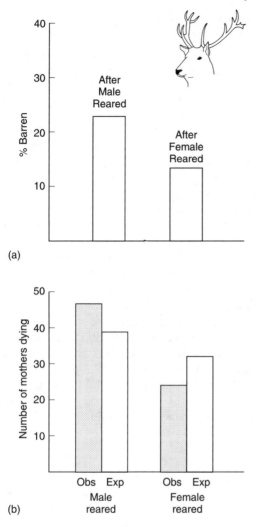

Fig. 6.1 The cost of producing male progeny to red deer hinds (after Clutton-Brock *et al.*, 1981). (a) The proportion of hinds that failed to conceive in the year that they gave birth to male or female calves. (b) The number of hinds that died within a year of giving birth to male and female calves. The expected numbers are based on the assumption that hinds that have produced sons and daughters have an equal likelihood of dying.

Female reproductive success is less dependent upon condition. Although a large female may well have a higher lifetime reproductive success than a small one, small females still have a significant fecundity. A female who can provision her developing offspring well, both pre- and post-natally, will increase her own long-term reproductive success considerably if she provisions a male, for a well-provisioned male will have higher lifetime reproductive success than a well-provisioned female. However, if the amount of provisioning that can be afforded is low, then the best allocation of the resources is, in selective terms, to a daughter, because a poorly provisioned female has, on average, higher lifetime reproductive success than a poorly provisioned male (Fig. 6.2).

Post-copulatory male competition

The competition between males described above is concerned with the monopolisation of access to females. However, in a large number of species (e.g. chimpanzees, birds, adders and many insects) females regularly mate with more than one male. When this occurs, there is selection on the male not just to maximise the number of matings he achieves, but also to ensure that his sperm are involved in fertilisation. This may lead to a phenomenon known as **sperm competition** (Parker, 1970).

Sexual selection favours males which promote their sperm over others already in the female. When competition between sperm from different males is intense, sexual selection favours males which produce more sperm. For instance, the relative testes size of the promiscuous chimpanzee (measured as testes : body mass ratio), is greater than that found in species of primates where females have one partner at a time (Harcourt *et al.*, 1981). In insects, mechanisms of promoting a male's own sperm above others are more subtle. Male dung flies (*Scatophaga stercoraria*) flush out the sperm of previous males, finely tuning the time spent doing this to their probability of a future mating (Parker and Simmons, 1994). Recently, studies have revealed that male fruit flies (*Drosophila melanogaster*) deposit chemicals in the fluid of the ejaculate (the

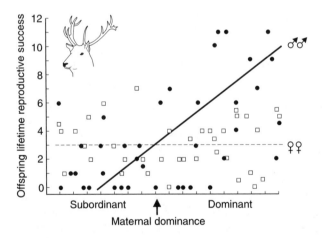

Fig. 6.2 Lifetime reproductive success of male (•) and female (□) progeny born to red deer hinds of known social rank. The data represented are taken from six different cohorts. (After Clutton-Brock *et al.*, 1984.)

accessory fluid) which kill or disable previously stored sperm, but leave their own ejaculate unharmed (Harshman and Prout, 1994). Whether this has anything to do with the fact that this species produces 'a most impressive male gamete' some 1.76 mm long (Cooper, 1950) remains to be seen.

Sexual selection also favours traits in males which safeguard paternity in the face of further males desirous of copulating with his partner. This selection is most intense in species with biparental care. Here, the male is protecting his investment. In many bird species where both sexes care for the offspring, females are observed to mate with males other than their partner (**extra-pair copulations** or **epcs**). In species where epcs are common, males guard their partners, and will force them to copulate with him if sexual infidelity is suspected (see Birkhead and Moller (1992) for a review). Mate guarding has also evolved commonly in insects. In many species of damsel fly (Odonata), the male will stay clasping onto a female until she has laid a clutch of eggs.

Other male strategies for protecting paternity have also evolved. In the domestic cat, the damage the male's barbed penis does to his partner's vagina may be interpreted as resulting from selection acting on males to lower the desire of his partner to mate again. Males of the South American butterfly *Heliconius erato* introduce an anti-aphrodisiac into their partners which repels further males for several weeks (Gilbert, 1976). In other insects, such as some grasshoppers and butterflies, the male inserts a physical barrier into his partner during copulation, in the form of a sperm plug (Parker and Smith, 1975; Ehrlich and Ehrlich, 1978). Finally, the male fruit flies of the species *Drosophila bifurca* produce giant sperm some 58 mm in length, or 20 times the length of their bodies. The reason for the production of this giant sperm is not known, but it seems inconceivable that some form of sperm competition has not played a part in the evolution of these gametes (Pitnick, 1995).

Sexual selection often goes unnoticed when the effect is on ensuring fertilisation rather than access to mates. However, selection to ensure paternity has produced, if anything, a more diverse range of adaptations in males than selection to gain access to mates. The principles are broadly the same, with perhaps one exception. Competition between males for access to mates often has little direct effect on the mates in question. For example, the competition between red deer stags has little effect on the hinds. Although females are sometimes innocent bystanders who get injured in the battle, the effects on them of the competition between males are generally limited. Competition between males to ensure paternity often has a much more tangible effect. Male ladybirds which mount for long periods of time represent an energetic cost to the female (Majerus, 1994). Male *Drosophila* introduce chemicals into the female's reproductive apparatus which damage her (Chapman *et al.*, 1995), so paralleling the tom cat's barbed penis. In an extreme case, male anthocorid bugs of the species *Xylocoris maculipennis* bypass the female's sperm storage mechanism (Carayon, 1974) by puncturing her cuticle and injecting sperm directly into her to seek the eggs. Clearly, the behaviour of males in maximising their access to fertilisation can injure their female partners. In these cases, male competition has led not just to a conflict of interests between males (over access to females and fertilisation), but also to a conflict of interests between males and females.

Male competition: future directions

Darwin's theory of sexual selection by male competition has had a successful career. Examples were, and are, easy to think of – so easy in fact that it has received relatively little careful and critical theoretical or experimental study. Furthermore, past studies may have missed many of the subtleties of this issue. This is because variance in male reproductive success has until now been hard to quantify. We can watch female and male behaviour, and record copulations, but we may miss certain copulations. More importantly, it has been difficult to ascertain paternity accurately when females mate multiply. Many studies have relied on the use of genetically marked laboratory-reared males to give insights into sperm competition.

Molecular genetic techniques with more variable genetic markers now allow us to assess reproductive success in the field, for they allow accurate ascertainment of paternity when females are promiscuous. We can now determine the winners and losers in the game of sperm competition. They are also giving us insights into the reproductive biology of species whose ecology renders them difficult to observe. For instance, we are for the first time developing a picture of the pattern of reproductive competition in aquatic mammals. It is to be expected that as our knowledge of paternity becomes more exact, and the number of systems we study becomes more diverse, new insights will be developed into the evolutionary biology of male competition.

Sexual selection by female choice

Whereas the antlers of a red deer stag clearly serve to combat other stags, many of the most extreme **secondary sexual characters** of males appear to have nothing to do with fighting. **Sexual dimorphisms** involving traits such as the extravagant plumes of male birds of paradise, ruffs and widow-birds, the iridescent colours of male birdwing butterflies, the colourful throat pouches of Anolis lizards and many others, are difficult if not impossible to explain in terms of male competition. Darwin (1859, 1871) recognised the problem that such traits pose. To explain their existence he proposed that females exercised choice between available males, mating preferentially with the most adorned males. Darwin did not provide a mechanism for the evolution of female choice, he simply took it as a premise to explain traits of this type.

While there has never been any doubt that males compete for females, Darwin's suggestion that females choose between males has been the subject of a long debate. Demonstrating female choice is often problematic because it is difficult to partition out the effects of female choice from those of male competition. It is easy to see that there are some males which gain more copulations, but it is more difficult to discern whether this is because they have warded off other males, or because they are more attractive to females. Recently, extensive efforts have been devoted to the enquiry of whether mating preferences (generally, but not exclusively, expressed by the female) influence the evolution of secondary sexual traits in the opposite sex, and to demonstrating that females do choose between males. Few evolutionary biologists would now deny that females of many species choose between males. Studies of species from widely disparate taxa have yielded unequivocal data on this point. It is also certainly true that female preferences can be genetic (see Bakker and Pomiankowski (1995) for a review).

The evolution and maintenance of female choice genes

Whilst it is easy to see why, in the presence of choosy females, males should have evolved traits that make females more likely to select them as mates, there is much debate as to why females should have evolved these mate preferences. There are three main types of hypothesis concerned with the spread and maintenance of female choice.

1. Female choice between males may have arisen not in the context of mate choice, but in some other ecological context. Choice is not selected because the partner obtained is in any way better, but arises as a by-product of selection on the sensory biology of the organism in some other context. This type of effect is termed the **sensory exploitation hypothesis**, for selection acts on males to exploit pre-existing biases in the sensory systems of the female.

2. Female choice between males may have spread and be maintained because of some direct benefit that the preference gives to the female. Here, the number of progeny raised in the lifetime of the female is increased by her choice, and choice is said to have spread under **direct selection**. The benefit received may be in the form of increasing her prospects for future survival and reproduction, or in aiding the prospects of her progeny. Crucially, the benefit does not derive from the genes inherited by her progeny.

3. Female choice between males may have arisen and be maintained because her preference, although it does not increase her lifetime reproductive success directly, increases the lifetime reproductive success of her progeny. The female choice trait spreads and is maintained because the preference is for the genetically fittest males. The preference means her progeny have a genetic constitution which gives them higher fitness. These progeny also inherit the choice trait, and the correlation between preference and genes producing high fitness may result in the spread and maintenance of the preference. Mate choice is said to have spread and be maintained due to **indirect selection**.

Sensory exploitation: pre-existing female preferences?

The sensory exploitation hypothesis, as outlined above, suggests female choice may spread not because choice itself is beneficial, but because of biases in the sensory biology that have evolved in contexts other than mate choice. A good example of how such biases may occur is provided by Heather Proctor's study of the water mite, *Neumania papillator* (Proctor 1991, 1992). In this species, individuals detect prey by adopting a 'net stance' posture (Fig. 6.3), in which they rest on their hind four legs and hold their front four legs in the water column to detect their swimming copepod prey by the vibrations they produce in the water. On sensing these vibrations, the female water mite orientates towards the prey and clutches at the source of the vibrations. Males have evolved to make use of this predatory response. They mimic prey by vibrating their first and second legs near the female. This often results in the female treating the male as if it were prey, and clutching him between her forelegs. However, she does not eat him and may subsequently take up a spermatophore deposited by him. The match here, between male and prey behaviour, and the similarity of the response of the female to these sensory inputs, leads to the conclusion that the male trait has evolved to exploit the predatory responses of the female. Female behaviour is adaptive in the context of foraging, but has also resulted in sexual selection in the male. The female choice is a by-product.

Female choice is likely to be common whenever it is the females that approach males in courtship, rather than the reverse. When this occurs, we can expect selection upon males for traits which match the sensory biology of the female, and therefore attract her to him. Where vision is important, males will be the colours that females see most easily; where sound is important, males will emit

Fig. 6.3 The net stance of a female water mite, *Neumania papillator*.

sounds of the pitch that females hear most easily; and where olfaction is important, males will emit those odours that females detect most easily.

In the examples above, there is a clear link between the species' sensory
biology and the female preference. However, other instances can be found
where female choice is more cryptic, with no obvious sensory link. Here we
need to ask which came first, the trait or the preference. Did the male trait
evolve to exploit a pre-existing preference or did males always have some form
of the trait and females subsequently evolved their preference? Only in the case
of the latter, if female choice is recent, can we argue that it brings direct
benefits. Otherwise we must conclude that mate choice is merely a by-product.
In order to sort out such circularity it is useful to construct a **phylogenetic tree**
of the group in question, assess each member of the group for the presence of
preference and preferred character, and from this infer when the female preference and male trait arose (Fig. 6.4). A recent analysis of female preferences
in platy and swordtail fish (genus *Xiphophorus*) has suggested that female preferences for what are quite extreme male traits may be pre-existing, and that
the male traits may be an evolutionary response which exploits the sensory
biology of the female (Box 6.1).

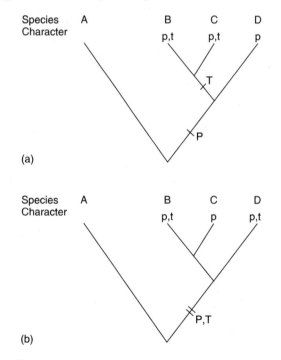

Fig. 6.4 Two phylogenetic scenarios for the evolution of mating preferences and preferred traits. (a) Sensory exploitation: trait evolves after preference. (b) Coevolution of
trait and preference. A, B, C and D are different species. p and t indicate the presence of
the mating preference and the preferred trait respectively in current species. P and T are
the inferred times that the mating preference and preferred trait evolved.

Box 6.1: Sensory exploitation: the case of the female preference for male swordtail fish in species where the males lack swords

The genus *Xiphophorus* contains species which bear sword-like extensions to the caudal fin (the swordtail fish), and species without this extension (the platy fish). In the swordtails, there is a strong female preference for males bearing the sword-like extension. What is interesting is that when Alexandra Basolo examined the sexual preferences of female platy fish, she found they also responded preferentially to male swordtails which are adorned with a sword over males of their own species (which are naturally swordless) (Basolo, 1990).

How could such a strange situation have arisen? There are two possible explanations. First, the sworded condition and female choice of sworded males could have coevolved, and that platy fish, for some reason, have lost the sword but maintained the choice. Alternatively, the female preference for sworded males could be ancestral, and the swords have evolved later. This would be to say that the female choice evolved in the absence of the male trait and that the choice did not spread because of the trait (it is a by-product). Basolo constructed a phylogeny of the genus based on their morphology (Fig. a). This suggested that the non-sworded condition is ancestral in the genus. She therefore concluded that the preference of females to mate with a sworded male arose before any males had swords, and the preference for sworded males is a by-product.

The story became controversial when Axel Meyer and co-workers constructed a phylogenetic tree of 22 species of *Xiphophorus*, based on analysis of both mitochondrial and nuclear DNA sequences (Fig. b). This evolutionary tree differed very significantly from the morphology-based tree used by Basolo. In particular, it indicated that the sworded condition has originated and been lost repeatedly, and it suggested that the ancestor of the genus is more likely to have been sworded than not (Meyer *et al.*, 1994). There is no evidence from phylogenetic considerations of this genus for the conclusion that the preferences for sworded males existed prior to the sword itself, and thus no evidence for Basolo's thesis that the female choice is a by-product, later exploited by males in the swordfish.

The sensory exploitation interpretation of female choice has, however, been revived with Basolo's observation that the preference for the sword is also present in the sister genus of the platy fish, which do not have swords. She argues that even if the sword is ancestral in the genus *Xiphophorus*, it is not ancestral to the sister genus and has never evolved here (Basolo, 1995). She concludes that in spite of the evidence of Meyer *et al.*, the preference for swords pre-dates the swords, and the preference is a case of a sensory bias.

continued

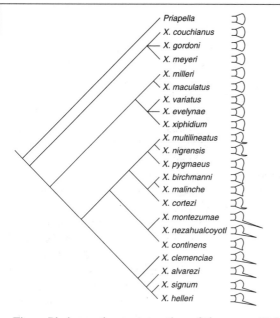

Fig. a Phylogenetic reconstruction of the genus *Xiphophorus* based on morphological analysis (after Basolo, 1990).

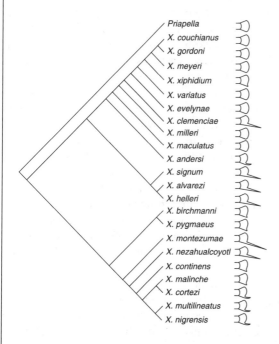

Fig. b Phylogenetic reconstruction of the genus *Xiphophorus* based on mitochondrial and nuclear sequence analysis (after Meyer *et al.*, 1994).

How do sensory biases arise?

It is possible that 'by-product' female choices, such as that which Basolo suspects in the genus *Xiphophorus* (Box 6.1), have arisen more frequently than previously thought. This possibility has been raised as a result of recent theoretical treatments of the problem of signal recognition.

Signal recognition is easy in a world where the signals are constant in appearance. We can develop computer systems which can, with suitable sensory inputs, correctly identify different shapes, as long as conditions of light and distance are fixed. However, as those who have been attempting to construct such systems have found, recognition of variable signals is much more difficult for a sensory system to manage. Because the light conditions which illuminate the subject, the background the subject is on, and the distance of the subject from the perceiver are all variable, recognition systems in nature have to be designed to cope with variation in the image received. Magnus Enquist and Anthony Arak argue that in animals, the need to cope with variation produces selection for neural nets which can generalise sensory inputs, that is to say, classify them into different types. Neural networks can be designed which manage the required generalisation, but these nets give rise, by chance, to responses to unrelated signals. These 'by-product' responses may result in mate preferences. Proponents of this type of hypothesis have suggested that female preferences for both exaggerated tails and for symmetry may be a result of the way females have solved the problem of recognising variable objects (Arak and Enquist, 1993; Enquist and Arak, 1993, 1994; Johnstone, 1994).

The most common signals a female is likely to use are those associated with species identity (Enquist and Arak, 1993). When two species come into regular contact but are genetically isolated due to **post-zygotic reproductive isolation mechanisms** (hybrids between them are inviable, sterile or severely maladapted), selection will favour females which can exercise a choice to mate with members of their own species. This choice will spread, and either one or both species will develop a signal used as a cue in mating. The mate preference and the signal will each go to fixation. The preference does so because individuals which exercise this choice gain benefit in avoiding the energy wastage that would be incurred if they mated with non-conspecifics. The signal goes to fixation because individuals expressing the signal will find more willing conspecific partners. Of course, although once a specific signal has become fixed in a species it cannot be the basis for choice between conspecifics, the choice that has evolved may, either in the short or long term, produce other mate preferences. Selection acts to promote the first variant that arises that increases fitness. In this case, any conspecific mate recognition gene(s), however it is expressed behaviourally (choice of specific signal, most expressed signal, super-normal stimulus, etc.), would be beneficial initially because it would reduce matings with non-conspecifics. This means that in some instances, extreme exaggeration of a signal may evolve, not because there is any advantage to this extreme exaggeration, but merely as a consequence of the behavioural expression of conspecific mate recognition genes. Furthermore, if the selective pressure to maintain particular conspecific recognition systems is relaxed at some future time, by, for example, the loss of sympatry with similar species, the selective constraints on mate preferences would also be relaxed with the consequence that these genes would be free to evolve in new directions. It is thus possible that female mating preferences may turn out to be by-products of the problem of signal and species recognition.

Direct selection on female preference

It is clear that, in the majority of examples of female choice, preferences have spread and are maintained in populations because they bring direct benefits to choosy females. Direct benefits may be easily assessable by females. For instance, in hanging flies (*Hylobittacus apicalis*) females exhibit a preference for males which provide the greatest offering of resources, in the form of a captured insect. Here, female reproductive success is strongly related to the size of these 'nuptial gifts', for a female's ability to produce eggs relies on her nutritional status (Thornhill, 1976) (Fig. 6.5), and it is easy to see that genes conferring female choice of males that bring the largest gift will spread.

Similarly, the territory quality of a partner is a parameter of key importance in raising progeny in many species of bird, and females select males accordingly. In other cases, the benefit of choice is less easy to see. In some cases, the

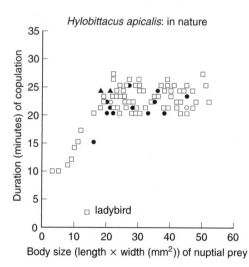

Fig. 6.5 The effect of the size (and palatability) of nuptial prey on the duration of copulation in the hanging fly, *Hylobittacus apicalis*. The symbols represent different numbers of similar observations as follows: □ = 1, ● = 2, ○ = 3, ▲ = 4 (after Thornhill, 1976).

choice may be advantageous because it reduces the costs females incur in mating (Maynard Smith, 1991). For instance, female choice for males with loud calls (or other highly apparent traits) may have evolved because it lowers the time, and perhaps the danger, involved with finding a mate. A loud call is easiest to follow when there is noise interference in the environment, and selecting a male with the loudest call will reduce the cost of searching. Alternatively, mating costs may be reduced if, by exercising choice, a female can lower her probability of mating with a parasitised male who might infect her during copulation or other reproductive interactions (Borgia and Collis, 1989).

All of the above benefits could be obtained by the female, even though the signal may only be an indicator. However, in species with biparental care, direct benefits may also accrue from choosing a male who will make a large future contribution to rearing offspring – in other words, a good provider. Unfortunately, there is a problem in that poor-quality males could potentially 'cheat' by displaying whatever trait is used as a quality indicator. Females must therefore be careful. For example, female common terns (*Sterna fuscata*) will break off interactions with males who fail to bring in large amounts of food during courtship. The ability of a male to bring in food during courtship reflects his ability to bring in food when the chicks have hatched, and the female choice is for a current indicator of a future desired ability (Nisbet, 1977). A better strategy is for females to choose according to an aspect of male biology which cannot lie, some direct measure of quality. For example, the preferred trait might be costly, such that it can be produced only by a healthy, well-fed male. Female choice for males with such 'handicaps' will spread because they ensure greater 'honesty' on the part of the male.

Indirect selection of female preferences

The above examples of the spread and maintenance of female choice rely on the direct benefits that the choice produces. These are easy to understand, and most authors consider that direct benefits underlie the evolution of most cases of female choice (e.g. Fisher, 1930; Maynard Smith, 1991). However, some cases of female choice are difficult to explain in such terms. Some polygynous birds, such as the peacock, and mammals, such as the Uganda kob, exhibit a lek mating system. In a lek mating system, males congregate in one area and defend small territories virtually devoid of resources. Females migrate to these areas solely for the purpose of mating; they select males, and in doing so, they show preferences for certain males. No nutrient transfer appears to take place, no male care of progeny is observed, and the female vacates the male's territory shortly after mating. Females receive only gametes from their mate. Many of the cases of extra-pair copulation in birds also seem impossible to explain in terms of direct benefit to the female. Here, the females appear to obtain no resources from the male, just an extra copulation (Birkhead and Møller, 1992).

In cases such as these, where there appear to be no direct benefits, female choice can only have spread through the indirect benefits associated with being present in fitter progeny. The choice spreads because it is found in genetically fitter progeny. Much theoretical work has been devoted to the nature of these hypothetical genes which increase fitness. Two basic hypotheses have emerged. First, these genes may confer increased viability and are therefore favoured by natural selection. This hypothesis is known as the **'good genes' model of sexual selection**. Alternatively, female choice may select males which are fitter because they are attractive to females, such that they produce sons who are themselves attractive, so conferring an advantage through sexual selection. This second scenario is referred to as the **'sexy sons'** model or the **Fisherian mechanism of sexual selection**.

The spread and maintenance of female choice

Good genes   A female choice may spread if the choice is for a male trait which is an indicator of good genetic quality. This is because her progeny will tend to bear both good genes and the preferred trait. The choice generates a correlation between good genes and preferred trait which causes the trait to increase. However, if the characteristic of males chosen, although initially indicating good genes, is also achievable by males with bad genes, then we expect selection due to female choice to favour males with poor genes which mimic the preferred trait. When this occurs, female choice becomes either unproductive or even costly. Female choice will thus only spread if the trait females choose is a reliable indicator of fitness, that is to say, it can only be produced by males of genuinely good genetic quality. This need for honesty led Amotz Zahavi to propose the **handicap principle**. Females should exhibit preferences only for costly traits (handicaps) which, because of their cost, are incorruptible indicators of genetic quality (Zahavi, 1975, 1977). Only a male with good genes can bear a handicap, and thus females can use them as reliable indicators of genetic quality.

As first proposed, the handicap principle was flawed. The problem was a lack of correlation between the handicap and good genes. Females which mate with handicapped males would produce, in addition to attractive (handicapped) sons with good genes, unattractive sons with good genes and attractive sons with poor genes (Maynard Smith, 1976b). These mismatches negate the benefit to female choice. However, the model can work if the expression of the handicap is modulated by the quality of the individual bearing it. The size of the handicap must bear a direct relation to the quality of the male; it must be a **condition-dependent handicap**. Under these circumstances, poor-quality sons of choosy females do not bear the full cost of the handicap, and the choice may spread. Bill Hamilton and Marlene Zuk (1982), for instance, suggest that traits indicating parasite resistance might be condition-dependent.

Males with parasites would be weakened, and would not have sufficient energy to dedicate to the development of bright coloration or a long tail.

A problem which 'good genes' models have confronted is that consistent female preference for males of good genetic quality should lead to a rapid loss of genetic variation for quality. When females select males with good genes, they are selecting out bad genes from the population. At some point, female choice no longer confers a benefit because all the bad genes have been selected out of the population and all the males are thus of the same quality. At this point, the female choice will stop spreading. If the choice is costly because, for example, it puts females at risk, wastes foraging time or leads to later breeding, the choice may start to decrease in the population.

For female preferences to spread and be maintained, therefore, it is necessary for there to be something continually generating differences in fitness between individuals. Two responses to the problem of diminishing genetic variation have been proposed. First, it has been argued that females may choose males which have good genes at very many loci, with variation in fitness being produced by environmental variation and mutation. Second, it has been argued that genetic variation in fitness at either one or a few loci may be maintained because of interactions with coevolving enemies. This argument, elegantly encapsulated by Hamilton and Zuk, points to the role of parasites in maintaining diversity at a particular locus.

Consider a situation where parasite resistance is encoded at a single locus in the host, with resistance alleles being complementary to different parasite virulence alleles. When one particular host resistance allele is common, then parasites bearing virulence alleles which evade the resistance are favoured and become more common. Selection will then produce changes in the frequency of host resistance alleles, and so on. Coevolution means one genetic type is never always best, but there is always genetic variation in viability between individuals, and any female choice for a trait whose condition indicates parasite resistance may spread and be maintained (Hamilton and Zuk, 1982).

It is hard to test the hypothesis that female preference results in progeny with better genes and thus improved survivorship. To assess the importance of good genes benefits, it is necessary to measure the fitness of progeny. The environment strongly affects survivorship, making fitness hard to measure. Furthermore, the study must then control against maternal effects. For example, observations of fitter progeny from choosy females may only mean that choosy females place more resources into parental care.

Empirical evidence for female choice of males with good genes comes from a recent study by Marion Petrie on the evolution of female choice in peacocks, *Pavo cristatus*, in Whipsnade Park. Petrie's previous work on this lekking species had shown that peahens prefer to mate with peacocks with the most elaborate tails, measured by the number of eye-spots they contain (Petrie *et al.*, 1991). She followed this work by tracking the fate of progeny of females who had randomly been allocated males with different tails sporting different numbers of eye-spots. Progeny from these females were raised under controlled conditions, and their weights measured after 84 days. She then released the offspring into the park, and monitored their survival. She found both that progeny sired by males with high eye-spot numbers were heavier after 84 days, and had a greater chance of survival to their second birthday. In summary, progeny sired by males with the longest tails had consistently higher survival, and hence fitness, under a variety of measures (Petrie, 1994).

There is now evidence that female choice can give a 'good genes' benefit. Other work has shown that 'good genes' can be parasite-resistance genes. It is clear from work by Anders Møller (1990) that the conditions for Hamilton and Zuk's parasite-related hypothesis to occur are met, at least in part. Working on the barn swallow, *Hirundo rustica*, he showed that females prefer to mate with males with long tails, that these males have fewer parasites, and that they bear parasite-resistance genes. The latter he demonstrated elegantly by cross-fostering experiments, thereby ensuring that parasite load reflected their genetic quality, and was a product neither of parental phenotype (healthy parents work harder at rearing offspring and could therefore rear healthier progeny) nor parental parasitisation rate (progeny from healthy parents would have fewer parasites because they would not have contracted them from their parents). He found that, despite cross-fostering, progeny of females that mated with long-tailed males carried fewer parasites than those born as a result of matings involving short-tailed males. This is good evidence for a genetic component to parasite resistance. There is thus evidence in this system for a 'parasite resistance' centred benefit to female choice.

Fisher's runaway hypothesis   The 'good genes' model of sexual selection relies on the correlation between mate preference genes and genes which produce individuals of high viability and thus high fitness. However, fitness is a product of both natural selection and sexual selection. Correlations between female preference genes and genes which make male progeny more attractive may also cause female choice to spread. When a significant proportion of females in a population prefer to mate with a certain type of male, males carrying the preferred trait will be fitter, and the trait will increase in frequency. At the same time, these 'sexy sons' will tend also to carry the female preference gene(s). Consequently, both trait and preference form a feedback loop, both increasing in frequency at an accelerating rate.

The rate of increase by genetic correlation, termed **Fisherian runaway**, is initially extremely slow because, when the preference is at or around the mutation rate, so few females are selecting males with the preferred trait. Fisher (1930) recognised this problem and solved it by proposing two selective influences on the evolution of the male trait:

1) an initial advantage not due to sexual preference, which advantage may be quite inconsiderable in magnitude, and
2) an additional advantage conferred by female preference, which will be proportional to the intensity of this preference.

Fisher was here proposing that female choice would only spread if, at first, the female was choosing males with a selectively advantageous trait. Fisher envisaged that correlation of gene(s) for this beneficial trait with the preference gene(s) would produce its initial spread. Thereafter, the spread of the two would occur primarily as a result of the reproductive advantage gained by the sons of choosy females who would have the preferred trait and would carry the preference gene(s). The expected outcome of this process is that the preferred trait either increases to a frequency in the population above that predicted under natural selection alone (if a single or limited number of genes are involved) or be elaborated to a greater extent than expected under natural selection (if many genes are involved) (O'Donald, 1967, 1980; Lande, 1980; Kirkpatrick, 1982). The process reaches equilibrium when the sexual selective advantage due to the preference is exactly balanced by the natural selective disadvantage of any further increase in the frequency or exaggeration of the preferred trait. The difference between the Fisherian mode of spread and good genes models is that Fisher did not envisage the presence of any 'good genes' effect at equilibrium. The only advantage envisaged is from sexual selection. Thus, when Fisherian runaway is important, females which choose should produce daughters which are no more fit than those produced by unchoosy females, and sons which, although less viable, are sexier. By contrast, in good genes models, daughters of choosy females should be more fit.

The major criticism levelled at the Fisherian process is that, although it is clear that it can cause the spread of a female choice and male trait, it is less clear why a female choice which has spread by Fisherian runaway should be maintained. This is because strong **directional selection** on the preferred male trait will rapidly deplete genetic variability for this trait, leaving all males equal. Any cost to female preferences will then select out the choosiness gene(s) because there are no longer any compensating benefits to choice. One possible way in which female preference could be maintained would be if mutation were biased against the preferred trait (Iwasa et al., 1991). If this is true, and it is perhaps likely, then choosiness would continue to pay and a low-cost preference could be maintained.

There is, as yet, no incontrovertible evidence that the Fisherian process produces the spread and maintenance of a choice gene to an extent that a naturally selected disadvantage results. To provide proof of Fisherian benefits, it is necessary to demonstrate a correlation between preference and preferred character, and show that the preferred character confers a sexually selected advantage. This scenario has not yet been found. This is largely because the level of genetic correlation between choice and preference is empirically difficult to measure. Males show the preferred trait phenotypically, but not the mate preference, and females show the mate preference, but, in most cases, not the preferred trait. It is hard to measure a correlation when an individual only expresses one of the characters at a time!

The best evidence for the importance of sexually selected benefits comes from Andre Gilburn, Simon Foster and Tom Day's collaborative study of the seaweed fly, *Coelopa frigida*. Female seaweed flies prefer large males (Gilburn *et al.*, 1992). Male size is largely encoded for by genes present in an inversion segment of one of the chromosomes. Females can therefore be assessed directly for both the preference, through a choice test, and the preferred character, cytologically. A correlation between the presence of the preferred male character and preference is observed (Gilburn *et al.*, 1993). What is as yet unclear is the importance of sexually selected benefits in the evolution of female choice in *C. frigida*. Genetic correlations between preference and preferred character are expected not only under Fisherian models, but also under 'good genes' models.

Indirect benefits: assessing the evidence

The work by Petrie on peacocks and Møller on barn swallows provides evidence for 'good genes' benefits to female choice. The investigations of seaweed flies by Gilburn *et al.* have shown that the most important feature of the Fisherian mechanism, genetic correlation, occurs in nature, and thus sexually selected benefits to female choice are also likely to exist. These and other case studies are important in that they confirm that the assumptions behind the various models of indirect sexual selection can be met in nature, and thus confirm the hypotheses as tenable explanations for female choice. However, they do not, on their own, constitute proof that the advantage in question is the cause of the spread and maintenance of choice either in a particular case, or generally. As Mark Kirkpatrick and Michael Ryan (1991) note, finding evidence for one benefit in a particular species does not mean that this benefit is the only cause of the maintenance of the choice, nor does it imply that this mechanism operates in other species.

Kirkpatrick and Ryan (1991) illustrate this principle by considering Møller's study of the barn swallow, *Hirundo rustica*. Female barn swallows exhibit a preference for males with parasite-resistance genes. However, it may not in fact be this benefit that maintains the choice. In order to prove the importance of parasite-resistance genes or any other benefit in the evolution of female choice, it is necessary not only to show that they gain a benefit from this source, but also that the evolution of the preference is not amenable to other interpretations. In the case of the barn swallow, where male parental care exists, female choice of males which are in good condition because they are unparasitised may have spread because of direct benefits to the choice. Unparasitised males may be capable of more effective care and are likely to pass fewer parasites on to her and on to her progeny. It may be that parasite-resistance genes constitute too weak a force to produce the spread of a preference in themselves, and that although they correlate with choice, they did not cause the choice, and the choice might even be maintained in the absence of heritable resistance to parasites.

The next step, therefore, is to examine not only whether the assumptions behind models are tenable, but also to examine individual cases for all benefits. With such an assessment, detailed models can be developed for the species being examined to partition out the advantages, and discover which of them were, and are, important in the spread and maintenance of female mating preferences: direct benefits, good genes benefits, Fisherian benefits, or a combination of all three.

A further complication to female choice: evolution and maintenance of preferences may have different causes

The passages above clearly show that female choice is an inherently difficult subject to study empirically. With so many potential factors, it is hard to dissect out the particular forces maintaining a female preference. To add to this complex situation, there is one more problem. The study of female choice in an extant population tells us about the current factors maintaining female choice, but says less about the factors which produced the initial spread of the trait. This point is exemplified in Michael Ryan's study of the tungara frog, *Physalemus pustulosus*.

In tungara frogs, females are attracted to the calls of males, and have a preference for calls of certain frequencies. Tests on females of two related species, *Physalemus pustulosus* and *P. coloradorum*, showed that both were attracted towards calls including a low-frequency chuck element, despite the fact that *P. coloradorum* males do not 'chuck'. The attraction of females towards chuck sounds appears to pre-date the chuck, and, therefore, may be best interpreted as a by-product of an auditory bias of the female ear. Chucking appears to be a case of sensory exploitation (Ryan *et al.*, 1990). However, previous work by Ryan and co-workers showed a strong correlation between chuck frequency and male size. Preferred males, those with low-frequency chucks, were also larger. Large males produce more sperm than small males, and therefore fertilise a higher proportion of a female's eggs. There is, therefore, a direct benefit to female choice (Ryan 1983, 1985). Thus, despite the fact that the initial evolution of the female preference appears to have little to do with direct benefits (it is ancestral), the female preference now gives the female a direct benefit, and this may be construed as maintaining the preference. So it appears that, in this case, the reason for the evolution of the male trait and the reason for its maintenance are different.

Female choice: future directions

The mechanisms which may maintain female choice genes in nature are beginning to become understood. Future research is likely to be centred on the relative importance of direct and indirect benefits to female choice and in attempting to collect unequivocal evidence of Fisherian selection or indicators of viability in the spread and maintenance of female choice. This is a formidable task to perform in any species. As has been seen, providing such evidence for the role of good genes and Fisherian benefits is hard. Assessment of the relative importance of these factors requires that many such studies are carried out.

Two possible factors may make this task easier. First, there is evidence that some female preferences are encoded by only a single gene. This was shown originally for two-spot ladybirds, which have a genetically determined preference to mate with black rather than red males (Majerus *et al.*, 1986) (Box 6.2), and has been shown in other organisms (Bakker and Pomiankowski, 1995). In such cases, genetic crosses, combined with technology for rapid gene mapping, may allow the preference gene to be identified, cloned and characterised. Insight into one system may lead rapidly to a better understanding of others.

Second, Fisherian runaway makes a strong prediction. If and when it occurs it will tend to result in the rapid fixation of whatever chromosome(s) carry the preference and trait genes. Such a process would constitute a selective sweep, effectively eliminating neutral genetic variability from chromosomal regions flanking the key genes. With the number of available genetic markers and methodologies for quick genetic typing increasing rapidly, perhaps these chromosomal regions with low variability may soon be identified. At the centre of each bald patch should lie the gene.

Another issue which is increasingly receiving attention is that of whether females choose between sperm from different male partners. Our understanding of female choice is, at present, biased towards considerations of female choice of the partners with whom they mate. However, it is now recognised that, just as male competition may occur in the form of sperm competition in species where females mate multiply, so female choice may also occur after copulation. Females may choose males by choice of sperm rather than mating partners. That post-copulatory effects may be important is suggested by a recent study of the adder, *Vipera berus*. Here, females which multiply mate produce offspring with higher viability than females which do not (Madsen *et al.*, 1992). The implication is either that competition between sperm resulted in the sperm from fitter males fertilising the female, or that the females chose the sperm from fitter males. Future research will reveal whether females encourage sperm competition to get the best sperm, or whether they can even bias fertilisations towards sperm with good genes.

Outside of these extensions to our current understanding of the operation of female choice, other novel areas are emerging as foci of research. Most studies of sexual selection have dealt with animal subjects, but plants may also be subject to sexual selection. The process of dispersing pollen to other plants, and choice of pollen grains by the recipient plant when they land on the stigma, are examples of male competition and female choice. Research is currently

Box 6.2: Female mating preference in the two-spot ladybird is genetic

The two-spot ladybird, *Adalia bipunctata*, is highly polymorphic for the colour patterns on its elytra. In many populations the commonest forms are the non-melanic f. *typica* (red with two black spots) and the melanic f. *quadrimaculata* (black with four red spots). In some of these populations the proportion of melanic males in mating pairs exceeds the frequency of melanic males in the population as a whole.

By means of a selection experiment, in which only females which had mated with melanic males were allowed to produce eggs for the next generation, the excess of melanic males among mating pairs was increased significantly over three generations of selection. In the control line, all females, irrespective of the males that they mated with, were retained to produce eggs for the next generation, and this line showed no change in the excess of melanic males among mating pairs (Table a).

Table a: The excess of melanic males amongst mating two-spot ladybirds has a genetic basis. The number of ladybirds used and the matings observed for four generations of the selected and control lines, together with the estimate of female mating preference for melanic males (assuming the excess of melanic males amongst mating pairs is due to female choice). For the full analysis, see Majerus *et al.* (1982).

Generation	Number of ladybirds used		Matings (males first)				Estimate of female preferences for *quad.* males
	typ.	*quad.*	$Q \times Q$	$Q \times T$	$T \times Q$	$T \times T$	
Selected line							
1	224	96	21	31	24	46	0.180
2	112	48	13	18	9	14	0.392
3	84	36	12	14	6	10	0.456
4	140	60	18	21	5	12	0.566
Control line							
1	98	42	13	15	8	21	0.273
2	70	30	7	11	7	12	0.266
3	112	48	10	18	10	18	0.286
4	112	48	11	22	13	28	0.209

The excess of melanic males could have been due to a greater competitiveness of melanic males compared to non-melanic males in securing mates, or to females preferring to mate with melanic males. The issue was resolved by reciprocal tests using either males or females from the selected stocks (i.e. the stocks with higher excesses of melanic males in

continued

mating pairs) respectively with females or males from the original wild population. The results (Table b) showed the level of the excess of melanic males in mating pairs to be characteristic of the stock from which the females were drawn. The obvious conclusion was that the excess of melanic males in mating pairs was due to female choice, not male competition.

Table b: Formal proof of female choice in the two-spot ladybird
Seventy *typ.* and 30 *quad.* from specific stocks (equal numbers of males and females of each form), were placed in cages and matings were recorded. 'Selected' means that the ladybirds were drawn from the progeny of the fourth generation of the selected line (Table a) with a level of preference in excess of 50 per cent, while 'Keele' means that the ladybirds were drawn from unselected Keele stocks characterised by a level of preference of around 20 per cent. The observed matings and estimate of level of female preference for melanic males are given. In each case the level of preference is characteristic of the females used in the experiment.

Stocks and sexes of ladybirds used	Numbers of matings (males first)				Estimate of female preferences for *quad.* males
	$Q \times Q$	$Q \times T$	$T \times Q$	$T \times T$	
Selected ♂♂ × Keele ♀♀	11	28	17	34	0.191
Keele ♂♂ × Selected ♀♀	16	28	6	15	0.539
Keele ♂♂ × Keele ♀♀	18	35	21	42	0.224

Analysis of mating tests on F1 progeny from isofemale lines drawn from a high preference line then showed that the female preference for melanic males is controlled by a single dominant gene with incomplete penetrance (Table c). Some females have the preference allele and mate preferentially with melanic males; others do not and mate randomly. Those females that do contain the preference allele show a level of preference of about 85 per cent when equal numbers of melanic and non-melanic males are offered in choice tests. That the maximum level of preference is not 100 per cent makes evolutionary sense. It would be unrealistic to expect that a female with a preference to mate with melanic males, who continually encountered non-melanic males, would go on rejecting these throughout her life. More probably, she would eventually become so frustrated that she would accept any male as a mate, irrespective of his phenotype. Indeed, sexual frustration, shown by female two-spots that have been denied mating opportunities, has been used to break down rejection of non-conspecific males in order to produce interspecific hybrid matings (Ireland *et al.*, 1986).

continued

Table c The genetics of female choice

Results of formal mating tests in which daughters of 21 families of two-spot ladybird, chosen at random from the tenth generation of a line selected for high female preference for melanic males, were offered equal numbers of melanic and non-melanic males. The number of daughters mating with males of each form and the estimate of preference for melanic males are given for each family. The families are divided into four groups on the basis of the levels of preference they show. For the full analysis, see Majerus *et al.* (1986).

Group	Family	Matings with males		Estimate of female preference for melanic males
		quad.	*typ.*	
1	Z4	51	48	0.033
	Y4	218	204	0.038
	Z33	49	45	0.046
	Y12	57	52	0.050
	Totals	**375**	**349**	**Mean 0.041**
2	Y18	134	73	0.324
	Y14	149	74	0.367
	Z36	141	69	0.375
	Y13	223	107	0.380
	Z17	212	102	0.381
	Z34	141	66	0.393
	Z20	68	33	0.398
	Y15	154	70	0.411
	Z32	74	32	0.430
	Totals	**1296**	**626**	**Mean 0.381**
3	Z13	218	79	0.501
	Y9	150	50	0.537
	Z16	80	26	0.544
	Y3	153	36	0.654
	Totals	**601**	**191**	**Mean 0.553**
4	Y19	272	34	0.801
	Z37	91	6	0.887
	Z40	102	5	0.918
	Z35	203	7	0.942
	Totals	**668**	**52**	**Mean 0.872**

focusing on the differences and similarities between the operation of sexual selection in plants and animals.

Sexual selection was first proposed to account for some (but not all) secondary sexual traits. However, the finding that some female two-spot ladybirds, *Adalia bipunctata*, have a single dominant gene that gives them a preference to mate with black rather than red males, and that the preferred trait is not **sex-limited**, has led to a consideration of how widespread mating preferences may be (Majerus, 1986, 1994). Female choice has also been implicated in the maintenance of non-sex-limited polymorphism in the scarlet tiger-moth, *Callimorpha dominula* (Sheppard, 1952b), the arctic skua, *Stercorarius* *parasiticus* (O'Donald, 1983), the snow goose, *Anser coerulescens* (Findlay *et al.*, 1985) and the four-spotted milkweed beetle, *Tetraopes tetraophthalmus* (Eanes *et al.*, 1977). It is clear then that mate preferences do not have to be for traits that are sex-limited. Consequently, attention is turning to the role of mating preferences, expressed by one or possibly both of the sexes, in the evolution of non-sex-limited traits, including elaborate courtship behaviours.

Summary

Sexual selection is selection for traits which increase an individual's lifetime reproductive success by increasing the number or quality of mates they obtain. Sexual selection may involve traits that increase access to mates by excluding competitor individuals (intrasexual selection) or it may involve the spread of traits that increase attractiveness to mates (intersexual selection). The rationale and evolutionary consequences of the former mode is well understood. Principally, selection favours allocation of resources away from maintenance to weaponry and increased body size. Intersexual selection is not so well understood, the question centring on why one sex finds certain types of mate more attractive than others. Although it is easily understood in cases where the choice of mate is based upon the direct benefits that choice produces (such as resources, or avoidance of copulations with non-conspecifics), other cases, where there appears to be no direct benefit to choice, are more difficult to understand. Two groups of hypotheses, sensory exploitation hypotheses and indirect benefit hypotheses, are currently being evaluated. In the former, choice of mate is held to be the by-product of selection on some other aspect of sensory biology. In the latter, the spread and maintenance of mate choice is thought to be associated with obtaining mates with genes which either make the progeny more attractive to females (spread of mate choice and chosen trait by Fisherian runaway) or which make the progeny more viable in natural selective terms (good genes models).

Suggested reading

BALMFORD, A. and READ, A. F. (1991). Testing alternative models of sexual selection through female choice. *Trends in Ecology and Evolution*, **6**, 274–276.
  A commentary upon empirical difficulties encountered when attempting to discover the factors important in the spread and maintenance of female choice.
JOHNSTONE, R. A. (1995). Sexual selection, honest advertisement and the handicap principle: reviewing the evidence. *Biological Reviews*, **70**, 1–65.
  Comprehensive review of issues surrounding the handicap principle, and a good general introductory section dealing with the problems of female choice and honest signalling.

KIRKPATRICK, M. and RYAN, M. J. (1991). The paradox of the lek and the evolution of mating preferences. *Nature*, **350**, 33–38.
  An interesting discussion of the problems inherent in the evolution of mate preferences in the absence of direct benefits.

MAYNARD SMITH, J. (1991). Theories of sexual selection. *Trends in Ecology and Evolution*, **6**, 146–151.
  A good (if quite hard) short review of theories of the evolution of female choice.

Chapter 7

Evolution and interspecific interactions

Life and death in a nettle bed

A leisurely wander through the countryside, using one's eyes and ears, can reveal many fundamental truths about ecology and evolution. On sunny May afternoons in Cambridge, it has not been uncommon, over the last few years, to observe scientists sitting in the middle of beds of stinging nettles watching the activities of ladybirds. Consider the notes of one of these scientists' observations for one such afternoon. Some of the ladybirds were mating. Some of the females were laying eggs. Others were involved in interactions with other species.

One seven-spot ladybird had a rather large pale yellow maggot emerging from its abdomen. This larva wove silk around the legs of the ladybird, rendering it immobile, and then formed a silken cocoon between the ladybird's legs in which it pupated. Elsewhere, an adult insect emerged from a similar cocoon. The insect that emerged was a parasitic wasp, *Dinocampus coccinellae*. The wasp (a female) expanded and dried her wings. She was already mature, and, as these wasps reproduce parthenogenetically (males of *D. coccinellae* are not known), by the end of the same afternoon she was seeking out other seven-spot ladybirds in which to lay her eggs.

The two-spot ladybirds which were not mating (and some of those which were!) roamed over the surface of the nettles. When they encountered an aphid, they attempted to stop it, subdue it and eat it. Some ladybirds were successful, but in other cases the aphid escaped. Some of the aphids appeared to move away or jump off the plant before a ladybird reached them. They seemed to detect the approach of a predator at distance.

On some nettle leaves, batches of ladybird eggs were hatching. Many batches gave the impression that they were overflowing with minute larvae, each egg having successfully hatched. But a few batches looked different. Close inspection revealed that only about half of the eggs had hatched, and that the newly emerged larvae were eating the unhatched eggs. In the margin of the observer's field-book is a note of two possible causes of the half hatch-rates in these batches of eggs. One possibility is that the females that laid these clutches were infected with a cytoplasmic bacterium that kills male embryos, but not female ones. If so, the eggs that hatch would bear just female individuals. Alternatively, the females that laid them may have mated with a close relative, the low hatch rate being a consequence of inbreeding depression.

From these few observations, it is obvious that interactions between members of different species are often as important to an organism's survival as those that occur between members of the same species. In the nettle bed, a chain of interactions between species may be followed. Sucrose, which the nettle has produced through photosynthesis, is sucked up from the nettle's phloem by nettle aphids. These in turn are eaten by ladybirds, hoverfly larvae, lacewings and ants. The ladybirds are also eaten by a variety of predators, including spiders, beetles, bugs and a few birds, and they are heavily parasitised by flies and wasps. Micro-organisms live inside both predators and prey alike. The ladybirds may be infected with male-killing bacteria, while the aphids contain an odd internal structure. This is the mycetome, which contains a bacterium, *Buchnera*. This bacterium is passed between generations, and 'contributes' amino acids to its host. This is of significant benefit to an insect which feeds on fluids from plant phloem vessels, for this fluid is poor in amino acids.

The fitness of an individual is clearly not just a function of its 'suitability' to the abiotic environment and its ability to procure mates and food in competition with conspecifics. Its 'suitability' in interactions with individuals of other species is also critical. A *D. coccinellae* female that can locate a seven-spot ladybird in which to lay an egg more rapidly and efficiently than others will leave more progeny. If this increased ability is genetic, then in the next generation, such wasps will, on average, find more oviposition sites more rapidly. As the population of *D. coccinellae* improves in efficiency, the proportion of seven-spot ladybirds that are parasitised increases. This imposes a gradually intensifying selection pressure on the seven-spots. Ladybirds that have resistance to parasitism by this wasp will be favoured, as appears to be the case with the two-spot ladybird, which is attacked but not parasitised successfully by this wasp.

When species interact, evolutionary change in one species, say a parasite, may have a 'knock-on' effect on the environment of others, for example, its host. The study of the evolutionary ramifications of interspecific interaction is often described by the term coevolution. In its broadest sense, coevolution may be said to occur when the direct or indirect interaction between two or more evolving units produces an evolutionary response in each (Van Valen, 1983). The important feature of coevolution is that it involves reciprocity of selection pressure. Both interacting 'units' evolve. When the units in question are species, and the interaction either mutualistic or antagonistic, evolutionary biologists frequently consider two patterns of coevolution, based on the number of parties involved in the interaction. The first type involves interaction between just two species. Considered by Dan Janzen (1980) to be the

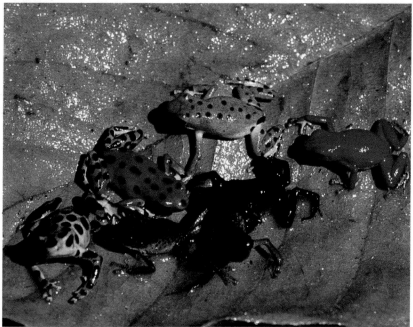

Plate 1 Diversity among different species of poison arrow frogs. The bright colour patterns all serve the same purpose, warning potential predators of the toxicity of these frogs. (Photograph courtesy of Dr Kyle Summers.)

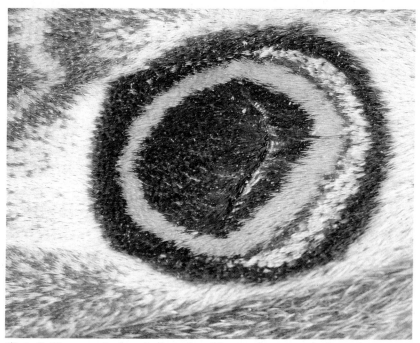

Plate 2 The complexity of adaptation: The startle eye-spot of a male emperor moth *(Saturnia pavonia)*. (Photograph courtesy of Mr Gerald Burgess.)

Plate 3 Different life-history stages may employ different defensive morphologies and behaviours. (**a - top**) The larvae of the buff-tip moth *(Phalera bucephala)* are distasteful and warningly coloured. Members of a clutch forage together. (**b - bottom**) The adults of the buff-tip are highly cryptic, habitually resting on the ends of broken twigs.

Plate 4 Adaptations may result from natural or sexual selection. (**a - top**) A mantid uses camouflage to avoid predation and to get close to prey without detection. (**b - bottom**) A merveille du jour moth *(Dichonia aprilina)* uses cryptic colouration, hiding amongst lichens (see over for plate 4c).

Plate 4(c) A male soldier beetle *(Oedemera nobilis)* with enlarged hind femurs for clinging to reluctant females.

Plate 5 A parasitoid and its host: A larva of the wasp *Dinocampus coccinellae* exits from an adult seven spot ladybird *(Coccinella 7- punctata),* before making its silken cocoon between the legs of its paralysed host.

Table 7.1 The major classes of interspecific interaction.

Type of interaction	Name	Example
Reproductive success of individual from species A increased by interaction with B, and vice versa	Mutualism	Vesicular arbuscular mycorhizal fungus, *Glomus mosseae*, associated with clover *Trifolium subterraneum*. Mineral salts (especially phosphate and zinc) move from fungus to plant, photosynthate moves from plant to fungus
Reproductive success of individual from species A increased by interaction with B; reproductive success of individual of species B decreased by interaction	Predation	Thrushes feeding on the snail *Cepaea nemoralis*
	Parasitism	Downy mildew, *Erysiphe graminis*, growing on barley, with hyphae of the fungus extending into the leaf cells, drawing out photosynthate
Fitness of individual from species B decreased by interaction with A; individual from species B also suffers decreased reproductive success because of the interaction	Competition	Seed-feeding *Geospiza* finches in the Galapagos Islands, competing for a limited seed resource. Plants in tropical rain forest competing for access to mineral nutrients in the soil

'true' coevolution, we term this 'tight coevolution'. In this type of interaction, members of species A provide the selection pressure for change in species B, and the ensuing change in members of species B has a reciprocal effect on the evolution of species A. This may be contrasted with interactions involving many parties. Interactions of this type, where one group of species is the motive force for change in another group and vice versa, we term 'diffuse coevolution'.

The purpose of this chapter is to consider the evolutionary products of interspecific interactions. Three main types of interaction will be considered; **interspecific competition**, interactions between predators and prey or parasites and hosts, and those between mutualists (Table 7.1). In cases of interspecific competition, the lifetime reproductive success of both interacting individuals is decreased, because they both utilise the same resource that is in limited supply. Here, the removal of individuals of one species increases the survival/reproductive rate of individuals of the others. In predator–prey and host–parasite interactions the lifetime reproductive success of one of the interactants (the predator/parasite) is increased by the interaction, but that of the other (the prey/host) is decreased. Put simply, the former exploits the latter. In contrast, **mutualistic interactions** result in an increase in the lifetime reproductive success of both participants. Removal of one of the parties reduces the lifetime reproductive success of the other, and in extreme cases causes death.

Interspecific competition

There are few laws in ecology. One of these few is Gause's dictum, otherwise known as the **competitive exclusion principle**, which states that coexisting species must differ in their use of limiting resources. As Gause (1934) wrote:

It is admitted that as a result of competition two similar species scarcely ever occupy similar niches, but displace each other in such a manner that each takes possession of certain peculiar kinds of food and modes of life in which it has an advantage over its competitor.

This quotation is concerned with the effects of competition without evolution. Without evolutionary change, interspecific competition may produce the loss of less well adapted species in particular areas as a result of competitive exclusion. It also suggests an effect over evolutionary time. In the 1940s and 1950s two of the founding 'evolutionary ecologists', David Lack and G. Evelyn Hutchinson, put forward the view that interspecific competition was the motive force behind the evolution of niche specialisation (Lack, 1947; Hutchinson, 1959).

In the founding study, Lack was concerned with explaining the reasons for differences in the niches of different species of bird. Focusing on the radiation of Darwin's finches (*Geospiza* spp.) on the Galapagos Islands, Lack examined the beak sizes of the various species on the various islands (Fig. 7.1). Where the species of finch occurred on the same island, their beak sizes were different, and appeared to be regularly spaced. However, within a species, the beak sizes

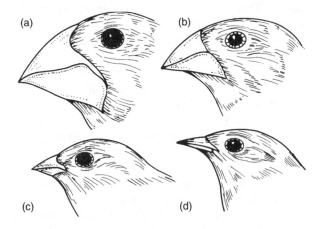

Fig. 7.1 Beak morphology in four of Darwin's finches on the Galapagos Islands. (a) *Geospiza magnirostris*; (b) *G. fortis*; (c) *G. parvula*; (d) *Certhidea olivacea*. Within an island, beaks vary such that no two species utilise exactly the same seed resource. However, within a species, beak morphology varies between islands. The ecological similarities of the islands prompted Lack to suggest that beak morphology changes were driven by selection to avoid interspecific competition.

differed between islands. Believing the islands to be broadly similar ecologically, Lack concluded that divergence was unlikely to be related purely to the abiotic environment and that the most parsimonious explanation for the differences in beak size between islands was that selection had acted to minimise interspecific competition, producing differences in beak size (and therefore niches) between the species where they coexist (see also p. 236). This change in beak size and niche resulting from selection to avoid interspecific competition is termed **character displacement**.

Beak size is just one feature of adaptation to a niche that may be the product of selection to minimise interspecific competition. Hutchinson focused on the size of different species. In a classic paper entitled 'Homage to Santa Rosalia, or why are there so many kinds of animals?', Hutchinson put forward his observations and insights gained from watching the beetles in a pool at a monastery in Santa Rosalia. Hutchinson developed the idea that niche differentiation could be associated with size.

He observed that within any niche different species varied in size. Size differences appeared to Hutchinson to be non-random. Within a niche, coexisting species were regularly spaced in terms of size. Arranging the species in order of size, he observed a regular increase in size between species in the ratio of around 1.2–1.3×, measured in one dimension (e.g. length). He held that this regularity is one of the hallmarks of competition. Animals should diverge over evolutionary time such that niches become different, but not so far that any part of the resource axis would be left untapped: the principle of **limiting similarity**. Selection will favour animals that are dissimilar, but not wastefully so.

Empirical evidence for character displacement driven by interspecific competition

The evolution of character displacement is difficult to demonstrate empirically because it is an end state. It is expected that during the early stages of the process there will be strong selection for a reduction in interspecific competition. However, by definition, by the time character displacement has been achieved, the underlying selective pressure will have fallen away, possibly to the extent that it can no longer be measured. The best evidence has therefore come from studies involving artificially engineered interactions, where putative early stages in the evolutionary process are recreated by mixing populations with overlapping phenotypes. One of the best cases is Dolph Schluter's study of character displacement in the three-spined stickleback complex, *Gasterosteus aculeatus*.

Previous work by Schluter had shown that the lakes of British Columbia contained three distinct types of three-spined stickleback. Where two species of stickleback occurred together, one species was adapted to feeding upon plankton in the body of the water (it is small and slender, with many gill rakers to filter out particles), and the other was adapted to feed upon larger benthic invertebrates (it is large with a wide gape and few gill rakers). Where there was only one species of stickleback in a lake, the species was intermediate between the two forms that were found together. This pattern suggested to Schluter that the species found together had diverged through competition-driven character displacement (Fig. 7.2) (Schluter and McPhail, 1992).

To test his thesis he set up two artificial ponds, each of which he divided into two. In one half of each pond he placed two species of stickleback, the slender, specialist plankton-feeding form, and the intermediate form which is naturally found when only one species is found in a lake. In the other half of the pond, he placed just the intermediate form on its own. His thesis was that if interspecific competition was important in driving character displacement, then selection should promote forms of the intermediate morph which were akin to the benthic morph when the plankton feeder was present, but not when it was absent.

Schluter's observations were consistent with this thesis. Forms of the intermediate species which were most similar to the plankton-feeding species had reduced growth rates in the pond where the plankton feeder was present, but this depression was not observed in isolation. Selection thus appeared to be acting against the intermediate species when its habits came into direct competition with the specialist plankton-feeder. As a natural corollary, the selection pressure in this artificial set-up was for the intermediate stickleback to evolve towards the benthic feeders in morphology. Interspecific competition thus produces selection for divergence of coexisting species, as predicted by Lack and Hutchinson (Schluter, 1994).

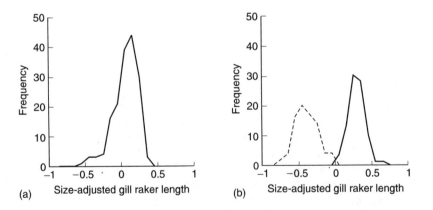

Fig. 7.2 Gill raker length in the three-spined stickleback, *Gasterosteus aculeatus*. When two species are present in the same lake (right-hand graph), there is considerable difference in their gill raker lengths. One species is a benthic feeder and has small gill rakers, whereas the other has large gill rakers and is a filter feeder. When one species is present on its own (left-hand graph), mean gill raker length is intermediate and more variable. The presence of two diverged morphs of this fish where two species coexist suggests character displacement has occurred. (After Schluter and McPhail, 1992.)

Is the stickleback model generally applicable?

The idea that differences in the biology and ecology of different species are driven by selection to avoid interspecific competition is intuitively appealing, and the stickleback case study shows that interspecific competition certainly can be an important evolutionary force. Furthermore, nature is full of instances where related species differ in size. It is tempting to extrapolate to the conclusion that interspecific competition is a major force in evolution. However, this may be overstating the case. All animals are different in some way, but clearly only some of these differences are the result of selection to minimise competition.

For most organisms it will not be possible to conduct experiments of the sort Schluter used on his sticklebacks. Thus, the problem of how to test for the role of competition in producing niche differences remains difficult, requiring evidence of competition occurring in the past. Joseph Connell (1980) recognised this problem in a seminal paper entitled 'Diversity and the coevolution of competitors, or the ghost of competition past'. Here he argues that, although species differences are regularly ascribed to 'the ghost of competition past', empirical evidence to back this assertion is difficult to obtain. He gave three criteria that would need to be fulfilled for complete proof. First, that current differences between species are genetic. Second, that at some time in the past, the species were similar. Third, that competition drove the divergence of the species. By these criteria, the complete set of evidence will frequently be impossible to obtain, for although current differences may be demonstrably genetic, and molecular genetic and palaeontological analyses may show species to have diverged, it is highly improbable that past ecological interactions can be unearthed in any but the most recently diverged species.

Identifying current examples of interspecific competition

The previous section paints a bleak picture for the possible analysis of the importance of competition in evolution. Only when phylogenetic history is known, and ecological history can be inferred, as in the Galapagos finches or the stickleback, can the role of interspecific competition in evolutionary design begin to be assessed with any hope of accuracy. However, we may begin to assess the general evolutionary importance of competition by appraising the role of competition in dictating the niches of organisms on a day-to-day basis. If interspecific competition is not important in this time-scale, then it is also unlikely that it is an important evolutionary force.

Three types of evidence can be used to infer interspecific competition. The first of these involves examining the effect of removing one of a pair of potentially competing species. If this produces an increase in the area occupied by the other, then it is likely that the two species previously competed for resources. For example, in a classic study, Connell (1961) looked at two species of barnacle living on British shores. He noted that one, *Chthamalus stellatus*, regularly occupied a zone higher up the shore than the other, *Balanus balanoides*. This was despite the fact that large numbers of *Chthamalus* larvae settled in both zones. In order to examine whether the absence of *Chthamalus* from the lower shore was due to competition, *Chthamalus* larvae were allowed to settle in this region and then monitored. For some, neighbouring *Balanus* were removed. The results were clear, showing that the reason *Chthamalus* do not grow lower down the shore is due to competition with *Balanus*. In the absence of *Balanus*, *Chthamalus* grew happily; in their presence, the young *Chthamalus* were undercut, smothered out or crushed.

An alternative test for interspecific competition is to examine natural variation in niche breadth. The expectation is that if competition is an important force in structuring communities, then two species which sometimes coexist should have narrower niches when they are found together than when found on their own. Jared Diamond (1975), for instance, examined the niches of doves in New Guinea. He looked at three species: *Chalcophaps indica*, *C. stephani* and *Gallicolumba rufigula*. These doves coexist on the main island, where each species occupies a separate niche, either coastal scrub (*C. indica*), light forest (*C. stephani*) or inland rain forest (*G. rufigula*). However, the same species are also to be found living separately on the smaller islands. In several instances, a species living separately was found to use more than one habitat type. Their niche is wider. This pattern is clearly compatible with the notion that competition on the mainland restricts each species to a smaller niche.

A third approach for investigating the pervasiveness of competition is to examine the spacing of niches. If competition is important, species should be spaced at regular intervals along the resource axis such that no two species living in the same location should use the same resource in the same way. Tests of this notion have produced varying results, depending on the taxonomic group concerned. In a review of the evidence for size regularity, Simberloff and Boecklen concluded that in many groups no evidence for regularity was found. In other groups, some regularity was found, but not always in the precise 1.2–1.3× ratio suggested by Hutchinson (Simberloff and Boecklen, 1981).

More recent evidence remains equivocal but suggests a pattern. Optimal species packing has been found in a study of gramniverous (seed-eating) rodents (Bowers and Brown, 1982). Within a community, the species are regularly spaced with respect to body size and, conversely, species which are similarly sized do not live in the same community (Fig. 7.3). However, among leaf-eating insects, size/niche relationships appear to be rare or absent. The probable explanation was identified by Donald Strong, John Lawton and Richard Southwood (1984) in their extensive review of the ecology of this community. They argue that, to a leaf-eating insect, resources are not limiting. To them, the world is green, and population size is, in the main, controlled by predators and parasites. One only has to look up at a hillside of bracken in summer to realise quite how much food there is for everyone, a point

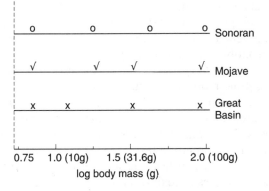

Fig. 7.3 Optimal species packing in rodents: the size of rodents in three different desert communities (Sonoran, Mojave and Great Basin). Each community contains four species. In each of the communities, each character ($\sqrt{}$, x, o) represents the mean mass of individuals of one of the species present in it. It is notable that the masses of the species present in each community are regularly spaced, and are not random. (After Bowers and Brown, 1982.)

emphasised by the apparent abundance of vacant niches in this habitat (Lawton, 1984).

Summary: is competition an important evolutionary force?

The evolutionary importance of interspecific competition is difficult to assess. To obtain evidence of character displacement, Connell has correctly noted the need for historical knowledge, which is unavailable in most instances. However, the idea of character displacement is intuitively sensible, and the potential for competition to drive character displacement is supported by empirical studies. The question is now one of how frequently and in what groups interspecific competition is important in driving evolution. Removal experiments and evidence from the non-random sizes of coexisting species do suggest that, in some groups, interspecific competition is occurring. However, interspecific competition requires a limiting resource. When species are controlled by other factors, as seems the case for leaf-eating insects, interspecific competition, at least for food, is probably of little importance.

Predation and parasitism

There is probably no species on the Earth that suffers neither predation nor parasitism. Top predators may suffer only the problem of parasites, but individuals of most other species are probably affected by both curses. Parasitism and predation can be defined in terms of the effect of encounters on the reproductive success of each party. The host/prey loses and the parasite/predator gains reproductive potential because of the interaction. Generally speaking, we can also define predation and parasitism by examining population level effects. The removal of predator or parasite individuals usually leads to an increase in the average lifetime reproductive success of members of the prey or host species (but see Washburn *et al.* (1991) for a cautionary note).

The adverse effects of predators and parasites on their prey/hosts imposes strong selection for genes which produce prey/hosts less likely to succumb to these fates. Few things promote adaptation faster than the prospect of becoming something else's lunch: the gazelle is fast, the snail resides in a shell, many plants and moths produce or sequester toxins, puffer fish blow up, hedgehogs curl up, rabbits and prairie dogs live in burrows, many species are cryptically coloured or camouflaged, birds can fly and often nest in trees. The list of anti-predator adaptations is endless. Similarly, adaptations for the avoidance of parasitism are common. Many animals have immune systems which selectively destroy non-self matter, plants isolate areas of damage or parasitism, many animals allot time to grooming and some animals have a fever response to infection to kill off invaders. Beyond this, as discussed in the last chapter, mate choice may have evolved to avoid matings with parasitised mates, or to select mates with a higher likelihood of bearing genes for parasite resistance.

With all these adaptations of host and prey, it is a wonder that the predators and parasites survive. But predator and parasite are not passive. Predators will evolve mechanisms of detoxification of unpalatable prey and mechanisms for locating cryptic prey. Parasites will adapt to evade host defences and they will evolve levels of virulence which maximise their own transmission. Paradoxically, the more successful prey and hosts are at avoiding one set of predators or parasites, the more liable they are to come under attack from others. Species of prey or hosts that are successful 'escapees' will increase in number. Any predator or parasite that can breach the defences of these common species will be strongly selected. The pay-off for overcoming these defences is high. Even some of those species that appear to be best defended have had their defences breached by some exploiters. For example, many of the most toxic plants have their herbivores. Milkweed is eaten by larvae of the monarch butterfly (*Danaus plexippus*) and by the milkweed beetle (*Tetraopes tetraopthalamus*), spurges (*Euphorbiaceae*) are eaten by larvae of the spurge hawk-moth (*Hyles euphorbiae*), and ragwort (*Senecio vulgaris*), which is sometimes called mad cow weed because of its adverse effect on cattle that eat it, is voraciously consumed by caterpillars of the cinnabar moth (*Tyria jacobaeae*). In a world where competition is rife, genes that allow individuals to use a common, under-utilised resource will spread.

Clearly, there exist continuous 'evolutionary battles' between predators and prey and between parasites and hosts. These continuing evolutionary battles led Leigh van Valen (1973) to formulate what he termed 'a new evolutionary law', commonly now referred to as the **Red Queen** effect. This derives from the beliefs of the Red Queen in Lewis Carroll's *Alice Through the Looking Glass*. She states of her country that:

. . . here, you see, it takes all the running you can do to keep in the same place. If you want to get somewhere else, you must run at least twice as fast as that.

Van Valen's point is that because of antagonistic (+ −) interactions such as predation and parasitism, evolution would continue in the absence of environmental change. It is 'necessary' to change merely in order to survive. The fact that this interaction often involves round after round of novel attack followed by improved defence has led to these cases of coevolution being referred to as **arms races**.

The evolutionary biology of parasite–host interactions

The feminising bacterium of the pill bug, *Armadillidium vulgare*, the pill bug, is an odd creature. Ancestrally, this species has a simple system of chromosomal sex determination with females being heterogametic. Individuals are male unless masculinisation is blocked. In certain populations there exist females which give rise only to daughters. This phenomenon has been extensively studied by Pierre Juchault, Thierry Rigaud and co-workers. They found that these females give birth to roughly the same number of progeny as others, indicating that the sex ratio effect is not caused by the preferential death of zygotes of one sex. Cytological analysis showed that the bias is not caused by alterations in the segregation of the sex chromosomes in the female. Bizarrely, they found that exposure of these females to elevated temperatures caused the sex ratio trait of their progeny to disappear. Furthermore, injection of macerate from affected females into a normal one causes that female to produce a biased sex ratio. The biased sex ratio is caused by the presence of a bacterium of the genus *Wolbachia* (Rousset *et al.*, 1992). This bacterium lives inside the cells of the host and is passed between generations. Any individual which inherits it develops into a female, because either it has the female sex chromosome complement, or it has the male complement but has the masculinisation pathway blocked by the *Wolbachia*.

The behaviour of *Wolbachia* makes good evolutionary sense. The bacterium inhabits the cytoplasm of its host's cells. In *A. vulgare*, as in most other animals, female gametes are large, consisting of both nucleus and cytoplasm, while sperm are small and have little cytoplasm. This means that while the *Wolbachia* may be passed from a female to her progeny in the cytoplasm of her large gametes, it cannot be passed on from males. In a male, the bacterium is at an evolutionary 'dead end'. Consequently, any means by which an ancestral *Wolbachia* could enhance its chances of being in a female would be selected. Of a number of possibilities, *Wolbachia* has found the most direct: it simply short-circuits the sex-determining pathway to prevent the expression of maleness.

The phenomenon of maternally inherited elements biasing the sex ratio towards females is not restricted to *Armadillidium*. The parasitoid jewel wasp, *Nasonia vitripennis*, bears a maternally inherited element (as yet uncharacterised) which increases the proportion of females born to affected females (Skinner, 1982). Other insects carry bacteria which kill male, but not female embryos (Hurst, 1991). The two-spot ladybird, *Adalia bipunctata*, provides a good example (Hurst *et al.*, 1992, 1993; Werren *et al.*, 1994).

In Cambridge populations, 7 per cent of females bear a cytoplasmic bacterium of the genus Rickettsia which kills the male progeny of the females it infects. In many species, this action would be of little consequence to the bacteria's fitness. However, hatchling ladybirds are cannibalistic and immediately devour any unhatched eggs in the clutch. This behaviour helps to make sense of why the bacteria commit suicide. All the bacteria in a clutch are related but half find themselves in male eggs and half in female eggs. Those in male eggs are already at an evolutionary dead end, so they do themselves no greater harm by killing their hosts. By contrast, those bacteria in female eggs get a free meal and a flying start in life. The net result is that the bacteria improve the quality of individuals in which they live. Put another way, male killing effectively diverts the mother's resources from the sex in which the bacteria has no future to the one in which it does (Hurst and Majerus, 1993). This type of behaviour is not infrequent in insects.

These cases show how the conflicting interests of parasite and host can result in strange, but explicable, parasite adaptations. They also illustrate the strength of the selective forces that can be imposed upon a host by one particular parasite. Not surprisingly, the host is not evolutionarily passive. For example, when the bacterium in *Armadillidium* increases to high levels, males can become scarce. As Fisher (1930) showed, natural selection perennially favours a 1 : 1 population sex ratio (see Chapter 4), therefore any mutation in the host which increases the frequency of males will be favoured. Vindicating this prediction, Rigaud and Juchault (1992) report genes which prevent the transmission of *Wolbachia* by female *A. vulgare*, thereby allowing production of males. The strange consequence of this is that *Wolbachia* has been partially co-opted into the sex determination pathway, in some populations of pill bugs, sex is determined chromosomally. In others it is regulated by genes which control the transmission of a parasite (Fig. 7.4).

General perspectives

The interaction between sex ratio distorting bacteria and their hosts is perhaps the scenario where tight coevolution is most likely to occur. The key feature is a strong one-on-one interaction. Such specialisation of parasites upon hosts is relatively common, driven (in evolutionary terms) by the need to match many aspects of host biology over a protracted period for successful completion of the life history. Specialisation is expected whenever parasite individuals spend

(i) Sex determined by constitution with respect to sex chromosomes. Female heterogametic (ZW).

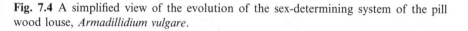

(ii) Invasion of cytoplasm bacterium which feminises male individuals. ZZ individuals with bacterium are female; those without are male.

(iii) Population becomes female-biased.

(iv) Invasion of nuclear gene which prevents transmission of the bacterium, because this gene effects production of males, the rare sex.

Fig. 7.4 A simplified view of the evolution of the sex-determining system of the pill wood louse, *Armadillidium vulgare*.

large periods of their life history interacting with a host. Coevolution is likely to follow specialisation, selection favouring host types which are less prone to infection or damage by the parasite. Although the host may suffer other risks in life, the inherent pervasiveness and potency of the parasite make it a very important selective factor. This said, tight coevolution can occur without complete specialisation. The interactions between the cuckoo and its various hosts are a good example. Here, the cuckoo is involved with separate arms races with a number of hosts. The situation still perhaps represents something of a quandary (Box 7.1).

The coevolutionary relationship between a parasite and its host can also be very long-lived. The tendency to find correspondence between host and parasite taxonomy was anecdotally recorded in the 1940s and was thought to be due to the long-term interaction of specialist parasites with their host producing the tendency for them to speciate with the host. Phylogenies of lice parasites of chewing gophers, constructed from comparisons of protein and DNA variation, have borne out the assessment of taxonomists (Hafner and Nadler, 1988; Hafner *et al.*, 1994). Although movement of lice species between host gopher species is known, co-speciation is the rule in this interaction. For every split in the host lineage, the parasite lineage splits (Fig. 7.5). We must therefore conclude that the interaction between host and parasite is indeed an ancient one. One reason for the extreme longevity of some of these interactions may be the nature of the arms race. If parasite numbers decline due to host adaptation, then any resistance genes which have a cost to an unparasitised individual will decline in the population. The benefit gained from resistance is rare, but the cost is ubiquitous. The parasite may thus be given an 'evolutionary breathing space' in which to effect recovery.

Box 7.1:   Coevolution between cuckoos and their hosts

Brood parasites are individuals which do not care for their own offspring but instead lay their eggs in the nests of other birds in the hope that these other birds will care for the eggs and resultant nestlings. One of the best known brood parasites is the European cuckoo (*Cuculus canorus*), which has been studied extensively by Nick Davies and Michael Brooke. Female cuckoos lay eggs in the nests of other, generally smaller, birds. The female will watch prospective nest owners, and when they are foraging away from the nest she will fly in, remove one of the eggs already in the nest and lay one of her own. The whole sequence takes only seconds. The cuckoo embryo develops unusually rapidly and, on hatching, the blind baby cuckoo ejects all other unhatched eggs and/or young chicks from the nest. It thus monopolises the food that its foster-parents bring. This behaviour clearly places a selection pressure on birds to avoid parasitism.

The cuckoo is a brood parasite of many species of bird, but it usually lays eggs which mimic those of its host. Placement of a non-mimetic egg in the nests of these hosts results in the egg being rejected. However, cuckoos that lay in the nests of dunnocks, *Prunella modularis*, produce non-mimetic eggs. Such eggs are rarely rejected by dunnocks. This is interpreted by Davies and Brooke as being evidence of an arms race between the parasite and the host. Following the initiation of parasitism of a bird by the cuckoo, genes which produce hosts able to detect obviously foreign eggs spread. Genes in the cuckoo which produce egg mimicry will then be favoured. The dunnock–cuckoo interaction is in an early stage, before selection has produced an increase in the discrimination of the host which would in turn impose a selection pressure on the cuckoo to evolve effective egg mimicry (Davies and Brooke, 1989a, b).

Two pieces of evidence support this interpretation. First, where a host species inhabits a region that lacks cuckoos, the tendency to reject non-mimetic eggs is less intense (Soler and Møller, 1990). Meadow pipits in Iceland (where there are no cuckoos) are less discriminatory than their British counterparts. Second, study of hosts that are known to have been utilised only recently by cuckoos, such as the azure-winged magpie, *Cyanopica cyana*, in Japan, have shown that new hosts do not discriminate. This endorses the notion that the discrimination is an evolved response to the cuckoo (Yamagishi and Fujioka, 1986; Nakamura, 1990).

It seems clear that there is an arms race between the cuckoo and each species it parasitises. There is tight coevolution, despite the face that the cuckoo has multiple hosts. The question that must be raised is how the cuckoo manages to specialise on many species of bird at one time.

continued

Cuckoos are to be commonly found parasitising four species of bird in the British Isles. With the exception of the dunnock, the cuckoos produce eggs mimetic to the host egg. But the pattern produced differs between the species of host bird in which the egg is laid. It is this, perhaps, that makes the cuckoo so interesting. The species is a specialist on four rather different species at once. In fact, each female cuckoo is a specialist parasite on one species, and possesses both host preference and adaptations to that specific host. The question that remains to be answered is how host preference and host adaptations are co-inherited.

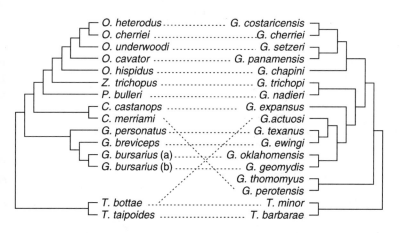

Fig. 7.5 Co-speciation of pocket gophers and their chewing lice, as illustrated by comparisons of phylogeny of gophers and lice with current associations. Dotted lines between hosts and parasites indicate current associations. Tree constructed from maximum likelihood analysis of the sequence of cytochrome oxidase I gene of each species. (After Hafner *et al.*, 1994.)

One important caveat of the above discussion of parasite–host coevolution is that the interactants do not inevitably descend into an arms race. If resistance to different parasite types is encoded by alleles at one locus, and parasite virulence for particular host types is encoded for at a complementary locus, then cyclical change in the frequency of host resistance and parasite virulence alleles may result. Effectively, the four forms chase each other continuously round an allele frequency box. Whenever one host morph is rare it gains an advantage and increases, in turn favouring the parasite morph which is adapted to it (Fig. 7.6). Polymorphism results and is maintained (Barrett, 1988). This is still tight coevolution. However, unlike in an arms race, there is no 'progress' in terms of parasite and host adaptation. In an arms race, a particular parasite genotype is fittest against all hosts, and spreads to fixation.

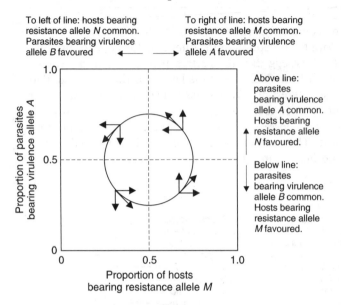

To left of line: hosts bearing resistance allele N common. Parasites bearing virulence allele B favoured ← → To right of line: hosts bearing resistance allele M common. Parasites bearing virulence allele A favoured

Proportion of parasites bearing virulence allele A

Proportion of hosts bearing resistance allele M

Above line: parasites bearing virulence allele A common. Hosts bearing resistance allele N favoured.

Below line: parasites bearing virulence allele B common. Hosts bearing resistance allele M favoured.

Fig. 7.6 Gene-for-gene interaction between host and parasite may result in cyclically varying allele frequencies, instead of an arms race. In the phase diagram, we jointly consider changes in the frequency of parasite virulence and host resistance alleles. Parasites bear one of two different virulence alleles, A and B. Host susceptibility to parasites is determined at a complementary locus, with two alleles M and N. Hosts bearing allele M succumb most readily to attack by parasites with virulence allele A, whereas hosts bearing allele N succumb most easily to parasites bearing allele B. Now, when hosts of type N are common, parasites bearing allele B will be at a selective advantage. The B allele will increase in frequency in the parasite population. This will result in greater parasitisation of hosts bearing allele N. Allele N will therefore decrease in frequency relative to M. As M increases, so parasites bearing allele A gain an advantage. The system varies cyclically. (After Barrett, 1988.)

In the gene-for-gene situation, each genotype is only ever fittest against one particular host type, producing a powerful density-dependent restoring force which prevents it from reaching fixation.

Predator–prey and parasite–host interactions contrasted

The evolutionary biology of predator–prey interactions does not in general follow the same patterns as observed between parasite and host. Predators and prey are less likely to enter into tight coevolution than parasites and hosts because the interaction is usually less precise. Whereas parasites regularly become specialised upon hosts, predators often hunt several prey species. This rather more diffuse interaction acts to spread the evolutionary load. This can be seen by consideration of a single predator species which eats a wide variety of other prey. If any one of these prey species evolves a mechanism to reduce predation, it is clear that the selective pressure on the predator to evolve a counter measure will be much less than if this were its only prey. If adaptation to the one prey species disrupts adaptation to others, then change would not be favoured. Dawkins and Krebs (1979) have dubbed this type of scenario the 'rare enemy effect'. This kind of interaction is exemplified by the predator hunting by a search image, discussed in Chapter 3, but with multiple prey species. Here, the innate bias against the commonest prey form drives the evolution of prey polymorphism, each new form gaining an advantage by being rarer than other potential prey items. The interaction is thus one-way: the prey evolves in response to the predator but the predator responds merely by switching attention to a different prey.

Predator–prey interactions also differ in evolutionary terms from those between parasites and hosts because there tends to be an asymmetry inherent in the predator–prey relationship. Whilst a predator which fails in a particular capture will often lose only its lunch, a prey species which fails to evade a predator will lose its life. The selective pressure on the prey for avoidance of the predator is much higher than the selection pressure on the predator for increased capture efficiency. Dawkins and Krebs have termed this the 'life–dinner principle'. Tight coevolution is unlikely to occur because within the interaction there is a far greater selective pressure for the prey to adapt than the predator. This contrasts with the situation in parasite–host interactions. The life of the parasite is as much in the hands of the host as the life of the host is in the hands of the parasite.

These two principles perhaps explain why predator–prey and parasite–host interactions differ. Parasites are often specialised and depend for their existence on their hosts. They are thus likely to exhibit tight coevolution. The rare enemy effect and the life–dinner principle mean individual predator and prey species are unlikely to enter into a mutual arms race. Rather, predators and prey are more frequently involved in diffuse coevolution. Predator biology is an adaptive response to the biology of a range of prey species and prey biology is an adaptive response to a range of predator species. Although predators and prey can be said to be involved in an arms race, the arms race is diffused because it involves suites of predators and suites of prey.

This picture of diffuse relationships between predators and prey works well when both predators and prey are animals. However, when the prey is a plant (i.e. the interaction is herbivory) specialisation is more frequently observed. Many species of phytophagous insect are monophagous. This is presumably allowed by the sessile and more constant nature of their plant prey. If we add to this the realisation that part of one host plant can be sufficient for a herbivorous insect to complete its development, the evolution of specialisation becomes less remarkable. The relationship between plant and insect in these cases is more like a parasite–host relationship than a 'classical' predator–prey relationship. This similarity is confirmed by recent analysis of damage by leaf-mining insects in fossil leaves. Different species of leaf-mining insects produce different patterns of damage in leaves. The patterns are so characteristic that a specialist can tell the identity of a mining insect from the mine alone. They do not need to extract the insect from it. When damage patterns of fossil leaves were compared to those produced by extant species of leaf-miner on the descendant plants, an excellent match was found. This correspondence of damage pattern suggests that members of the same leaf-miner taxon have been associated with the same host plant clade for at least 97 million years (Labandeira *et al.*, 1994). It is therefore likely that tight coevolution will occur in at least some insect–plant interactions.

Mutualistic interactions

Some of the most interesting ecological interactions are those where members of different species work in harmony. This occurs where members of two different species interact in such a way that the lifetime reproductive success of both is increased. Perhaps one of the most beautiful examples of coevolved mutualism is the association between the ant *Pseudomyrmex ferruginea* and the bull's horn acacia, *Acacia cornigera*, first described in detail by Dan Janzen.

The bull's horn acacia occurs in Central America, and the acacia is associated with ants which nest in the hollow thorns of the acacia. The acacia appears to culture the ants. It provides them with nutrients in the form of sucrose from extra-floral nectaries (areas of nectar secretion not associated with flowers), and protein from Beltian bodies at the tips of leaves (Janzen, 1966, 1967) (Fig. 7.7). Conversely, the acacia benefits from the presence of ants. The ants are extremely aggressive and drive away intruding herbivores, reducing damage and increasing the seed set of the acacia. In addition, the ants also prune neighbouring plants, giving the host acacia an advantage in the competition for light and space.

Acacias that harbour ants lack the cyanogenic chemical defences in their leaves present in related species of acacia that are not associated with ants (Rehr *et al.*, 1973). Two contradictory explanations may be constructed from this observation. Either selection for the association with ants was stronger in the bull's horn acacia because this acacia lacked chemical defences. Alternatively, the acacia has recently lost its chemical defences as a direct consequence of its association with ants. By offering an alternative form of defence, the ants reduced the need to expend metabolic-energy-making toxic chemicals, which in turn became an unjustified cost and fell into disuse. It is difficult to demonstrate which of these evolutionary pathways actually occurred, but either way, the result is a system of coevolution between ant and plant.

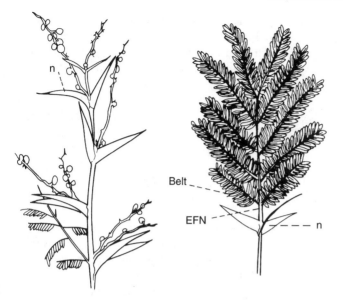

Fig. 7.7 The interaction between bull's horn acacia and *Pseudomyrmex* ants. The acacia (dry and wet season forms shown) provides hollow thorns in which the ants nest (n), and food from Beltian bodies (Belt) and extra-floral nectaries (EFN). The ants prune competitor plants and remove insect herbivores.

Pollination ecology: a case study in the evolutionary biology of mutualistic interactions

Tightly coevolved mutualisms such as that between ant and acacia are rare. Why is this? The reasons are perhaps best discovered by examining interactions that may be mutualistic, but in which the parties rarely specialise and coevolve. Study of the interaction between pollinating insects and plants is useful in this context. Plants need pollinators, and pollinators in general are rewarded by plants. The interaction is frequently mutualistic. Additionally, there are cases where the interaction between pollinator and flower is highly specialised and the parties have obviously coevolved. However, although many interactions in pollination biology are mutualistic, most plants and pollinators are not specialised, and thus show few signs of tight coevolution.

Plants, being more or less immobile, frequently rely on attracting insects to disperse their gametes. Many adaptations to prompt insect pollination exist. Flower colour allows identification by insects, nectar is produced as a gift of food, anther and stigma position are optimised with respect to the transfer of pollen. The interaction is often a mutualistic one, with the insect benefiting in terms of nectar and pollen collection, and the plant benefiting from active pollen dispersal. Diffuse coevolution occurs. However, specialisation and tight coevolution have rarely been described. One example is the interaction between the Madagascar star orchid, *Angraecum sesquipedale*, and the hawk-moths of the genus *Xanthopan*. The star orchid flower is remarkably beautiful, with a corolla tube of some 30 cm in length. This corolla excludes most insects. Darwin recorded this flower but not the moth, yet he was sufficiently astute to predict that pollination must be achieved primarily by a moth with a very long proboscis (Darwin, 1862), a prediction that was confirmed by observations over a century later (Nilsson *et al.*, 1985; Nilsson, 1988). It is clear that *Xanthopan* proboscis and plant corolla tube have coevolved, for it is hard to see a moth with such a long proboscis existing without the flower, and vice versa. At some point in the past, both moth and flower would have had less exaggerated features which, through coevolution, have subsequently become enlarged.

But why are examples of specialisation and tight coevolution not more common? There are two principal reasons. First, plant–pollinator interactions do not lend themselves to specific one-to-one relationships. To begin with, a plant species which became exquisitely adapted to one species of pollinator would gain little, and might have everything to lose if that pollinator failed to appear in a particular year. The converse is also true. Pollinators which can feed from a variety of plants do better than those which are tied to one, because they are likely to find food distributed more widely in both space and time. A pollinator which is foraging for nectar is unlikely to gain from fidelity in the face of higher rewards being offered by other plants. It is notable that many examples of tightly coevolved mutualisms (not just pollinator–plant interactions) derive from tropical regions. Undisturbed areas of the tropics tend to show ecological stability. Here, long continued presence of given species in the same locality provides a stage on which specialisation can occur.

The second reason that specialisation may be rare is the tendency of mutualistic relationships to break down. Just as co-operation between members of the same species may be prevented because of the possibility of cheating (Chapter 5), so too will the formation of stable mutualisms. Pollination interactions frequently become parasitic. Plants, by producing both flowers without nectar and false baits, attract insects without incurring the cost of producing a reward. A beautiful set of examples are the bee orchids. These orchids have flowers that resemble female bees or wasps, and attract males which attempt to copulate with the flowers, and effect pollination in the process. In one species of Australian orchid, the deception by the flower is so compelling that male wasps actually go through the whole process of copulation and ejaculation, leaving spermatophores behind in the cup-shaped lower petals of flowers. The reverse is also true. Insects may cheat on the plant. Individual insects will be selected to gain the nectar reward in the fastest fashion, irrespective of the pollinating they do in the process. The fastest route may not necessarily be down the corolla, via anther and stamen. Inouye (1980) distinguishes two types of insect which gather food without pollinating on the basis of their behaviour, these being described as either thieves or robbers. Thieves steal nectar without causing physical harm to the plant; they merely miss the anthers/stigma on the way to the nectaries. By contrast, robbers actually damage the plant. For instance, short-tongued bumblebees bite a hole in the base of the field bean flower and steal nectar.

Reliance on one organism is dangerous in two ways. Either it may fail to appear in a given year, or it may evolve cheating behaviour. Either way, it is unsurprising that specialisation is the exception and not the rule. The question that occurs is why any examples of tight coevolution exist. What drives the evolution of specialisation? The benefit to a plant of having a specialist pollinator may lie in increased receipt of conspecific pollen, and increased probability of transmission of its pollen to a conspecific. If one species of pollinator is particularly efficient at pollinating, selection may favour exclusion of others. Once this has occurred, there may be selection for specialisation by the insect. The specialist plant offers increased assurance of reward. Perhaps the environmental stability of the tropics may promote mutualism, not just by producing continuity of interaction through climatic predictability, but also by increasing the intensity of competition, providing an advantage to specialisation and an assured reward.

Specialisation may also be a historical legacy. For instance, female moths of the genus *Tegeticula* not only pollinate their host *Yucca* plant, but also lay eggs in the ovule of the plant. The larvae which hatch eat some of the developing seeds. As with all parasitic interactions, success demands a good match between moth and host-plant biology. Interestingly, other moths in this family act purely as parasites, feeding off host-plant seeds without pollinating the plant, and species in the sibling genus *Parategiticula* pollinate but do not parasitise the plant. The question then arises as to which drove the specialisation, pollination biology or parasitism. After constructing a phylogenetic tree of the family, Pellmyr and Thompson (1992) concluded that pollination is a recent occurrence, and that seed parasitism is ancestral. The interaction began with specialisation because of the need for a good match of parasite to host, and later evolved into a mutualism, following pollination activity by the moth. In the genus *Parategiticula*, parasitism was then subsequently lost.

Dependence and mutualism

The cases of mutualism we have so far discussed involve instances where selection acting upon two independently living species has favoured closer interaction. The interaction has become specialised, and the parties have coevolved. Many cases of the most intimate symbioses, for example between bacterial symbionts and their hosts, or between fungus and alga within a lichen, appear at first examination to parallel these. The reproductive success of one party depends upon that of the other. In the case of bacterial symbionts of insects, the host has evolved mechanisms for the transmission and maintenance of the bacterium, while the bacterium may have lost certain metabolic pathways, have reduced DNA content, and may have lost the capacity to grow in isolation. Often, the interaction is one where the parties cannot live apart. However, it must be questioned whether there has ever been any mutual selection for specialisation in these cases.

The central issue is whether the initial intimacy of the interaction was a product of selection upon both sides to tighten the relationship, or whether it was the product of selection upon one side only. As Angela Douglas and Douglas Smith (1989) argue, it is important not to mistake the dependence of a bacterium on its host as indicating the interaction was, at the time of initiation, one of mutualism. It may equally be that the interaction was parasitic at first, with the host capturing and maintaining a bacterium which serves a metabolic role within it, despite resistance from the bacterium. Only following this capture would the parties coevolve. Possibly the most ancient interaction of this type is between the mitochondrion and its host eukaryotic cell. Neither can live apart. However, it is not clear whether this relationship was originally exploitative, parasitic or mutualistic. What now appears as a coevolved mutualism may have evolved not from two harmonious partners, but from a parasitic bacterium making the best of its lot.

Discussion: an assessment of tight coevolution

The above discussions have suggested that tight coevolution is relatively rare outside of parasite–host and insect–herbivore interactions. Recently, John Thompson (1994) has presented evidence that this view may be over-pessimistic. Tight coevolution is more important than we think. He contends that much coevolution is restricted in space and/or time. Consequently, although many mutualistic (and indeed other) interactions involve specialisation and coevolution, in order to detect these relationships it is necessary to examine the interaction carefully, over small geographical ranges and over short periods of evolutionary time. Single species may coevolve with different partners in different regions, even when a more general view perceives a diffuse coevolutionary interaction. Judging an interaction to be tight or diffuse across a range of areas is to blur the evolutionary picture. In doing so, we miss much of the local detail. Rather, we should examine the interactions in each area for evidence of tight coevolution.

Further to this, Thompson discusses the general importance of tight coevolution. In order to appreciate the importance of coevolution in framing the biology of the species, he argues we should combine our knowledge of local adaptation with knowledge of gene flow between locally adapted populations. There is what he terms a geographic mosaic of coevolution, with local coevolutionary events interacting through gene flow to shape the biology of the species. Thompson's thesis is thus that tight coevolution is a much more important evolutionary phenomenon than previously thought, with local coevolutionary events interacting to frame the biology of the species.

Summary

Interactions between species can be complex. Interspecific competition may lead to the promotion of biological changes that result in the ecological divergence of the parties. Mutualisms may become tightly coevolved, though it is unclear how often they do so. Many, but not all, parasites become embroiled in evolutionary arms races with their hosts. In contrast, predators and their prey generally experience a more diffuse relationship with no one species either depending on, or at the mercy of, another. The expected outcome of interspecific interactions depends on many features. Ecology is important, because the frequency of the interaction over space and time dictates the selection pressure that the interactants place upon one another. Genetics is also important, because this governs how selective pressures are translated into evolutionary change. From the range of examples we discuss it becomes clear that many of the selective forces an organism experiences derive not from the abiotic environment or from intraspecific competition, but from interactions with individuals of other species. Together, these interactions ensure that evolution will occur even in the absence of environmental change.

Suggested reading

DAWKINS, R. and KREBS, J. R. (1979). Arms-races between and within species. *Proceedings of the Royal Society of London B*, **205**, 489–511.
A paper dealing with the evolutionary biology of arms races and coevolution.

DOUGLAS, A.E. and SMITH, D. C. (1989). Are endosymbioses mutualistic? *Trends in Ecology and Evolution*, **4**, 350–352.
Short review of the nature of mutualism and dependence.

FUTUYAMA, D. J. and SLATKIN, M. (1983). *Coevolution*. Sinauer, Massachusetts.
Many papers on a variety of themes on the subject of coevolution. Contains chapters on pollination ecology, host–parasite associations, and the evolutionary role of competition.

THOMPSON, J. N. (1986). Constraints on arms races and coevolution. *Trends in Ecology and Evolution*, **1**, 105–107.

THOMPSON, J. N. (1989). Concepts of coevolution. *Trends in Ecology and Evolution*, **4**, 179–183.
Two papers dealing with the evolutionary biology of arms races and coevolution.

SCHLUTER, D. and MCPHAIL, J. D. (1993). Character displacement and replicate adaptive radiation. *Trends in Ecology and Evolution*, **8**, 197–200.
Discusses the evidence for divergence being driven by the advantage of avoiding interspecific competition.

Chapter 8

Conflict within the individual: genes that break Mendel's laws

Genetics effectively began with Mendel's laws of inheritance, which he deduced from his famous series of observations on pea plants. These laws describe how inherited characters segregate among progeny and how each chromosome pair assorts independently. Phrased in more functional terms, each chromosome in a pair has a 50 per cent chance of ending up in any particular gamete. When this simple rule was combined with Darwin's concept of natural selection, the foundations were laid for the new synthesis of evolutionary biology that had population and ecological genetics at its core.

For the first half of this century, only a handful of data sets in any way undermined the Mendelian bedrock of this synthesis. Correns' (1909) observation of **uniparental** inheritance of chloroplasts in the four o'clock plant, *Mirabilis jalapa*, showed that certain characters were regularly inherited in a non-Mendelian fashion. Further interesting anomalies included observations of female-biased sex ratios in several species of *Drosophila* (e.g. Morgan *et al.*, 1925) (see below) and the biased inheritance of alleles at the autosomal t-locus of mice (Chesley and Dunn, 1936). Evolutionary theorists of the day did not consider these few exceptions to be sufficiently important to warrant amendment of the way evolution was viewed. They were seen as rare exceptions to an otherwise good rule. The assumption that genes are inherited according to Mendel's first law appeared, to all intents and purposes, a reliable one.

However, since 1950, the number of exceptions to Mendel's laws has increased. A range of different phenomena have now been characterised, all of which involve a bias in favour of the transmission of one allele during meiosis. Furthermore, it has since become clear that the spread of these non-Mendelian elements is not some freak of nature, but can be explained by the action of natural selection. In the process of elucidation, we have been forced to alter fundamentally our thinking about the way in which selection acts.

There are two clear ways in which an allele can increase its frequency between generations. Long ago Darwin pointed out how inherited characters which improve fitness will be favoured. Since then it has been an implicit assumption that the products of Mendelian segregation are represented equally, and that the race for improved fitness begins only after fertilisation has occurred. This is the first way. The second path is intuitively less obvious and involves creating a bias before natural selection gets a chance to act on the zygote. This bias is caused by one allele promoting its own transmission.

Consider a hypothetical mutation that creates an allele which causes sperm to swim faster. In the race to the egg, compared with its homologue, a sperm carrying such a mutation is likely to win. The number of progeny produced will be the same as normal, but the vast majority will carry this new allele, allowing it to sweep through the population to fixation. This is an example of sperm competition within a single ejaculate. An alternative scenario would be if the new allele guaranteed its future, not by swimming faster, but by somehow actually killing sperm carrying its homologue. Under these circumstances, there is likely to be a cost to the organism in terms of reduced sperm production, but given that sperm are usually produced in large excess, this cost will be low.

The final case to consider is that in which a mutant gene kills its homologue using a delayed time fuse, causing death to occur after fertilisation. Now the costs and benefits are less clear-cut. From the point of view of the organism, such a gene reduces fitness to half, since half the offspring of a carrier die. From the point of view of the gene, the advantage gained is close to zero, since its representation in the next generation is unchanged. In fact, the mutant does gain a slight advantage because it reduces the fitness of its homologue. In large populations this advantage will be negligible. However, in a small population such a gene might spread because its effect is negatively proportional to population size. There are cases where the gene could spread even through a large population. Consider an organism in which the resources a pregnant female can contribute to her load of offspring are limited. Progeny of a female carrying half the normal load in her womb will gain more resources than average for the species, thereby gaining a flying start in life. Now the killing action of the gene does improve the survival of carriers of the gene. Although the number of offspring which carry it has not altered, it is now preferentially carried by individuals that have high fitness because they have been abnormally well nourished by their mothers.

These examples show clearly how hypothetical mutations could promote their own spread in a way which breaks Mendel's first law. They are able to do this, even if they lower the fitness of the individual which carries them, a point noted as long ago as 1932 by J. B. S. Haldane:

Clearly a higher plant species is at the mercy of its pollen grains. A gene which greatly accelerates pollen tube growth will spread through a species even if it causes moderately disadvantageous changes in the adult plant (Haldane, 1932).

Underpinning this observation is the key concept that individual fitness is usually shared equally between two homologues because meiotic segregation is fair. Thus, for an organism which produces ten progeny, any particular allele will be passed on to only five on average. It is this difference between the expectations of the gene and the individual which provides room for conflict. In the example given, any gene which manages regularly to bias segregation, such that it is found in more than five progeny on average, will have increased fitness. Such a gene will spread by natural selection, even if its action causes fewer progeny to be produced overall. Battles between homologues make good evolutionary sense. However, there is a powerful restriction. Fighting can only be carried out effectively at a distance, when the homologues are in different cells, otherwise the outcome must be mutual suicide.

In this chapter we shall consider a variety of real examples where Mendel's first law is broken, and which have been interpreted as showing selection acting on particular genes or chromosomes, below the level of the individual.

Violations of Mendel's first law

The first observations of a pattern of inheritance where alleles in a heterozygous parent were not equally represented in progeny were from three groups of workers investigating the sex ratio of fruit flies during the late 1920s and early 1930s. In each of three species, occasional crosses were found to produce female-biased sex ratios. Genetic analysis revealed the same pattern of inheritance in each case (Morgan *et al.*, 1925; Gershenson, 1928; Sturtevant and Dobzhansky, 1936). Although daughters from the distorted brood produced an equal number of males and females, half of their sons produced female-biased sex ratios, whatever the origin of their mates. The remaining grandsons produced normal sex ratios and were shown to have lost the trait.

Further analysis of this trait showed the sex ratio bias to be present at fertilisation. No difference could be found between the viability of eggs derived from pairs which produced a biased sex ratio (parents with the SR trait) and normal pairs. **Parthenogenesis** could be ruled out since the trait was associated with the action of a factor in the male line. The lack of intersexual characteristics in any of the female offspring suggested that there is not simply a conversion of genetic males to females. By exclusion, the authors concluded that the SR trait was the product of unequal participation of X- and Y-bearing gametes at fertilisation. That is to say, the sex chromosomes were breaking Mendel's first law.

The cause of this unequal participation of gametes was eventually discovered by comparing the pattern of spermatogenesis in SR and normal males. Using electron microscopy, Policansky and Ellison (1970) showed that SR males produce approximately half the number of sperm per bundle as normal males, implying that the female-biased sex ratio is caused by heavy mortality among sperm bearing the Y chromosome. Given the genetic analysis described above, which showed how the SR trait is passed down through the female line, it became clear that the effect was carried by the X chromosome. In other words, SR is caused by a gene on the X chromosome which preferentially kills sperm bearing a Y chromosome.

Meiotic abnormalities in which the two types of gamete from a heterozygote regularly occur at unequal frequencies among the zygotes fall under the general headings of either **meiotic drive** (Sandler and Novitski, 1958), or **segregation**

distortion. Of these, the latter term is perhaps more accurate, given that gametic death occurs after meiosis. Often, meiotic drive is shortened simply to 'drive', implying a positive process in which a gene's 'aim' is to do better than the 50 per cent transmission rate offered by Mendelian segregation. Examples of meiotic drive of sex chromosomes are now known in twelve species of *Drosophila*, two species of mosquito (*Aedes aegypti* and *Culex quinquefasciatus*), at least one butterfly, and the wood lemming, *Myopus schisticolor* (see Lyttle (1991) for a review). Meiotic drive of sex chromosomes is also seen in white campion, *Silene alba* (Taylor, 1994).

Interestingly, the nature of meiotic drive is not the same in all instances. In the two mosquito species, it is the Y chromosome that eliminates the X, so producing a male-biased sex ratio. In the African butterfly, *Acraea encedon*, again the Y chromosome kills the X, but since females are the heterogametic sex in Lepidoptera, this results in a female-biased sex ratio. To the extent that any rule can be constructed from so few cases, it seems that sex chromosome drive always occurs in the heterogametic sex, but that either sex chromosome may be driven.

Meiotic drive is not restricted to sex chromosomes. Driving autosomes have been observed in two species of mice (*Mus musculus* and *M. domesticus*), the fruit-fly *Drosophila melanogaster*, and two Ascomycete fungi of the genus *Neurospora* (Agulnik *et al.* 1991; Lyttle, 1991). However, autosomal drive appears to occur less frequently than drive on sex chromosomes. Whether this is a true reflection of its rarity or whether this an artefact attributable to the fact that sex chromosome drive is more conspicuous, is unclear. Certainly, the presence of a strongly biased sex ratio is much more likely to be noticed than most of the effects which a driving autosome is likely to produce. On the other hand, there are good reasons for believing that driving autosomes are less likely to evolve in this way. With the sex chromosomes, conflict is restricted to individuals carrying one X and one Y, and one chromosome attacks the other. The battle is clearly defined. By contrast, with driving autosomes it is possible to have an individual who is homozygous for the driving chromosome. In such cases, both chromosomes have the power to fight, potentially causing a decrease in fitness of the individual concerned.

Cases of autosomal drive have given us most of our information concerning the mechanism of drive. Analysis of the genetics of the segregation distorter system of *D. melanogaster* has shown that the ability to drive is encoded by a single locus, *Sd*, on chromosome 2. However, the action of this gene is modulated by a second gene, also on chromosome 2, known as *Responder*. If the homologue bears the sensitive allele, Rsp^s, then it will be driven. If it carries the insensitive allele, Rsp^i, then it is protected from drive and segregation occurs normally. Why does *Sd* not drive itself? The answer is because the chromosome on which it is found also carries the insensitive allele at the responder locus, Rsp^i. This chromosome thus carries both 'poison' and 'antidote'. Furthermore, this advantageous pairing cannot be broken down, because both genes lie within the same inverted section, preventing recombination (Fig. 8.1).

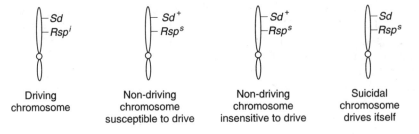

Fig. 8.1 Autosomal drive of chromosome 2 *Drosophila melanogaster*: chromosomal types.

The population genetics of driving sex chromosomes

The reason for the spread of a meiotically driving sex chromosome is easy to comprehend. If a male produces an excess of sperm, such that there is competition between gametes of a given male for access to the egg, then a chromosome in a male that drives out sperm bearing its homologue will remove competitors and therefore gain increased transmission to the next generation. It will thus spread through the population. In a seminal paper, Bill Hamilton (1967) modelled the dynamics of driving X and Y chromosomes, and their effect on population size and sex ratios (Fig. 8.2). Hamilton noted that the spread of a driving sex chromosome was potentially disastrous to the species which bore it, and could cause extinction. This scenario is perhaps the case in the butterfly *Acraea encedon*. In some populations of this species, drive is so frequent that the sex ratio has decreased to fewer than three males in every 100 individuals (Owen, 1965; Owen *et al.*, 1973). However, *A. encedon* may be the exception and not the rule. In many cases, driving chromosomes have settled at an equilibrium level in the population.

How can driving sex chromosomes come to an equilibrium? As a driving sex chromosome spreads, the population sex ratio becomes increasingly biased. This has two effects. First, it will produce selection for genes unlinked to the driving gene which resist drive. Second, the very act of reducing the frequency of one sex tends to lower competition among the gametes produced by that sex. This can have the knock-on effect of reducing the transmission advantage gained by the driving chromosome.

The first of these is easy to understand. As the driving chromosome spreads and the sex ratio becomes more biased, mutations on chromosomes other than the driver which prevent the drive will spread under natural selection, for these will result in the production of the rare sex into which they will pass. Members of the rare sex will, on average, have higher lifetime reproductive success. The drive resistance gene will thus spread in accordance with Fisher's principle (Fisher, 1930; see p. 84), and restore the sex ratio to equality or equilibrium.

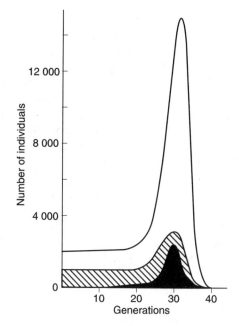

Fig. 8.2 The dynamics of X chromosome drive. In this simulation, a driving X chromosome was introduced into one male in a population containing 1000 males and 1000 females. Each male can fertilise a maximum of two females, and females in the population may be unproductive after the 27th generation due to a dearth of males. The curves track total population size (top line), the population size of males (middle line) and the population size of males bearing the driving X chromosome. (After Hamilton, 1967.)

The lessening of the transmission advantage to driving chromosomes as they spread is more difficult to comprehend, and is best brought across by analogy. Consider each adult male to be a school, preparing its pupils (both girls, X-carrying sperm, and boys, Y-carrying sperm) for outside employment (fertilisation of female gametes). Each adult female thus represents an employer providing jobs (female gametes to be fertilised by sperm) for the school leavers. Each school gets a quota of jobs for its pupils. The quota of jobs available to pupils at most schools is much lower than the number of pupils seeking jobs, just as the number of sperm produced by males usually far exceeds the number of gametes produced by females. The competition for jobs among pupils in a school is thus akin to intra-individual sperm competition. Pupils that fail to secure employment perish. Now consider that in some 'schools' girl pupils have drive, manifested by the fact that to improve their own chances of gaining employment, they assassinate the boys. This cut-throat behaviour is initially successful. Girls from schools in which there is drive are more likely to get jobs than any particular pupil from a school lacking pupils with drive. Gradually the ratio of employers to schools, and therefore jobs to pupils, alters. The number of schools (adult males) diminishes, but the capacity for employment (adult females) increases. Competition between pupils for jobs becomes progressively less severe. At some time, as the ratio of employers to schools increases, the number of jobs available will exactly equal the number of school leavers. Now there is no benefit to girls who assassinate their male class mates for they are guaranteed a job anyway. The advantage to drive has been lost. In the case of a driving X chromosome, this situation represents a stable equilibrium. The spread of the driver has altered the sex ratio, which in turn has reduced and finally removed the transmission advantage of the driving chromosome.

There is evidence that both of these processes maintain driving chromosomes at equilibrium levels in the population. Genes which prevent drive, known as restorer genes, have been found in the mosquito, *Aedes aegypti*, and in some species of *Drosophila*. In *A. aegypti*, X chromosomes differ in their suscept-ibility to Y chromosome drive (Hickey and Craig, 1966; see Wood and Newton, 1991, for a recent review). In *Drosophila*, where drive is a property of the X chromosome, Y chromosome and autosomal resistance to drive have been observed (Stalker, 1961; Voelker, 1972; De Carvalho *et al.*, 1989; De Carvalho and Klaczko, 1993). However, not all species of *Drosophila* where drive is observed show evidence of drive repression systems. In *D. pseudoobs-cura*, an extensive search failed to detect any evidence of repressor genes (Policansky and Dempsey, 1978). A similar search in *D. neotestacea* has also proved fruitless (James and Jaenike, 1990). In these species, it is possible that the transmission advantage of the driving chromosome is limited by the fall in transmission advantage discussed above.

The population genetics of driving autosomes

Meiotically driving autosomes do not necessarily endanger the species in which they are found. If they have no phenotypic effect in the homozygote, then they can increase to fixation, at which point drive is no longer seen. However, there are at least two examples where homozygotes show reduced fitness. In *Drosophila*, homozygotes suffer reduced viability and in the mouse they are lethal. Because of this, there is a degree of frequency dependence in the dynamics of these driving chromosomes. When the driving chromosome is at low frequency in the population, it gains a transmission advantage and will spread. However, as the driver increases in frequency, homozygotes become increasingly common, causing direct selection against the driver. At some point, there is a balance between transmission advantage in the heterozygote and decreased viability in the homozygous state. This equilibrium point is stable and the population stays polymorphic for the driving autosome.

In some populations of the house mouse, around 25 per cent of individuals are heterozygous for the driving t-complex. This continual presence of a dele-terious allele has had an interesting evolutionary consequence. Sarah Lenington (1983) tested the mate preferences of male and female mice. She found a marked preference to mate with partners which did not bear the meiotically driving chromosome. This makes good sense because males hetero-zygous for the t-complex have reduced fertility, and homozygotes are inviable. In the terms of mate choice discussed in Chapter 6, the mate preference for individuals bearing the wild-type chromosome is a choice of partners with 'good genes', and is a clear case of indirect selection for mating preference. This mate choice may also be considered a mechanism of host resistance to a driving element, like insensitivity to drive.

Meiotic drive and concept of ultra-selfish genes

Meiotic drive illustrates an important principle in evolutionary genetics. Alleles may spread through a population despite causing damage to the individual which bears them. Breaking Mendel's rules by killing half of an individual's gametes will produce a decrease in the fertility of that individual, but, per-versely, will favour the gene which gains the transmission advantage. Darwin's concept of natural selection applies not just to struggle between genes within different individuals, but also to genes within an individual.

Genes which spread despite, or rather because of the damage they cause to their host have been termed **ultra-selfish genes** (Crow, 1988). 'Ultra-selfish' is here used to differentiate the reasons for the spread of these types of genes from genes which spread because they increase inclusive fitness (selfish genes). Clearly, kin selection may produce the spread of genes which cause damage to the individual which bears them so long as the action of these genes pro-duces a sufficient increase in the reproductive success of relatives (Chapter 5). Ultra-selfish genes differ from these in that the damage to the individual bear-ing them is not compensated for by the increased survival of relatives. They spread despite being deleterious to the individual in terms of inclusive fitness.

Meiotic drive is one example of ultra-selfish behaviour. Two further exam-ples are provided by B-chromosomes and a newly discovered group of genes termed *Medea* genes.

B-chromosomes

Individuals of most diploid organisms contain an even number of chromo-somes, because each chromosome exists as one of a homologous pair. However, a wide variety of species of animal and plant are found in which there are chromosomes which are not necessary for the function of the organ-ism, and which are not necessarily found in a pair. Some estimates suggest that over 15 per cent of eukaryotic species bear these extra chromosomes (Beukeboom, 1994), which are often referred to as **B-chromosomes**, to differentiate them from chromosomes required for organismal function (A-chromosomes). Another name is supernumerary chromosomes, a title which implies that they are excess to requirements.

How are B-chromosomes maintained? Some B-chromosomes are maintained because, although not essential for individual survival and reproduction, they do increase the fitness of individuals which bear them. For instance, in the chive, *Allium shoenoprasum*, a B-chromosome improves the probability of seed germination (Plowman and Bougourd, 1993). In other cases, however, B-chromosomes appear to be maintained by ultra-selfish behaviour. They have no beneficial effects, and are merely a cost to the organism in terms of using resources. These chromosomes have invaded and are maintained because, when present singly, they have a probability of greater than one-half of being present in the gamete. Various methods by which they manage to increase their transmission efficiency are known, one example of which is given in Box 8.1.

B-chromosomes can be thought of as genomic parasites, maintained because they gain transmission above Mendelian expectation. B-chromosomes which are ultra-selfish produce selection for repression just as meiotically driving sex chromosomes do. In the mottled grasshopper, *Myrmeleotettix maculatus*, for instance, genes in the 'regular genome' of A-chromosomes counteract the accumulation of B-chromosomes (Shaw, 1984; Shaw and Hewitt, 1990).

Box 8.1: B-chromosome transmission: paternal sex ratio in *Nasonia vitripennis*

One classic case of a B-chromosome which is a genomic parasite was revealed by Jack Werren, Uzi Nur, Leo Beukeboom and collaborators during their studies of the parasitoid jewel wasp, *Nasonia vitripennis*. Some male jewel wasps display a trait termed the paternal sex ratio (PSR) where they produce few daughters (Werren *et al.*, 1981). In the jewel wasp, females are diploid, being the product of sex, and males are haploid, usually being derived from unfertilised eggs. Fertilisation ordinarily leads to the production of a diploid individual which is therefore female. However, when the sperm from a PSR male fertilises the egg, the zygote formed loses all the paternal chromosomes, becoming haploid and therefore developing into a male (Nur *et al.*, 1988). The factor which produces PSR is a B-chromosome which causes the male genome (with the exception of itself) to condense and disappear. The behaviour of the PSR-producing B-chromosome makes sense when it is considered that it will be transmitted with high probability through males (being haploid, males pass all chromosomes on), whereas they have a high probability of loss from females (where meiotic segregation means the loss of chromosomes in around half of the gametes). The chromosome is maintained because manipulation of the sex ratio of its host away from the host's optimum increases its own transmission (see Werren (1991) for a review). It is a classic 'ultra-selfish' gene.

Medea *genes*

It is frequently observed that when individuals from different geographic races are crossed, the hybrids show lowered viability, and often sterility, in either one or both sexes (Haldane, 1922). This phenomenon is known as **hybrid dysgenesis**. A variant of this pattern was recently observed by Beeman and co-workers in the flour beetle, *Tribolium castaneum*, and involves an ultra-selfish gene named *Medea* (Beeman *et al.*, 1992). In a search for hybrid dysgenesis between geographically separated races of *T. castaneum*, hybrids were created by mating between strains of the beetle from Singapore and the USA. Hybrids of both sexes were then back-crossed to the parent strains. Three of the four crosses were fully fertile, but one was not. Hybrid females mated to USA males were semi-sterile, with half of the eggs laid failing to hatch. Sterility was also found among hybrids mated to each other, although the proportion of eggs which failed to hatch was lower (Table 8.1).

The genetic causes of the increased egg inviability were analysed by Beeman and his co-workers. Genetic analysis showed the cause of the increased egg inviability to be an autosomal dominant gene, *Medea*, from the Singapore population. *Medea* (short for maternal-effect dominant embryonic arrest) is expressed by females, and results in a chemical being placed in the eggs which disrupts embryonic development. However, the effect of this toxin is nullified in embryos which inherit at least one copy of the *Medea* gene, either by *Medea* itself, or by a gene which is extremely closely linked to it. Thus, homozygote females are all fully fertile, since all eggs receive either no toxin (female is $+/+$) or both toxin and antidote (female is M/M). Heterozygote females $(M/+)$ poison all their eggs, of which half are then rescued by maternal antidote. The fate of the remaining embryos depends on whether or not they get rescued by a paternal chromosome carrying *Medea* (Fig. 8.3).

The results from Beeman's crosses may be understood when it is realised that beetles from the Singapore population are homozygous for this gene, and those from the USA lack the gene. The Singapore × USA cross is fully viable because all progeny inherit a copy of *Medea* from their Singapore parent.

Table 8.1 Semi-sterility in crosses of *Tribolium castaneum* from the USA and Singapore. (Data adapted from Beeman *et al.*, 1992.)

Female parent	Male parent	Average fertility
Singapore	USA	78%
USA	Singapore	86%
F1 female (hybrid)	Singapore stock	85%
F1 female (hybrid)	USA stock	27%
F1 female (hybrid)	F1 male (hybrid)	50%
USA	F1 male (hybrid)	95%
Singapore	F1 male (hybrid)	82%

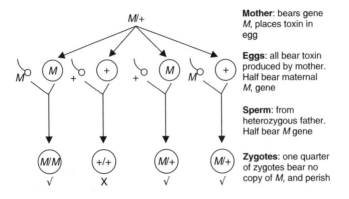

Fig. 8.3 The fate of embryos from a female heterozygous for *Medea* mated to a male heterozygous for the gene.

However, hybrid females are all heterozygous ($M/+$), and thus poison all their eggs but rescue only half. Matings of these females to Singapore males are fully viable, because all the offspring inherit a paternal copy of *Medea*. On the other hand, USA males are unable to rescue embryos because they do not carry *Medea* and thus when they are mated to hybrid females the result is semi-sterility. Matings among hybrids fare somewhat better, for, in accordance with Mendelian ratios, three out of every four embryos will inherit at least one copy of *Medea* and will be saved. Of course, since it is the females who poison their eggs, all crosses involving USA females are fully fertile.

Medea can thus be seen to be self-selecting at a cost to its host. Embryos are killed, but only those which do not carry *Medea*. However, this behaviour confers only the most marginal advantages, measured in terms of the number of non-*Medea* chromosomes which are killed. In order to spread rapidly, *Medea* must increase its own fitness. One plausible mechanism would be if limited embryonic death increases survival among the remaining eggs (Bull *et al.*, 1992). It is therefore interesting to note that sibling egg cannibalism exists in *T. castaneum*, allowing *Medea*-carrying progeny to gain resources by eating their dead non-*Medea*-bearing siblings. In this way the increase of the gene to fixation is inevitable.

How widespread are *Medea*-type genes? Beeman *et al.* (1992) report their incidence in another flour beetle, *Tribolium confusum*, and it would be surprising if they were not to be found in other species. Laurence Hurst has pointed out another possible example, the $Scat^+$ gene of mice (Hurst, 1993). $Scat^+$ is an autosomal dominant gene which is responsible for the disease murine severe combined anaemia and thrombocytopenia. Just as with *Medea*, the only individuals which are affected by the disease are wild-type progeny of heterozygous mothers (Peters and Barker, 1993). Mothers who do not carry a $Scat^+$ chromosome produce viable progeny whoever they mate with.

Ultra-selfish behaviour of cytoplasmic genes: cytoplasmic male sterility

The cases of ultra-selfish behaviour so far discussed are cases where an allele (or whole chromosome) in the nucleus spreads because it gains transmission to more than one-half of the progeny. Ultra-selfish behaviour is not confined to nuclear genes. It is also commonly observed in **cytoplasmic genes**, which are usually inherited in a 'non-Mendelian' fashion from the female parent only.

In Chapter 7, we briefly discussed how maternally inherited symbionts distort the sex ratio of their hosts. Such organisms promote the production of females by inducing parthenogenetic reproduction of females, by feminising genetically male hosts, and by killing males which inherit them. These observations are rationalised by the fact that maternally inherited genes find themselves at an evolutionary 'dead end' in males, since they are not transmitted through sperm. Consequently, it is in the evolutionary interests of cytoplasmic genes to force their host to produce and resource female rather than male progeny.

Mitochondria and chloroplasts are ancient intracellular symbionts, which, like cytoplasmic bacteria, are often inherited through the female line alone. Selection is expected to favour organelles which, like inherited bacteria, increase the output of female individuals. Although no case of mitochondria or chloroplasts causing a sex change has been recorded in dioecious plants (plants which exist in separate sexes), ultra-selfish cytoplasmic genes have been found in hermaphrodite plants. The trait found is known as **cytoplasmic male sterility**, or cms, and involves inhibition of the production of the male parts of the plant (see Hanson (1991) for a review). Its link to mitochondria was established by using restriction fragment length polymorphism analysis to distinguish between different lineages, cms being found associated only with particular mitochondrial **haplotypes** (Box 8.2). Species in which this form of male sterility has been observed include maize, chives (*Allium schoenopressum*), thyme (*Thymus vulgaris*), sugar beet (*Beta maritima*) and plantain (*Plantago maritima*), all of which are gynodioecious, with affected individuals being female and others being hermaphrodite.

Cytoplasmic genes which induce male sterility parallel male-killers in their evolutionary dynamics. Decreasing transmission of genes through the male line does not, in itself, increase the transmission of the cms gene. Cms-producing genes will spread only if there are beneficial 'knock-on' effects of the death of male parts. Death of the male germ-line must be compensated for by increased productivity of the female germ-line. In the case of cytoplasmic genes in plants, this means anther death must increase either the quantity or quality of the seeds which the individual produces. There are two ways in which seed quality or quantity can be increased. First, male sterility will tend to reduce the likelihood of self-fertilisation, thereby enhancing seed quality because it enforces out-crossing. Second, the death of anthers may result in resources being diverted away from the production of pollen, through which mitochondria cannot pass, towards the production of ovules and the resourcing of seeds (Lloyd, 1974, 1975, 1976; Charlesworth and Ganders, 1979).

Box 8.2: Restriction analysis of cytoplasmic genomes reveals correlation of mitochondrial variant with male sterility

The role of the mitochondrial genome in producing the male sterile trait was discovered by investigating restriction fragment length polymorphisms of mitochondrial DNA (**mtDNA**) and chloroplast DNA (**cpDNA**) of normal and male sterile plants. Mitochondria and chloroplasts were isolated from normal and male sterile plants by centrifugation. The DNA was then extracted, and digested with a variety of restriction enzymes. Restriction enzymes cut DNA at particular sequences, and mutations in DNA mean that these sites may be lost and gained (see Chapter 9). The restricted DNA can be run out on a gel, and the sizes of fragments observed. These fragment sizes indicate which restriction sites are present in the mtDNA. Different mitochondrial and chloroplast types can therefore be identified by cutting with restriction enzymes.

When RFLP analysis was performed on mtDNA and cpDNA from normal and male sterile maize plants, it was observed that one particular mtDNA restriction type was present in all male sterile plants. No correlation was observed between cpDNA types and male sterility (Levings and Pring, 1976).

Cases of purely cytoplasmically inherited gender have not been found. In all cases, sex is determined by a mix of cytoplasmic factors which tend to promote male sterility, and nuclear genes which tend to restore male function. This pattern has the same rationale as the evolution of autosomal and nuclear suppressors of meiotic drive. Just as meiotically driving chromosomes are detrimental to their hosts but still spread, so cytoplasmic genes which produce male sterility spread despite the adverse effects that they have on their carriers. Similarly, just as selection will tend to reduce the harmful effects of meiotic drive by favouring the evolution of drive resistance genes, the detrimental effects of male sterility genes may result in selection for nuclear genes which inhibit the expression of these cytoplasmic genes. This is one explanation for the ubiquity of nuclear genes which repress cms.

Intragenomic conflict and parasite–host conflict are similar evolutionary processes

The above examples illustrate how genes which break Mendel's first law can use a transmission advantage to spread, despite being detrimental to the individuals which bear them. Ultra-selfish behaviour is also known for genes which are always inherited in a non-Mendelian fashion, such as genes in the cytoplasm. Observations of the spread of ultra-selfish genes have augmented the 'gene-centred' view of natural selection that was initiated by observation of kin selection. Ultra-selfish behaviour reinforces the notion that the gene is the fundamental unit of selection.

In this model, each individual is seen as a collection of linkage groups. When the interests of all the linkage groups are identical, in terms of survival and reproduction, then selection maximises inclusive fitness (Dawkins, 1982, 1990). However, when the interests of linkage groups (e.g. a driving chromosome, its homologue, and the rest of the genome) differ, then there is said to be **intra-genomic conflict** (*sensu* Cosmides and Tooby, 1981), and the interests of the individual are no longer paramount. When we say 'interests' in this context, it is important not to see the gene as desiring an outcome, merely that selection would favour genes which did produce that outcome. Thus, to say that the interests of a driven Y chromosome are in not being driven is an alternative way of saying that selection will favour Y chromosomes that are insensitive to drive.

There exist strong parallels between how some genes interact within the genome and certain interspecies relationships. Both cytoplasmically inherited sex-ratio-distorting bacteria and mitochondria causing male sterility spread because their actions increase their passage down the female line. A meiotic driver may be considered to be parasitic and in conflict with its homologue. In this context, therefore, we may talk of host individuals and parasitic genes. Taking the analogy further, we may see the relationship between ultra-selfish genes and the rest of the genome as an example of coevolution in which the ultra-selfish 'parasite' gene is involved in an arms race with the rest of the genome. Just as parasites produce selection on the host for novel resistance genes, ultra-selfish genes have produced selection for unlinked genes which repress them (e.g. repressor genes exist for cms, meiotic drive and B-chromosomes). This similarity in evolutionary biology is reinforced by the observation that, just as parasite–host interactions are considered to produce the spread of mate choice for individuals with parasite-resistance genes (Hamilton and Zuk, 1982), the mouse t-complex has produced selection for mate choice for individuals which do not bear a parasitic chromosome.

Genomic conflict and the paradox of Mendelian segregation

Perhaps the most surprising fact is that the individual still survives despite all this genomic anarchy, and that genetic mutualism is the rule rather than the exception. Mendel's first law is true in the vast majority of cases, and cytoplasmic genes are generally well behaved. Of course, the anarchy within the genome does not deny that for many purposes the interests of the genes are the same. Genes which produce individuals that develop and compete efficiently will always be favoured. However, the question remains, why is anarchy not more rife? Cytoplasmic male sterility is not seen in all hermaphrodite plants, and chloroplasts generally appear benign. Mitochondria in animals do not produce obvious sex ratio biases, despite the fact that mitochondria which did so would spread. Meiotic drive appears to be rare. As yet, we have found few exceptions to Mendel's first law.

Within this context, Leigh (1971) argued that no part of the genome could be ultra-selfish for very long, for being outweighed in size by the rest of genome, selection would rapidly result in its suppression. The rest of the genome would act as a 'parliament of genes' and solve disputes. Interestingly, the debate has now moved on to consider whether ultra-selfish behaviours are rare because past evolution has favoured genomes and genetic systems in which the evolution of ultra-selfish genes is impeded. Those species in which ultra-selfish genes have arisen which have not evolved mechanisms to ameliorate their adverse effects may simply have been driven to extinction. This subject will be discussed more fully in Chapter 12.

Mobile elements

Among the many classes of DNA which litter the genome are sequences which possess the ability to jump from one site to another, and are variously called mobile elements, transposons, or jumping genes. Such sequences, by definition, break Mendel's laws and their very nature means they tend to increase in number, both in terms of the number present per genome, and the proportion of individuals which bear them. Although there are many variations on a theme, the P-element in *Drosophila* provides an excellent case with which to illustrate the process. Not only is it the best studied of all mobile elements, but also its origin can apparently be timed to within the last 50 years, giving insight into the population genetics of its spread (for a review, see Kidwell (1994)).

P-elements were first identified in *Drosophila melanogaster*, in which they sometimes cause hybrid dysgenesis, a phenomenon in which crosses between certain strains results in sons which suffer from elevated mutation rates, degenerate gonads and sterility. It was later shown that these symptoms reflect a bout of active jumping of one or more P-elements, triggered by the cross itself and restricted to germ-line tissue. The more jumping occurs, the greater the chromosomal disruption and the more complete the sterility.

On the basis of these crosses, flies could be divided into two primary classes, P strains, which carry P-elements, and M strains, which do not. Because P-elements repress their own activity, dysgenesis occurs only when P-elements are introduced into a genome which lacks them. Even then, sterility occurs only when the P-elements come from the male. The reason for this is that the eggs of a P-strain female carry repressor ability whereas male sperm do not. More recently, a third type of strain has been identified, M' strains, which lack active P-elements but instead carry many defective ones. M' flies do not cause dysgenesis but do benefit from much reduced sterility when they are crossed with P strains, apparently through the action of their defective elements.

P-elements in *D. melanogaster* appear to be recent in origin (Kidwell, 1994). All strains collected from the wild after 1978 contain P-elements, whereas collections made pre-1950 are almost invariably M strains (they lack P-elements). One possibility is that P-elements are ancient but that they are lost rapidly from laboratory strains. However, this explanation appears untenable given that the reverse trend is usually observed: mixed M and P strains almost invariably revert to pure P strains rather than vice versa. More plausible is the fact that *D. melanogaster* only recently came into contact with *D. willistoni*, in the USA (Engels, 1992). This second species, and all its closest relatives, carry P-elements virtually identical to those found in *D. melanogaster*. In the laboratory, P-elements can be introduced by micro-injection, converting an M strain to a P strain. It therefore seems reasonable to propose that *D. melanogaster* gained P-elements from *D. willistoni* by way of horizontal transfer, possibly mediated by the haemolymph-sucking mite, *Proctolaelaps regalis* (Houck *et al.*, 1991).

Intriguingly, the majority of naturally occurring fly strains in Europe, Asia and Australia are now M' strains. Genetic analysis reveals that these strains carry many copies of one particular defective element called the KP element (Black *et al.*, 1987). In laboratory crosses, flies carrying one or more KP elements show much reduced sterility in dysgenic crosses. It appears that the incidence of KP elements is increasing rapidly in the wake of the initial spread of P-elements, being selected because they reduce the detrimental effects of jumping (Rasmusson *et al.*, 1993).

The P-element story, here much simplified, illustrates a number of important points. First, the ability to replicate and move between chromosomes has clearly produced an increase in the mean number of elements borne by flies even in the face of their strong negative effects on individual fitness. Over a mere 50 years or so, P-elements appear to have invaded most of the world's populations of *D. melanogaster*. Of course, this spread has undoubtedly been aided by travel and scientific exchange, but its speed and extent are none the less impressive.

Second, when a mobile element's action has a significant cost to its host, the conditions are ripe for the invasion of any sequence which ameliorates the cost. Thus, the KP element was responsible for a secondary invasion, again rapid and presumably selected because it represses P activity. Perhaps we can expect additional waves of replacement by further derivative elements, each one tightening the P-element's molecular straight-jacket. Given what has gone on so far, it would come as little surprise if, over the next few centuries, P-mediated dysgenesis in the wild became a thing of the past.

Third, and relating to the last point, P-elements show how transient such evolutionary events can be. It is interesting to note that an initial argument against P-elements being of recent origin ran that the chance of finding such an unusual and rapid event was too remote! This would perhaps stand if it were not for the fact that human activity, introducing *D. melanogaster* to the USA where it could encounter *D. willistoni*, initiated the event. Our view of how commonly deleterious mobile elements invade populations is probably strongly coloured by our ability to detect it. Continuing the speculation from the previous paragraph, a scientist several centuries hence who came across *D. melanogaster* for the first time might well describe P-elements as passive passengers of the fruit-fly genome. Indeed, with their predicted lack of dysgenesis, this person might overlook them completely.

P-elements represent what Orgel and Crick (1979) termed 'selfish DNA'. A genetic element is selfish DNA if its sequence encodes a tendency to accumulate within the genome. The use of the term 'selfish' in this context is often confusing to biologists used to the concept of selfish genes. The spread of selfish (and indeed ultra-selfish) genes is one of directional increase in the proportion of individuals bearing the gene at that particular locus. In contrast, the spread of selfish DNA is characterised by an increase in the mean copy number of the element within the genome.

The spread of selfish DNA, like the P-element, may be deleterious to the genome, just as the spread of an ultra-selfish gene is deleterious to the genome. When it does cause damage, selection favours repression of the element (the spread of KP elements), in a similar manner to the way selection favours repression of ultra-selfish genes. However, there is one important difference between selfish DNA and ultra-selfish genes. The deleterious effect of selfish DNA, such as that of P-elements (hybrid dysgenesis), is incidental and not a necessary requirement for their spread. In contrast, the deleterious effect of ultra-selfish genes is an integral part of the spread of the gene.

The non-independence of mutations at different loci

The examples presented so far have one thing in common: they show how, by breaking Mendel's laws, individual genes and genetic elements can spread even when in doing so they reduce individual fitness. A second assumption of classical genetics is that every gene mutates independently, in a way which is not influenced by what happens at other loci. However, recent studies have shown this independence is also not an absolute truth. In some DNA sequences, notably gene families, mutations at one locus may be somehow copied from one member of a family to another.

A gene family is the name given to any sequence which is repeated within the genome. Examples include coding genes, such as the ribosomal RNA genes and the globin gene family, and satellite DNA sequences. The fundamental, non-Mendelian observation is that of concerted evolution (see p. 30). Repeat units within a population and within a species are more similar to each other than repeat units compared between species. Such a situation is unexpected if every repeat unit evolves completely independently. However, the occurrence of the same difference in 100 000 copies of a satellite between dophins and other cetaceans could hardly be interpreted otherwise (see p. 29).

Most repeated sequences exhibit concerted evolution, but the strength of the effect varies with the sequence involved. It appears to be strongest among sequences with high copy number, where the repeats are arranged in tandemly repeated blocks, and between sequences on the same rather than different chromosomes. However, although the observation is indisputable, the mechanisms which produce homogenisation of gene families are as yet not understood. Gabby Dover, who has championed the role of homogenisation in evolution, points to the large number of possible processes which cause 'turnover', the gain or loss and repair of repeat units (Dover, 1982). These include unequal exchange, gene-conversion events, slippage, transposition and a host of other molecular processes. Some or all of these may play a role, but exactly how they interact to do so remains unclear.

The processes causing turnover are collectively referred to as molecular drive. This term has always begged comparison with the processes of meiotic drive. However, although similar in name, in evolutionary terms the two processes are quite different. Meiotic drive is the spread of one chromosomal type because it irradicates gametes bearing homologues. We can predict the chromosomal type that will spread (the one which inflicts the damage!). Molecular drive, on the other hand, is not necessarily a property of the sequence which becomes homogenised. If we are shown two different members of a gene family, it is not possible for us to predict which (if either) of them will come to dominate within the family. There is not necessarily a drive, a predictable directional change, in the processes of homogenisation.

So far, the exact importance of homogenisation in evolution remains controversial. One beautiful case study of how concerted evolution and natural selection may interact is provided by Alan Templeton and co-workers and involves a study of the abnormal abdomen trait in the Hawaiian fruit-fly, *Drosophila mercatorum* (Templeton *et al.*, 1989). Abnormal abdomen is a phenotype where larval development is slow, some juvenile cuticle is retained into adulthood, and where females lay fewer eggs faster. Templeton realised that these characters were similar to those of flies having too few copies of the ribosomal RNA genes (rDNA).

In an elegant analysis, Templeton dissected this complex trait. He found that flies showing abnormal abdomen carried normal numbers of rDNA genes, but that many were made non-functional by the presence of a 5 Kb insert. However, not all flies with large numbers of insert-containing repeats show the trait. When the demands on protein production are greatest, fruit-flies have the ability to make somatic copies of their rDNA genes. Wild-type flies carry a gene which enables them somehow to amplify only functional copies, so nullifying the effect of having non-functional units. Only homozygous recessive flies, *aa*, which lack this ability, show the trait of abnormal abdomen. The system thus contains two parts. Flies are normal either if they carry few rDNA inserts or if they carry the wild-type allele at a second locus, abnormal abdomen, which allows them to under-replicate non-functional repeat units.

Life history analysis predicts that flies showing the abnormal abdomen trait will gain an advantage in populations with a young age structure, where laying eggs early is at a premium. Conversely, wild-type flies will be favoured in populations with an older age structure where longevity is the key determinant of reproductive success. On the Hawaiian Islands, extreme habitat diversity provided an opportunity to test and verify these predictions. In most seasons, the wild type is found towards the tops of the mountains, where high humidity maintains a limited but continuous supply of its larval food, rotting cactus leaves. Lower down the mountains, drier conditions produce patchy food resources and a young age structure favours abnormal abdomen. However, after a violent Kona storm, the pattern was reversed. The storm produced a short-term super-abundance of food high up the mountain, so selecting for early egg production, and the frequency of abnormal abdomen allele rose accordingly.

The story of *D. mercatorum* bears comparison with a second species, *D. hydeii*, living in the same habitat. *Drosophila hydeii* inhabits the same areas, and suffers the same selective pressures as *D. mercatorum*. Like *D. mercatorum*, a sizeable proportion of the 28S rDNA units are disrupted by inserts. The only difference is that *D. hydeii* does not have a gene which preferentially under-replicates non-functional repeat units. In this species, therefore, selection acts directly on the rDNA genes themselves, acting to modulate the number of insert-bearing repeats according to the selective regime (habitat).

Abnormal abdomen is a beautiful example of how the molecular processes of turnover interact with selection. Without homogenisation, insert-bearing repeat units would not have spread in the first place and the species would not have responded so sensitively to the demands of selection. Where this homogenisation produces an advantageous phenotype it is selected. In other circumstances, where the phenotype is deleterious, selection has favoured a gene which masks the effect. This latter case represents an example of what has been termed molecular coevolution, the spread of a novel gene in response to the homogenisation of a deleterious unit within a gene family. The situation in some ways resembles the response of the genome to ultra-selfish genes. A novel gene spreads, not because of selection from environmental change, but from selection to repress deleterious changes occurring within the genome.

Summary

Increasing the fitness of the individual organism is just one way in which an allele can spread. Other ways include the monopolisation of either gametic output or parental resources. The organism is not merely ignored, but suffers a significant drop in fitness. When this occurs the allele in question is termed ultra-selfish. In a similar fashion, the movement of transposable elements increases their representation within the genome, but their presence may harm the individual bearing them. Ultra-selfish genes and selfish DNA which disrupt function will lead to selection favouring mechanisms which repress their effect and restore order.

Ultra-selfish elements appear to be rather rare, even though their existence makes sound evolutionary sense. The reason for this may well lie in a combination of their inherent transience and the way in which they are detected. Mendelian segregation offers a full 100 per cent advantage to those genes which cheat effectively. Where successful, such genes may go to fixation extremely rapidly, or become nullified by the action of a second locus which is itself selected strongly. The power of the latent forces involved suggests that we will observe occasional violent battles which flare up rapidly but are just as soon resolved. One outcome may be extinction, for example if the sex ratio is pushed towards the elimination of one sex. Perhaps more commonly the genome will fight back, producing a silent conflict which will continue unseen until some event such as hybridisation upsets the balance.

Finally, the fact that alleles can spread despite causing harm to the individual, and that certain DNA motifs tend to accumulate within the genome, requires that we re-evaluate our ways of thinking about the evolution of individuals. Although natural selection may produce well-designed individuals, selfish genetic elements may equally act to disrupt that design. It is necessary to realise that selection acts at a variety of different levels, including individual genes, individual chromosomes, haplotypes, whole organisms, or groups. Depending on the genetics of the system in question, any or all of these may become units of selection.

Suggested readings

CROW, J. F. (1991). Why is Mendelian segregation so exact? *Bioessays*, **13**, 305–312.
A recent review of the evolutionary biology of meiotic drive.
DOVER, G. A. (1993). Genetic redundancy for advanced players. *Current Opinion in Genetics and Development*, **3**, 902–910.
LYTTLE, T. W. (1993). Cheaters sometimes prosper – distortion of Mendelian segregation by meiotic drive. *Trends in Genetics*, **9**, 205–210.
A recent review of the evolutionary biology of meiotic drive.
WERREN, J. H., NUR, U. and WU, C.-I. (1988). Selfish genetic elements. *Trends in Ecology and Evolution*, **3**, 297–302.
A wide review of the nature of ultra-selfish behaviour.

Chapter 9

Molecular ecology: a means of measurement

Parents and their offspring, populations within a species, species within a genus, orders within a phylum, and the kingdoms, are all linked to each other by genetic continuity. By harnessing methods which can describe genetic similarities and differences between individuals, it is possible to retrace the invisible path of evolution and so establish relationships between organisms. The spectrum of questions which can be answered covers a vast range. At one end we might want to establish the origin of certain cell types during development, or link progeny to their parents. At the other end we can strive to reconstruct the tree of life, relating each taxonomic grouping to every other. The methods which are being brought to bear to this end are now often grouped together under the heading 'molecular ecology'.

Molecular ecology can call on a wide range of laboratory protocols to examine DNA, and these may be applied to study a wide range of DNA sequences, each of which has its own peculiar properties. By tailoring the technique to the DNA sequence, most of the spectrum of possible questions can be approached. At first glance, a description of all these techniques appears far too large to be held by a single chapter. However, the basic principles, the ways in which the information we seek can be extracted, can be simplified down to the small number of DNA sequences which are studied most, and a few generic approaches. Each of these is discussed in turn, ending with a consideration of some possibilities for the future.

RFLP analysis: a generic tool

The study of DNA only became readily tractable following the discovery of **restriction enzymes**, a range of enzymes which cut DNA at specific sites. Each restriction enzyme recognises a different nucleotide sequence motif, usually either four or six nucleotides in length, and will cut the DNA wherever this sequence occurs. **Restriction fragment length polymorphism (RFLP)** analysis is the generic technique by which variability at the DNA level can be converted into patterns of bands on an electrophoretic gel. Contrary to a widespread misconception, RFLP analysis is not a single tool, but a description of a method which then has to be applied to a specific DNA sequence.

In concept, RFLP analysis is simple. Any particular restriction enzyme will cut total genomic DNA into smaller fragments, the size of which depends on how frequently cutting sites occur. Any particular fragment from a restriction digest is thus a piece of DNA which contains no cutting site in its middle, but has one at each end. If run on an electrophoretic gel and then highlighted specifically with a radioactive probe, such a fragment would appear as a single band. Imagine now that a mutation occurs such that one of the end restriction sites is deleted. The fragment will now be longer, having gained a neighbouring piece. Alternatively, a mutation in the middle might create a new restriction site, such that the fragment is now cut in two. Both these events will cause a change in length of the original fragment and can, therefore, be detected by electrophoresis. In other words, RFLP analysis uses restriction enzymes and electrophoresis to detect where mutations change the spacing of restriction enzyme cutting sites.

Generally, RFLP analysis is not a very powerful tool. Applied to a random section of non-coding DNA, RFLP analysis will reveal roughly the same level of polymorphism as found in **protein isozymes** (protein variants that are revealed by gel electrophoresis). The number of mutations which are detected can be increased by using several different restriction enzymes to cut the same fragment, but even so, the approach is inefficient. The problem is that only a small proportion of all possible mutations will create or delete a restriction enzyme cutting site, and so result in a change in length of the fragment of interest.

Restriction fragment length polymorphism analysis is most effective when it is targeted towards a specific sequence that is known to be polymorphic. A good example of this is DNA fingerprinting, where the simple RFLP approach is aimed at fragments of DNA which contain unstable sequences that show extreme length polymorphism. In this case, the ends of the fragment stay constant, but the middle mutates frequently by insertions and deletions (see below). A second possibility is where DNA sequence analysis has identified a polymorphic site of interest. Given the vast number of restriction enzymes that are currently available, it will usually be possible to design a simple RFLP test to reveal the alternative states. As a third option, when large numbers of probe sequences are required, some workers use a collection of randomly cloned 'anonymous' DNA fragments to generate RFLP patterns. This approach is quick, and may be useful, particularly for species where little is known about the genome. However, the recent development of more powerful methods should see a decline in the use of simple RFLPs.

Paternity and close kin analysis: DNA fingerprinting and its relatives

If two people were seen standing together at a bus stop, both with blond hair, dark brown eyes and standing over two metres tall, many passers by would think to themselves, 'relatives'. If the same pair of people were of average height and had mid-brown eyes and hair, few would give a second glance. The reason for this is a subconscious appreciation that rare traits are more likely to be shared by close relatives, and that the sharing of several rare traits is an even stronger indication. Rare traits thus provide a good method for identifying relatives, but, by definition, most people are 'normal'.

Close relatives can be identified routinely only if the genetic characters can be found which are so very highly polymorphic that everyone carries a rare trait. In 1985, a family of sequences showing just such levels of hypervariability was discovered by Alec Jeffreys, working in Leicester. These sequences are now referred to as minisatellites, and the technique which takes advantage of their extreme polymorphism to identify individuals and relatives has been christened DNA fingerprinting (Jeffreys *et al.*, 1985a).

Minisatellites are very short DNA sequences, 10–60 bp or so in length, which are found scattered throughout the genomes of higher organisms. At each chromosomal location, these motifs are arranged in long strings, head-to-tail in an uninterrupted line. Tens, hundreds, thousands, even tens of thousands of repeats of a motif may be present. It is this repetitive structure which forms the basis of minisatellite polymorphism. Mutations by which repeat units are either gained or lost occur at high frequency, resulting in high, often extreme levels of length variability (Jeffreys *et al.*, 1988a). As yet, there is no evidence to suggest that minisatellites are anything but selectively neutral.

When different minisatellites are compared, either within or between species, it is seen that they tend to share a conserved, GC-rich 'core' sequence. In his original paper, Jeffreys (Jeffreys *et al.*, 1985a) noted a strong homology between the core sequence and an archaic bacterial recombination signal which triggers chromosomal rearrangement. Later experiments demonstrated that this homology is not merely a coincidence, because a protein has been discovered which binds specifically to the core. It seems likely, therefore, that the instability which characterises minisatellite sequences is due to the activity of a sequence-specific protein (Collick and Jeffreys, 1990; Collick *et al.*, 1991).

The minisatellite core sequence is, by and large, conserved between loci, both within and between species. A radioactive probe for the core sequence, used at low **stringency**, will therefore detect many different minisatellite loci in almost any species which is studied (Beritashvili *et al.*, 1989; Burke, 1989; Amos and Pemberton, 1992; Blanchetot, 1992). In this way, a simple Southern blot analysis of restricted genomic DNA results in a complex ladder of bands resembling a bar code (Fig 9.1). Each band represents a fragment of DNA containing (usually) an entire minisatellite array, derived from a single genetic locus. The whole bar code is referred to as a DNA fingerprint and has three useful properties. First, as the term fingerprinting implies, and excluding monozygotic twins, each person's bar code is, to all intents and purposes, unique. Second, since each band is inherited by simple Mendelian rules (Jeffreys *et al.*, 1985b; Burke *et al.*, 1989), the bar code of a child is composed in equal parts of elements from each parent. Thus, if the bar code (**profile**) of a mother and her child are

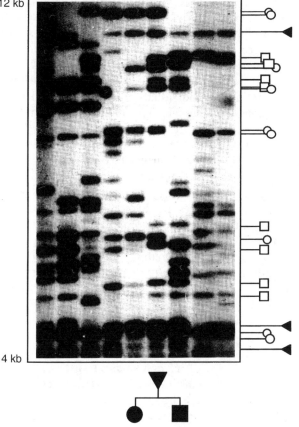

Fig. 9.1 DNA fingerprints prepared from grey seals. A mother, pup, father trio is indicated. Bands inherited from the mother are marked with circles, from the father with squares and from either/both with triangles. (After Amos *et al.*, 1993.)

compared, some bands will be shared and others will differ. Those bands in the offspring which are not in the mother must have been inherited from the father, and can be used in a genetic 'identity parade' (Jeffreys *et al.*, 1985b; Burke, 1989). Since the chance of a candidate father having any specific band is slim, the chance of a male matching every one of a child's paternal bands is vanishingly small, hence the power of the technique. Third, the similarity of two profiles, i.e. the proportion of bands which are shared, provides some measure of relatedness: the greater the number of bands shared, the more closely related are the two individuals concerned (Lynch, 1990). This measure is often called the **band-sharing coefficient**.

Use of DNA fingerprinting to establish individual identity is now routine in forensic science, where suspects can be matched to blood, semen or other tissue left at the scene of a crime (Honma *et al.*, 1989). For non-human applications, however, this is of limited use. One possible exception is a scenario where samples are collected from the study species by remote methods, where the sampled animal is not observed. For example, in a study of European bears, wire mesh cages were erected around the base of trees, so providing attractive rubbing posts. Hair caught in the wire then provided a source of DNA samples from individual bears (Taberlet and Bouvet, 1992), and DNA fingerprinting could have been used to identify individuals (Morin and Woodruff, 1992). In practice, there are logistic difficulties which make DNA fingerprinting an unattractive proposition for such applications. For example, DNA fingerprinting requires rather a lot of DNA for each analysis and the banding profiles are difficult to use in databases (discussed further below).

By far the most important use of DNA fingerprinting is to establish parental affiliation. In most higher organisms, the mother will be known, and hence we are dealing mostly with paternity testing. Indeed, given the energetic costs of reproduction which are born by the females of most species, through either egg provisioning or gestation and lactation, it is usually the males who have the greatest scope for evolving alternative mating strategies. In most cases, therefore, DNA fingerprinting is used as a tool to establish paternity. To date, studies have involved a wide diversity of organisms, from whales (Amos *et al.*, 1991) and lions (Packer *et al.*, 1991), through many different birds (Gibbs *et al.*, 1990; Rabenold *et al.*, 1990; Graves *et al.*, 1992) to stickleback fishes (Rico *et al.*, 1991), bees (Blanchetot, 1992) and apple trees (Nybom and Schaal, 1990).

It is no coincidence that many of the first studies to use DNA fingerprinting for paternity analysis in non-humans involved birds (Burke *et al.*, 1989). Quite apart from the fact that birds are honorary 'furry animals' and therefore research into their habits tends to attract more funding, birds have the important life history trait of building nests. A nest is ideal for DNA fingerprinting, since it contains an entire, usually fairly easily sampled family of chicks, along with the mother and, often, one or more attending males. Fertilisation and hatching occur within a single season. Thus, each nest serves as a focus which allows the identification and sampling of most, or all, key individuals. By fingerprinting the chicks and matching them to the adults it is easy to both ask and answer the biologically important questions of 'how many fathers?', 'do attending males match the clutch?' and occasionally even 'do all the offspring belong to the resident female?'

Other species present more of a problem for DNA fingerprinting. Most mammals are polygynous and in many, fertilisation and birth occur in different seasons. Usually, one is faced with a female and her offspring but no good clue as to who the father is. He could be almost any male in the population and, although behavioural observations made the previous season may indicate possible fathers, there is no guarantee that they will be correct. Indeed, where testing has been carried out successfully, behavioural observations have tended to be rather poor at identifying the father (Pemberton *et al.*, 1992; Amos *et al.*, 1993b). An important logistic problem is the ease with which comparisons can be made between samples analysed in different experiments (on different electrophoretic gels) (Amos, 1992). This will be discussed in more depth below.

The study of close relatives is really only an extension of paternity analysis. Within a DNA fingerprint, information relating to tens of highly variable characters is found. Relatives will tend to share some bands through inheritance and others by chance. The closer the relationship between individuals, the greater will be the proportion of bands which will be shared. With increasing genealogical separation, this sharing becomes more and more diluted and tends towards the level expected for non-relatives. What this means is that, in practice, if one calculates a measure of the degree of band-sharing for a group of non-relatives, and then repeats the performance for a set of mother–offspring pairs, the values obtained fall into two approximately normally distributed curves centred around different means.

Pioneering work by Lynch (1990) has examined the extent to which different degrees of relatedness can be told apart, purely on the number of fingerprint bands they share. He concluded that first-degree relatives can be told apart with some ease, that second-degree relatives can sometimes be told apart and that lower degrees of relatedness yield band-sharing values which are usually too close to those of non-relatives. A good example is the study by Packer *et al.* (1991) of the rules that govern which male lions help each other to take over a pride. They found that brothers will help each other even when some will later be excluded from mating, whereas unrelated males share the reproductive spoils of a pride take-over more fairly. On the basis of evidence of this sort, it appears that Lynch's theory is largely borne out.

In many ways DNA fingerprinting has acted as the flag-bearer for molecular techniques, leading the charge into ecology and behaviour. Its immense power caught the imagination of many who would otherwise not have contemplated using genetics, let alone molecular studies, to solve their problems. Here was a technique that held much promise. For example, for those interested in kin selection (see Chapter 5), the degree of relatedness of co-operators might be appraised accurately (Jones *et al.*, 1991), or, for those studying mating stategies (Chapter 6), data could be generated which showed with certainty which young were the progeny of which adults (Pemberton *et al.*, 1992).

However, within a few years of the discovery of DNA fingerprinting, there were already signs of discontent, as scientists grappled with several major technical hurdles. First, the technique itself is not straightforward, being rather slow and temperamental. Second, and more importantly, the very complexity of the ladders of bands being generated means that DNA fingerprints are extremely difficult to quantify. Individual bands can occur in an effectively infinite number of positions on each gel, and some, those whose affinity for the probe lies close to the hybridisation stringency used in probing (Fig 9.2), will come and go with minor fluctuations in experimental conditions. Third, DNA fingerprints are wasteful of information. Every band in a fingerprint profile is a band and nothing more. Somewhere in the profile there will be a

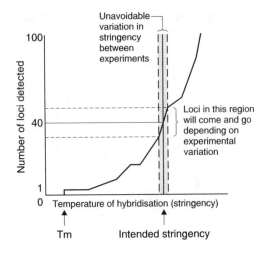

Fig. 9.2 Hypothetical representation of how stringency affects the number of bands (loci) in a DNA fingerprint. Tm is the melting temperature of a precise match between probe and sequence. The shaded area indicates the limits of inter-experiment consistency, such that detection of bands derived from loci in this region will vary between experiments.

partner from the same locus, but it is impossible to know where. Some bands will actually represent two bands, either because the locus in question is homozygous in that individual, or because a band from another locus has migrated to precisely the same point on the gel.

As a general rule, the problems encountered when trying to quantify DNA fingerprints are most severe when comparisons are attempted between samples typed on different gels. Within a gel, the hybridisation stringency will be uniform and bands can be compared readily by eye. Some people have attempted to write computer programs to convert DNA fingerprint information into a database format, and many sophisticated scanning devices have been brought in, but none has been particularly successful. In any case, each band must be stored as an estimate of its molecular weight, and will therefore always be associated with some error. Consequently, DNA fingerprinting is ideal for answering questions such a those posed by ornithologists, which require up to twenty or so samples (a nest full!) to be compared. When more samples are involved, such as in the study of polygynous species, work and frustration increase exponentially and other techniques are sought.

Single-locus minisatellite analysis

The first alternative to DNA fingerprinting was a logical extension. If a fingerprint is too complex, why not simplify it? When individual minisatellite loci are cloned and used as probes at high stringency, in place of the polycore probes used for normal fingerprinting, it is possible to highlight only those bands that derive from a single locus (Fig. 9.3). In one go, allelic information has been gained, problems over stringency have been lost and the complexity has been reduced down to only one (homozygote) or two (heterozygote) bands per individual. The feasibility of this approach was demonstrated both by probing with cloned human minisatellite loci and by using locus-specific PCR primers (Wong *et al.*, 1986; Jeffreys *et al.*, 1988a).

Single-locus minisatellite analysis appeared to present the ideal solution. However, several important problems emerged. First, each allele was still only quantified as an estimate of its molecular weight. Even using internal molecular weight markers, co-loaded with every sample, the errors in molecular weight estimation were sufficient to cause headaches when it came to data-basing the genotypes and for making comparisons between samples (Amos, 1992). Second, in order to use a marker it has first to be cloned. Unfortunately, the very quality which makes minisatellites so variable and hence useful is the bacterial recombination signal each repeat carries. Cloning is, therefore, a nightmare. Minisatellite loci tend just to fall to pieces inside bacterial hosts. Even using sickly strains of bacteria with severely deficient DNA repair systems, whose ability to cause problems has been spiked, the process remains inefficient and requires green cloning fingers. To date, even though human minisatellite loci are now used routinely in forensic studies and a few loci have been cloned from other species, by and large the technical difficulties have proved to be too much for most laboratories.

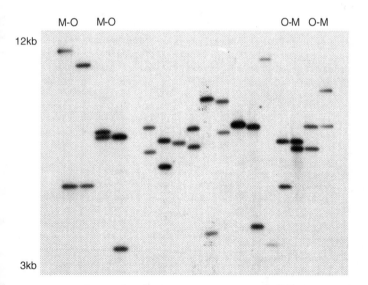

Fig. 9.3 Single-locus minisatellite analysis showing extreme variability. Mother–offspring pairs are marked by linked M and O symbols. Mendelian inheritance is shown by the fact that one band (allele) is shared between each mother–offspring pair.

Microsatellites: the answer?

The apparent answer to molecular ecologists' prayers arrived in 1989, when three papers were published simultaneously, drawing attention to an abundant class of polymorphic DNA sequences which became known as microsatellites (Litt and Luty, 1989; Tautz, 1989; Weber and May, 1989). Microsatellites are so-called because they are like minisatellites only more so. They comprise tandemly repeated strings of short elements, usually two, three or four nucleotides in length, and occur scattered widely throughout the genomes of all **eukaryotes**. Mammalian genomes contain up to several hundred thousand such markers, equivalent to one every 10–50 kb or so (Weber and May, 1989). Invertebrates and plants tend to have fewer, perhaps by an order of magnitude, but there are still plenty for population analysis (Condit and Hubbell, 1991; Estoup *et al.*, 1993).

Microsatellites are considered to be ideal markers for the following reasons. First, although less variable than the extremes shown by minisatellites, they are none the less highly polymorphic, with some two to ten alleles at most loci. Hypervariable loci, with heterozygosities reaching 95 per cent, have also been recorded (Amos *et al.*, 1993a). This polymorphism is thought to result from a process termed 'molecular slippage', whereby the two strands of the DNA molecule become misaligned during DNA replication (Schlötterer and Tautz, 1992) (see Chapter 2). Second, just like minisatellites, microsatellites are predominantly selectively neutral. If nothing else, natural selection cannot act simultaneously on such a large number of loci. Third, allele size is determined by PCR followed by electrophoresis on polyacrylamide 'sequencing' gels (see Fig. 9.4). Consequently, each allele is resolved unambiguously to single base pair resolution, and a sample typed in one laboratory will be assigned exactly the same genotype in another. There is no error associated with allele length determination.

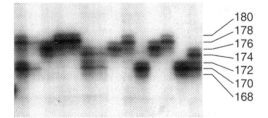

Fig. 9.4 Typical appearance of microsatellite alleles. Each individual carries either one (homozygote) or two (heterozygote) alleles. However, molecular slippage during the PCR reaction results in a number of ghost or stutter bands. Allele sizes are given in base pairs. Note that alleles differ from each other by small number multiples of two base pairs, as is expected if mutations primarily involve the gain or loss of one or two dinucleotide repeats.

The general acceptance that microsatellites are the genetic markers of choice is witnessed by the various genome mapping projects which have been initiated. Mapping requires large numbers of variable markers, such that the position of any particular trait can be located anywhere in the genome through its association with particular markers. The more markers that are used, the finer the scale of the map. For the human genome mapping project, a French organisation called Genethon was set up specifically to identify and generate large numbers of microsatellite markers. Although the method was discovered only in 1989, Genethon has already cloned nearly 8000 different human microsatellites (Weissenbach et al., 1992). Lagging behind, but still impressive, several hundred markers are now available for several other species, including mice (Love et al., 1990), sheep and cattle.

A particular advantage of microsatellites is that the technique is readily transferable between closely related species. The reason stems from the way in which their polymorphism is visualised. Each locus is first cloned and sequenced. Then PCR primers are designed to bind on either side of the microsatellite sequence itself. After amplification, therefore, the size of the product will reflect the number of repeats within the microsatellite array. Assuming the microsatellite locus itself is present in two species, the method will work for both, so long as neither PCR primer site has mutated. In practice, PCR primer sites appear to decay rather slowly, continuing to be useful over evolutionary periods of tens of millions of years. For example, the first four primer sets developed for pilot whales were found to work well on a diverse range of other cetaceans (Schlötterer et al., 1991), primers for cows have been found to work well in sheep (Moore et al., 1991), and human primers work well on other primates (Deka et al., 1994; Rubinsztein et al., 1995).

For population analysis, microsatellites can be used as Mendelian markers to answer exactly the same questions as DNA fingerprinting or single locus minisatellite analysis, with the proviso that the number of loci which must be typed is usually greater, and should be tailored to the level of genetic resolution required. To give an idea of the number of loci which should be screened, individual identity requires around five average loci; paternity analysis (mother known) can be achieved with around ten markers; paternity analysis (mother unknown) requires around fifteen loci, and with twenty loci or more one can begin to ask whether two animals have at least one parent in common.

In addition, new methods of analysis are likely to be developed to take advantage of the quality of data which microsatellites yield. Two examples which are already published come from a study on pilot whale group structure (Amos, 1993; Amos et al., 1993a). In the first, the degree of relatedness between group members meant that insufficient loci were available to identify individual mother–offspring relationships with confidence. However, a method was derived for estimating the proportion of all animals who were accompanied by their mothers, revealing a pattern consistent with individuals never leaving their natal pods (see Box 9.1). In the second case, a test was made to find out how often individuals disperse between social groups. For species in which

Box 9.1: Estimating how many pilot whales have mothers in the same pod

Consider a young pilot whale for which a number of older females may be identified as possible mothers and where all individuals may be screened for a number of reasonably variable genetic markers. Regardless of whether the true mother has been sampled, one or more females may match (share at least one allele at every locus) the young whale by chance. The number of such chance matches can be calculated based on the frequencies of all alleles carried by the youngster. On average, the number of chance matches we find (O) and the number we expect to find (E) should be equal. In other words, $\{O - E\} = 0$. However, when the mother has been sampled, she will match by force. Since all other parameters are changed little, $\{O - E\}$ will approximate to 1.

In any individual test, the value of $\{O - E\}$ is too variable to be of much use. However, when making a lot of comparisons, as one can in the case of a large social group such as the pilot whale pod, the mean value of $\{O - E\}$ provides a reasonable estimator of the probability that any one young whale's mother has been sampled. When this method was applied to a real pilot whale pod the results made biological sense. Among very young animals effectively all individuals are with their mothers (mean $\{O - E\} \approx 1$) and this proportion declines with increasing age (see figure). (After Amos, 1993.)

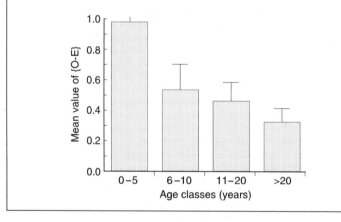

female offspring remain in the social group in which they were born, such groups will become enriched for alleles inherited through the female line. In other words, the progeny of successful females will inflate the frequency of their maternal alleles within their group. Consequently, a positive correlation is expected between group-specific allele frequency and age. Such a correlation was found in pilot whale pods and used to argue that neither sex disperses from their natal groups (Amos *et al.*, 1993a).

Microsatellites may well represent the best balance of properties of all currently known genetic markers, but they are not all sweetness and light. Most importantly, there is the question of null, or non-amplifying, alleles. Whenever a mutation occurs in one or other of the primer sites, amplification will no longer be effective. As a result, even though the microsatellite on that chromosome is still present, only one of the two alleles will be scored. In practice, this means that some apparent homozygotes will be null/normal heterozygotes and some individuals which fail to amplify (a common problem in PCR, particularly with samples collected under widely varying conditions) will in fact be null homozygotes. As yet, we do not know how frequently null alleles occur. Estimates based on a large number of human markers suggest that as many as 30 per cent of loci may carry nulls (Callen *et al.*, 1993), and preliminary studies have confirmed their presence in deer (Pemberton *et al.*, 1995) and damsel flies (Cooper, 1995).

How can null alleles be detected and what are their consequences? Common nulls (frequency > 10 per cent) are easy to detect, since they are manifest as a large excess of homozygotes and occasional non-amplification. Rare nulls are more problematic. In the extreme case, the presence of a single null in a large data set cannot be detected, but then its effect is likely to be unimportant. Null alleles are most dangerous when they occur at a frequency of around 1–5 per cent depending on the sample size. At such frequencies they are below easy detection but sufficiently frequent that they will interfere with subsequent analyses.

The most rigorous way to detect null alleles is to examine carefully a large number of parentage test cases. When a mother is heterozygous for a null, half her offspring will inherit this null and appear to be homozygous for the paternal allele, if this is visible. When the paternal allele differs from the normal maternal allele, the result is a mismatch between mother and offspring. Ignoring mutation, and assuming that all offspring have been assigned accurately to their mothers, this allows the frequency of null alleles to be estimated (Fig. 9.5).

Unfortunately, in many studies, large numbers of mother–offspring pairs are not available. In such cases, null alleles can be detected only though the apparent homozygote excess. The problem here is one of circularity. One of the commonest forms of genetic analysis which is performed on population data is the calculation of the so-called inbreeding coefficient, F_{IS} (see p. 58), which depends on the difference between the expected and observed frequency of homozygotes (Bowcock *et al.*, 1994; Roy *et al.*, 1994). Clearly, the presence of null alleles negates the use of F statistics, and other related methods which test for inbreeding through the detection of an excess number of homozygotes.

Perhaps the best way to circumvent the problems caused by null alleles is to use a twofold approach. First, several loci should be used together, as will be the case in most studies. Any loci which behave discordantly can then be

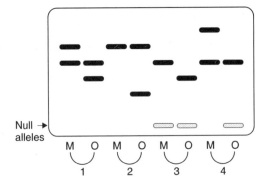

Fig. 9.5 Detection of null alleles in known mother–offspring pairs. In this hypothetical case, four mother–offspring pairs are shown, the mother being on the left in all cases. In case 1, both mother and offspring are heterozygous. In case 2, the mother is homozygous but shares her single allele with her offspring. In case 3, however, both mother and offspring are homozygous for different alleles. Such cases can be explained if the mother carries a null allele (indicated by the stippled allele) which is inherited by her offspring. In case 4, the offspring inherits a null allele from its father, but here a mismatch between mother and offspring is not recorded.

examined in greater detail or discarded. Second, awareness of the problem is half the battle. In paternity studies, false exclusions can only occur when both the putative father and the offspring are apparent homozygotes. Such cases are rare and can, for the sake of safety, be regarded as suspect. In due course, computer packages will become available which will be designed specifically to perform analyses based on microsatellite loci and which will undoubtedly incorporate modifications both to help to identify, and to compensate for the presence of, null alleles.

A second drawback associated with microsatellites is that every locus needs to be cloned and optimised. For most mammals, this is no longer a problem, since loci have been developed for use on a wide range of orders. In other taxa, there is still a need for more cloning. With time, sufficient microsatellites will be developed, such that adequate numbers of useful loci will be available for virtually any new species. For the moment, particularly amongst invertebrates and plants, the first step towards using microsatellites is to begin cloning.

Finally, evidence is now emerging that microsatellite mutations are not as simple as has so far been assumed. Until now, microsatellites have been assumed to gain repeat units as frequently as they lose them and to mutate at a rate which, at each locus, does not vary with allele length or between closely related species (Shriver *et al.*, 1993; Di Rienzo *et al.* 1994; Goldstein *et al.*, 1995; Slatkin, 1995). However, in a comparison of homologous loci between humans and chimpanzees, Rubinsztein *et al.*, (1995) found a startling tendency for mean allele length to be longer in humans. Such a consistent difference implies both directionality and a difference in average mutation rate. In other words, either microsatellites tend to lengthen, and in humans the mutation rate is greater than in chimpanzees, or there is a tendency for micro-satellites to shrink, and in humans the process is lagging behind. Furthermore, an early paper by Weber (1990) showed a relationship between mean allele length and heterozygosity, implying that the mutation rate varies with allele length.

Clearly, whilst microsatellites offer arguably the best range of properties yet found in a nuclear marker, there remains a need for caution. We must under-stand fully the rules which govern mutation before we place absolute faith in derived measures of genetic distance. Failure to do this may result in mislead-ing values, where, as appears to be the case here, the way in which we thought mutations occur differs greatly from what actually goes on in nature.

Mitochondrial DNA analysis

Mitochondrial DNA analysis is arguably the most widely used genetic method for studying animal populations (Moritz, 1994). The reason for this depends on a number of factors, among which its user-friendliness and unusual mode of inheritance are the most important. These attributes, discussed below, are enough to ensure that whatever developments take place in the study of nuclear DNA, mitochondrial DNA will remain a key tool for the study of population structure and taxonomy.

In the cytoplasm of eukaryotic cells are found small membrane-bounded bodies known as mitochondria. It is now accepted that these organelles are the remnants of ancient **endosymbiotic** bacteria. Every mitochondrion has its own degenerate genome, containing all the RNA coding genes required for ribosome function, along with a few genes coding for important proteins located in the inner mitochondrial membrane (Wilson *et al.*, 1985). The single chromosome is bacterial in character, being double-stranded, circular and with a single origin of DNA replication. In size, the animal mitochondrial genome is usually around 16 kb, and gene order tends to be conserved (Brown *et al.*, 1982).

Mitochondrial DNA has a number of attributes which make it useful for population studies. First, lying outside the cell's nucleus, in animals it tends to be inherited only through the female line. The consequence of strict maternal inheritance is a simple pattern of inheritance with no mixing between lineages, either through reassortment or through recombination. Any mutation in a female's mitochondrion can be passed on to her progeny and her's alone; the behaviour of males is irrelevant.

Until recently, strict maternal inheritance was regarded as an absolute rule. However, it is difficult to show categorically that paternal transmission does not occur, and examples have recently come to light among marine molluscs where up to 10 per cent of progeny inherit mitochondria from their father (Zouros *et al.*, 1992; Skibinski *et al.*, 1994). Other, less spectacular examples include mice and nematodes. None the less, although these instances serve as a potent warning, in the absence of further revelations it seems that paternal inheritance is the exception rather than the rule, and that for most purposes it can be ignored.

The second important feature of mitochondrial DNA is that it evolves rapidly. In mammals, most of the mitochondrial genome evolves five to ten times more quickly than equivalent nuclear counterparts, and one region, the so-called D-loop or control region, evolves up to ten times as rapidly again. Such rates of evolution mean that patterns can be discerned between, and even within, populations. Examples range from the identification of human ethnic groups (Cann *et al.*, 1987), through to humpback whale migration routes (Baker *et al.*, 1990) and nematode population structure (Blouin *et al.*, 1992). Mitochondrial DNA analysis is one of the most widely used techniques in molecular genetic analysis of population structure and fine-scale taxonomy.

Third, mitochondrial DNA appears to be selectively neutral, although it could be argued that this is more a hope than a statement. In terms of its use to examine population differentiation, this assumption is critical, but remains relatively untested. To begin with, we know for sure that the mitochondrial genome is under selective constraint, since third-position codons mutate more rapidly than first- and second-position codons, and there are large differences in the rate of evolution of different sections. Evidence of positive selection is limited, but has been reported from a study correlating oxygen uptake with mitochondrial genotype in athletes (Dionne *et al.*, 1991). There is also an observation that mitochondria evolve much faster in warm-blooded animals relative to fish, specifically sharks (Martin *et al.*, 1992). This observation could be explained if the evolution of warm-bloodedness altered greatly the optimum mitochondrial phenotype, resulting in strong selection for change, presumably towards increased energy output and larger numbers per cell. Perhaps the most rigorous study in this line is that of Ballard and Kreitman (1994) who compared sequences in related species of the fruit-fly *Drosophila*. They found that polymorphic sites which affect the final protein are much more likely to evolve to become fixed differences between species than are sites which do not affect the protein. Such examples are rare because selective effects in mitochondria are difficult to detect and, in most cases, have yet to be looked for. Until rigorous studies have been conducted on a variety of species, this slight doubt should not be ignored.

Fourth, mitochondrial DNA may provide a 'molecular clock'. If a DNA sequence accumulates changes at a constant rate, then there exists the possibility of using the degree of genetic divergence to estimate the time of separation. Mitochondrial DNA offers arguably the best combination of characteristics for this: it is thought to be predominantly selectively neutral, it is constained by size such that most mutations are point mutations, and its mutation rate is high enough to be informative over relatively short evolutionary periods (Hillis and Moritz, 1990). Since its inception, the concept of the molecular clock has formed the focus of heated debate. On the one hand, empirical data suggest that there are plenty of instances where a relationship between divergence and time exists (Hillis and Moritz, 1990), particularly if an adjustment is made for the existence of polymorphism within a species (Lynch and Jarrell, 1993). On the other hand, the clock has undoubted imperfections, in that rates of evolutionary change can vary greatly between taxa (Martin *et al.*, 1992), and one must always be aware of possible interference from natural selection.

Mitochondrial DNA is studied in any of a number of ways. Traditionally, it was subjected to simple RFLP analysis, either using Southern blots and probing, or by first purifying the mitochondrial DNA, then cutting and staining with ethidium bromide. Neither method can be regarded as perfect. Southern blot analysis is useful only for DNA fragments larger than around 300 bp, effectively precluding the use of four-base cutting enzymes. This approach is therefore rather uninformative. On the other hand, the purification of mitochondria is technically slow, tedious and requires large amounts of high-quality tissue. However, the rewards are in direct proportion. Freed from the need to identify mitochondrial DNA fragments specifically by probing, small fragments can be included in the analysis, resolved on polyacrylamide gels and detected by silver staining or by radioactive **end labelling**. Not only does this offer a route by which radioactive probing can be avoided, but it also increases greatly the number of restriction-enzyme cutting sites which can be screened. In both cases, genetic information is usually increased by repeating the analysis many times with different restriction enzymes.

With the advent of PCR, mitochondrial DNA analysis has been simplified greatly (Kocher *et al.*, 1989). The simplest method is to amplify the highly variable D-loop region and then use RFLP analysis (four-base cutter enzymes and acrylamide gels) as before. However, the sequencing of PCR products is becoming ever faster and easier, and provides more information than can ever be gleaned using RFLP analysis. At the time of writing, the numbers of studies using sequencing and restriction analysis are roughly equal, even if the sequencing studies have yet to make themselves felt fully in the literature. In the near future, however, we can expect to see the number of RFLP studies fall by the wayside, and rightly so.

Whichever method is used to assess mitochondrial DNA variability, the information gained is used in roughly similar ways. Through the application of one of a number of the generally available computer programs, individual samples are placed on a tree-like pattern reflecting their sequence similarity (see Chapter 11). Whether comparing populations, species or higher taxa, the branching pattern of the resulting tree provides a pictorial description of where divisions occur and how great they are. In the case of populations, a strong bifurcation which correlates with geographic or other data can be used to infer subdivision (Baker *et al.*, 1990). When species or other taxa are compared, the lengths of the branches are often used to infer which classes are most closely related to each other (Smith and Patton, 1991; Edwards, 1993).

In population studies, mitochondrial sequence data can provide information over and beyond mere differentiation. Since males are mitochondrial dead ends, the distribution of mitochondrial variation can provide clues about different dispersal patterns between the sexes. For example, if there are two hypothetical populations between which males move freely but females do not, the two populations will evolve different mitochondrial genotypes (usually referred to as haplotypes, because there is only one state per individual). However, whereas males of both types will be found in both populations, females will show complete segregation. It is a shame that so few mitochondrial studies take advantage of this by including the sex of the sampled organisms.

Mitochondrial DNA is also a good means by which to infer population bottlenecks. For nuclear genes, each mating involves four copies, all of which are equally likely to be carried by their progeny. In contrast, the same mating involves only two mitochondrial haplotypes, one in the male and one in the female, of which but one, the mother's, will be represented amongst the progeny. In any population, therefore, the effective size with respect to mitochondrial sequences is always smaller by a factor of four than for nuclear genes. Consequently, population bottlenecks will have a much greater effect in reducing mitochondrial variability than they will on nuclear sequences (Hoelzel *et al.*, 1993).

Population subdivision

The maternal inheritance of mitochondrial haplotypes makes mtDNA analysis a powerful tool in dissecting population substructure. However, it must not be forgotten that the picture obtained depends strongly both on the relative movements of males and females and on the sampling regime employed. This point is best made by considering some hypothetical examples:

1. Females do not disperse but males do; only females are sampled. Here a mitochondrial study will reveal a highly structured population with well-differentiated local subgroups, even though nuclear genes are experiencing more or less complete mixing.
2. Females do not disperse but males do; the sampling regime only catches dispersing males. Here, although the species could be the same as in case 1, by sampling the other sex, a completely different pattern emerges, indicating no substructure at all.
3. Females disperse; sampling concentrates on the males. Here again no substructure will be apparent, this time irrespective of male behaviour. The female movement ensures complete mixing.
4. Only non-breeding females disperse and these are greatly over-represented in the sample as compared to breeding females. Here the situation is particularly misleading. Reproductively successful females will tend to create substructure, but this will go unnoticed because a quirk of sampling means that these individuals are not being sampled.

These examples are summarised in Table 9.1. Together they show the importance of knowing the sex of the individuals being sampled, and of being able to relate this to the breeding system. It is therefore surprising to note how few published studies note the sex of each sample. In the past, for species such as whales which are difficult to sex in the field, this has been excusable. However, with the advent of molecular sexing methods, there seem to be fewer reasons why this important facet should not be incorporated.

In species which show strong female site fidelity, the degree of isolation between populations can be measured reliably only by looking at the distribution of neutral variability in the nuclear genome. The way in which this is usually done is to take advantage of two opposing forces. On the one hand, genetic isolation will allow mutation and genetic drift to promote divergence, and on the other, gene flow will redress the balance towards genetic homogeneity. At equilibrium, the degree of genetic differentiation observed provides a measure of the degree of isolation.

Virtually any polymorphic neutral genetic marker can be used to identify population subdivision. However, the sensitivity of the test correlates with the maximum potential rate of divergence, given complete isolation, and this, in turn, is directly proportional to the mutation rate at the loci concerned. Given this, it follows that the best markers for studying population subdivision are those which are most variable. Logically, this implies using microsatellites, although larger numbers of less variable markers can be substituted. It should be noted, though, that care must be taken not to fall into the null allele trap

Table 9.1 Influence of sampling regime and breeding system on inferences from mitochondrial DNA studies. In these hypothetical examples, four different sampling regimes are employed to study four species showing various breeding behaviours. Samples are not sexed (as in most current studies) and are treated as representative of the geographical region in which they were collected. Intersect boxes show what inferences would be drawn. Clearly, there are many instances in which a classification of samples by age and sex would prove informative.

	Random sampling of all individuals	Just males are sampled	Just females are sampled	Just migrants or mobile individuals are sampled
Females show maternal site fidelity, males disperse when adult	Weak substructure	No substructure	Strong substructure	No substructure
Both sexes show maternal site fidelity, non-breeding sub-adults are mobile	Fairly strong substructure	Fairly strong substructure	Fairly strong substructure	No substructure
Male site fidelity, females disperse when adult	No substructure	No substructure	No substructure	No substructure
Two breeding sites with unidirectional male migration from A to B (due to prevailing winds, ocean currents, river flow or other factor)	Two populations	Weak substructure	Two populations	One population of type A

(see pages 201–2) by using too few loci and forgetting that some homozygote individuals may be heterozygote null/normals.

The primary problem facing the study of population subdivision is less of finding suitable markers, and more of quantification. Given that polymorphism is found, a lack of difference between two groups implies the existence of current, or at least fairly recent, gene flow. Similarly, a significant difference implies that gene flow is in some way restricted. The problems come when trying to relate the magnitude of any genetic differences to migration rates linking the groups. To do this it is necessary to assume that the situation is at equilibrium, i.e. that neither the sizes of the groups concerned, nor the number of migrants per generation, has changed over a long period. Clearly, this is unlikely to be true in most natural populations. Indeed, it is difficult to think of many examples which have not been perturbed by human activities. Consequently, whereas large differences tend to indicate low migration rates and small differences tend to be associated with appreciable levels of interchange, it is difficult, perhaps even dangerous, to be more precise.

It is unfortunate that one of the most frequently asked questions in population genetics is one that is the most difficult to pin numbers to. In marine ecosystems the term used is 'stock identity', and this is regarded as the golden fleece in both the fishing industry and by the International Whaling Commission. Enormous research effort has been applied in the hope of being able to define stocks more precisely, yet satisfactory answers have yet to emerge. One of the best hopes for the future may lie with DNA sequences that are repeated many times in the genome, and this is dealt with at the end of this chapter. A further intriguing possibility is to use mitochondrial DNA.

There are two views of how to define a stock. In an ideal world it would be described as fully as possible, with information on migration rates, breeding system and population substructure. To a large extent, this view is still a pipe dream. Alternatively, one could aim for a more functional definition. For this purpose mitochondrial DNA has much potential, since its mode of inheritance automatically defines the most important individuals in the population, the females. If population management is based on female mitochondrial haplotypes, the scheme is fail-safe. Where females move, such that they are capable of recolonising depleted areas of their range, the population will be treated as panmictic. Conversely, when female movement is limited, individual subgroups will become more important, but this structure will show up in their haplotypes and each one should be protected by any management strategy.

RAPDs

The acronym **RAPD** stands for **random amplified polymorphic DNA**. In some ways this is the quickest and easiest way to visualise genetic variability. In other ways, there is a great irony. RAPDs tend to be used by people looking for a quick option, usually non-expert molecular biologists who appreciate neither the numerous pitfalls inherent in this technique, nor the ways in which these problems can be overcome. Any scientist who understands the basis of RAPDs, and so potentially could use them in an informative way, would almost certainly know enough to avoid them like the plague!

In any PCR reaction, a polymerase molecule trudges backwards and forwards over the same piece of DNA, making ever more copies. The section of DNA which gets copied is defined by the two primers, specific oligonucleotides which anneal to sites at either end of the fragment. At high stringency, and using long primers (> 20 bp), it is unlikely that anything other than a single product will result. However, anyone who has tried low-stringency PCR (PCR in which the annealing temperature is lowered so as to allow imprecise priming) will know that, whilst this may allow homologous products to be amplified across species, extraneous 'artefact' bands also appear. These extra bands are usually regarded as annoying background, and derive from sites where the primers have, by chance, found a place to bind.

The technique of RAPDs involves the use of low-stringency, non-specific PCR to generate many diverse artefact bands (Williams *et al.*, 1990). Short primer sequences, 8–12 bp long, are annealed so as to generate a ladder of bands resembling a DNA fingerprint. Profiles compared between individuals, either within or between species, often show differences. Polymorphism is thought to be due largely to the gain or loss of primer binding sites, making allele scoring asymmetric. Absence of a band indicates that neither chromosome produces a product, but presence of a band indicates that either one or both chromosomes give a product (Lynch and Milligan, 1994).

Random amplified polymorphic DNA analysis can be criticised on many grounds, ranging from reliability to interpretation. The first problem encountered is one of reproducibility. Anyone who has ever used PCR will know that artefact bands come and go with the day of the week. Small fluctuations in ambient temperature, humidity (affecting evaporation rates during setting up, and hence final concentrations), reagent batch number, PCR machine temperature calibration, DNA quality/consistency and a host of other factors can determine whether particular bands appear or not (Schierwater and Ender, 1993). In this context, it is particularly important to note that whereas different tissues may well contain nominally the same DNA, they may vary greatly in other components, such as glycogen. Depending on the method of purification, different tissues will tend to generate their own characteristic contaminants which, whilst they may not inhibit PCR, may be eminently capable of causing shifts in its dynamics. This is restrictive, for in many studies of natural populations, it is important to be able to use any tissue sample that becomes available.

The problem of variation in PCR stringency and dynamics is confounded by interdependency within each profile. The intensity of each band is not independent from the intensity of all others, since all are co-amplified in a single reaction. If one band is amplifying well, another may become fainter. As PCR is an exponential process, even slight changes in the starting balance can lead to dramatic differences in the final product. Certainly, the possibility should not be ruled out. These effects are illustrated well in papers such as the one by Schierwater and Ender (1993) who looked at how RAPD patterns change with different suppliers of Taq polymerase. To use RAPDs at all, standardisation of every conceivable aspect of the reaction has to be painfully rigorous. Comparisons between laboratories should be regarded as virtually impossible, and every gel must be festooned with standards.

A corollary of standardisation is the question of DNA quality. RAPDs are, by definition, completely non-specific. Thus, if DNA is extracted from a diseased organism, the genome of the bacteria, virus, protozoan or whatever will be at complete liberty to contribute spurious bands to the profiles being generated. Plant material may include epiphytic moulds and fungi, and many insects harbour symbiotic micro-organisms. Samples taken at different times of the year, or from different regions, may thus give reproducibly different patterns for reasons which have nothing to do with the organism's own behaviour, or, at least, reasons which merely reflect an unknown correlate.

The second problem is one of quantification. RAPDs are usually scored on agarose gels. This returns to the question of band size estimation which we first met with DNA fingerprinting. Once again, databasing is difficult, if not impossible, and most comparisons should be restricted to those within a gel. To some extent this problem can be overcome by loading each gel with multiple standards, but this is both wasteful of experimental effort and, if inconsistencies are detected, it is unclear how they should be treated.

Third, and most fundamental, is the question of heritability. We still have only a vague idea of the basis of most RAPD bands. Wherever parentage tests are carried out, there seems little problem in identifying bands in the parental profiles which are transmitted to the offspring. However, in one of the more rigorous studies so far, the authors report finding many non-parental bands (Riedy et al., 1992). Furthermore, where offspring are 'simulated' by mixing parental DNAs, these patterns appear different from true offspring (Ayliffe et al., 1994). Whatever the reasons for this, the signal is clear: shared RAPD bands provide support for the possibility of relatedness, but differences may be due to a variety of reasons and do not, in themselves, preclude relatedness.

Two uses come to mind in which RAPDs may be an acceptable approach. First, many plant species suffer from problems in extracting DNA suitable for restriction enzyme digest, yet the same DNA will PCR well. Given the lack of knowledge about plant genomes in general (with the notable exceptions of species such as *Arabidopsis*, barley, rice and the like), RAPDs offer a means by which to identify strain- or species-specific differences. The method involves generating RAPD patterns and then cloning the PCR bands which identify different groups of interest. By sequencing the fragment and then designing new, high-stringency primers for each informative band, more conventional markers can be developed with little experimental effort. Here, the RAPD technique is used simply as a convenient means by which to screen for polymorphic sequences.

The second use is for simple, one-family paternity analysis. The ideal situation occurs when a single female can produce many offspring and there are few possible fathers, all of whom can be sampled. Here the question is unambiguous, 'which male fits best for each offspring?' Each test can then comprise all possible parents and as many offspring as the gel can accept. Such a system controls for most of the defects inherent in the RAPD system, particularly inter-gel variation. A good example of an application that would be appropriate is the laboratory study of sperm competition in insects with multiple mating (Hadrys *et al.*, 1992).

Point mutation detection

Restriction fragment length polymorphism analysis is an extremely inefficient method for detecting mutations in DNA. Sequencing is much more powerful, but remains a labour-intensive option which cannot be used when sample sizes are large, although to some extent this is changing. The critical limitation is the fact that the electrophoresis of DNA fragments separates only by size. Thus, in order to uncover differences in sequence, such differences must first be converted to length differences. Recently, several methods have been developed which allow differences in sequence to be detected directly. Two methods stand out as being preferred, **single-strand conformation polymorphism** and **denaturing gradient gel electrophoresis**. Both suffer from long names, and are usually shortened to **SSCP** and **DGGE** respectively. This cannot be helped.

Of the two, SSCP is conceptually the simplest. In most applications, when DNA molecules are subjected to electrophoresis they lack secondary structure, being either double-stranded or single-stranded but fully denatured (in sequencing). In neither case is there an opportunity for nucleotide bases to pair within a strand. However, consider what happens if DNA is first denatured, to separate the strands, but then cooled too fast for complementary strands to find each other. Under these circumstances, the bases in the single-stranded molecules will still try to pair up, and will fold back upon themselves to create secondary structure. Such forms are akin to proteins, being three-dimensional and adopting shapes which depend critically on the precise sequence of bases.

The similarity to proteins goes further. During electrophoresis, there is no longer a strict length dependency. Instead, mobility now depends on the precise nature of the folding. Importantly, minute changes to the sequence can result in rather large changes to the three-dimensional form. Remarkably, as many as 90 per cent of all possible single base changes can be detected through the changes in mobility they cause, the method being most effective with fragments of around 150 bp in length (Sheffield *et al.*, 1993).

Although the sequence of each variant is not known, the technique offers a dramatic improvement in the efficiency with which genetic variability may be screened. For example, studies using mitochondrial DNA variability are often limited by sample size. If sections of the D-loop are amplified using PCR and then subjected to SSCP, each individual can be classified as one of a number of different variants, representatives of each then being sequenced to determine their relationship to one another. In this way, it becomes feasible to screen literally thousands of samples, instead of the tens or, at most, hundreds that are currently dealt with. Medical studies frequently use SSCP analysis to identify the region of a gene which is defective and hence associated with a particular disease (Nigro *et al.*, 1992).

Denaturing gradient gel electrophoresis offers an alternative approach to identify molecules which differ slightly in their sequence (Fischer and Lerman, 1979, 1983). In the double-stranded DNA molecule, the two strands are held together by the attraction of opposing bases, G to C and A to T. Of these, GC pairs are considerably stronger than ATs. As a result, the precise temperature at which a sequence melts depends strongly on the proportion of its bases which are Gs and Cs as opposed to As and Ts. More subtly, the precise melting temperature is also affected by the distribution of the bases along any fragment: where Gs and Cs are grouped the melting temperature is higher than in regions where all bases are dispersed evenly.

Denaturing gradient gel electrophoresis takes advantage of these differences between molecules. In DGGE, DNA fragments are made to migrate into a gradient of either increasing temperature or increasing denaturants (urea and formamide). At some point, dependent on the sequence of the fragment, one or other end will begin to 'fray', unzipping to form a Y-shape. When this happens, the mobility of the molecule decreases greatly, effectively halting the fragment's progress. Consequently, a short fragment which frays early will be overtaken by longer, more stable fragments. As with SSCP, the method can resolve small changes in sequences between molecules of the same length (Sheffield *et al.*, 1993).

Fraying does not cause a complete cessation of migration, merely a drastic lowering. If the fragment moves far enough to fray completely, its mobility will increase again and it will be free to move off the end of the gel. A simple modification can reduce this problem greatly. For fragments which are generated by PCR, one of the primers can be modified by the addition of a long GC-rich tail, which effectively locks one end (Sheffield *et al.*, 1989). In this way, electrophoresis can be continued until equilibrium is reached, without the risk of some fragments being lost. Sensitivity can be increased further by altering the gradient, making it shallower or steeper. Ultimately, DGGE is somewhat more laborious that SSCP, and may have a slightly lower resolution, but may be more reproducible and its bands are crisper and easier to interpret.

A miscellany and the future

The methods and study sequences described above encompass the vast majority of those that are used in molecular ecology today. However, to a large extent they reflect our current understanding of evolution: for example, DNA fingerprinting is based around a family of highly mutable genes and lineage analysis depends largely on the quirks of mitochondrial inheritance. Should new molecular evolutionary processes be discovered, it is reasonable to predict that they, in turn, will be harnessed as new tools for studying evolution.

A recent discovery which can bring great benefit to studies involving mammals is the testes-determining gene, *Sry* (Berta *et al.*, 1990). This gene is found only on the Y chromosome, and is diagnostic of males. Using a simple PCR test, tissue samples can be typed quickly and easily for sex (Palsbøll *et al.*, 1992). This may have little relevance to those species which are clearly sexually dimorphic at all ages, but where the sexes are hard to tell apart, for example in many species of whale, or where forensic-type samples are collected, for example faeces or hair, such that the target animal is not seen, it could be important to be able to determine sex molecularly.

Mitochondrial DNA has revolutionised our understanding of population structure and, in particular, patterns of female dispersal. However, in most mammals and some birds, it is probably the males that do most dispersing. If a male equivalent of mitochondrial DNA could be developed, this would be highly desirable. The obvious avenue to follow is to look for polymorphic regions of the Y chromosome. Unfortunately, as yet, the Y chromosome seems to be sadly monomorphic, at least in humans, where most experimental effort has been directed. On the bright side, both minisatellite and microsatellite sequences have been identified on the Y chromosome (Jobling *et al.*, 1994). Therefore, it seems likely that, in the near future, Y-specific markers will become available.

One of the greatest practical assets of mitochondrial DNA is its accessibility. Conserved regions provide PCR footholds such that virtually any organism can be studied with little lead-in time (Kocher *et al.*, 1989). This observation goes some way towards explaining the immediate attractiveness of RAPDs. What is needed is a class of nuclear DNA sequences which show high levels of variability, but for which PCR primers can be designed without the need for cloning and sequencing. One good class of candidate sequences are introns. The majority of genes in higher organisms have the strange property of being interrupted, the actual coding regions containing one or (usually) more non-coding sections. Each gene is therefore a patchwork of alternating coding (**exon**) and non-coding (intron) sections. Sequence analysis reveals that the introns are for the most part functionless 'junk', which evolve rapidly.

To the molecular ecologist, introns offer the exciting possibility of generating a wide range of rapidly evolving nuclear markers. Individual introns can be amplified by simply designing PCR primers to the relatively conserved coding regions on either side (Fig. 9.6) (Palumbi and Baker, 1994). The amplified product can then be screened for variability either using RFLP analysis or, more powerfully, by using one of the point mutation detection systems such as SSCP. Subsequent analysis is not straighforward, because the PCR reaction

Fig. 9.6 Amplification of intronic sequences. Most genes in eukaryotes contain alternating regions of coding and non-coding sequences. The coding regions tend to be conserved, whereas the non-coding, intronic regions evolve rapidly. Consequently, PCR primers designed against coding sequences but facing each other across an intron provide an easy way to access DNA sequences which are variable and likely to be selectively neutral (though this may not be true if the gene itself comes under selection). In the diagram, coding exonic regions are white and intronic regions are filled. Primers A and B could be designed to conserved exon sequences and then be used to amplify the left-most intron.

will amplify both alleles and these may differ from each other. However, technology is moving forward so rapidly that one can reasonably expect this problem to be overcome in the near future.

Little attention has so far been paid to the phenomenon of concerted evolution (see Chapter 2). This is the name given to the strange observation that members of a gene family tend to be more similar to each other within a species than they are between species (Dover, 1982). Although the mechanisms remain obscure, it appears that new mutations either get eliminated or increase in number at the expense of other versions, a process which seems to be too fast to be the result of neutral drift-like processes alone. The net result is that gene families within a population, or species, tend to evolve as a single unit, but when gene flow ceases, divergence takes place. For those interested in defining population boundaries, the advantage of this is twofold. First, the evolutionary process is actually dependent on, and therefore reflects, the process of genetic isolation. Second, being cohesive, with all repeat units within an individual and within a population being similar, a single individual tends to be representative of its population. Compared with other methods, where many individuals must be sampled in order to compare allele frequency differences, a method based on repeated sequences need, in theory, require only a single sample.

As yet, repeated sequences are treated by many with a good pinch of salt and considerable caution. Although this may be unfair, some scepticism is a good thing. As yet, we know precious little about the mechanism by which homogenisation is achieved, nor the DNA sequences in which it will act fastest or slowest. In some instances there is evidence of large changes occurring with extreme speed. For example, all cetaceans share a highly repeated satellite sequence which is around 1.5 kb in length and may have 100 000 copies per genome. All baleen whales differ from all delphinids by a small, 200 bp deletion (Árnason *et al.*, 1988). No intermediates are found. What sort of process can cause an identical change in 100 000 repeat units over a million generations or so? In other instances evolution appears to be less striking. Thus, whilst repeated sequences are a powerful tool for identifying differences between groups of organisms, no attempt should be made to relate these differences to time.

A further class of DNA sequences which one might speculate could be informative if their behaviour could be defined more precisely comprises the so-called jumping genes. A number of sequences are known which have the facility to jump, that is, to move from one site in the genome to another. Sometimes the process is duplicative, a further copy being produced; sometimes it is not. Examples of such sequences are the P-element and *copia* elements in *Drosophila*. In the best understood of these, the P-element system, when a bout of jumping takes place the result is genome disruption and sterility. This phenomenon is known as hybrid dysgenesis, since it occurs usually when two strains are crossed. The reason is simple. Given the detrimental effects of jumping, the genome has evolved a repressor system, such that little jumping goes on in a strain where all flies carry P-elements. The repressor is passed on from mother to offspring, such that it is only when P-element-bearing males mate with non-P females that dysgenesis occurs (see Chapters 8 and 10).

Jumping genes thus behave almost like infectious diseases, where the infection leaves behind a calling card. In theory, the spread of elements such as the P-element could allow highly sensitive measures of gene flow between populations to be developed. This would be most useful in situations where neutral genetic drift lacks either the power or the speed to cause detectable changes. This process of spread has already been monitored, when the P-element, having come over from the USA, began spreading across Europe. This may not be a one-off example. Jumping genes are common and there are undoubtedly many that have yet to be discovered. In time, and as we learn more about them, the possibility of harnessing them to monitor dispersal may be realised. However, as a cautionary note, the fact that they appear to have the ability to transfer horizontally between species may restrict their value.

Horses for courses: the right method

At present, the range of techniques available for studying DNA sequence polymorphism is large and increasing. The methods are becoming ever more powerful. In parallel, our knowledge of the diverse forms of sequence within the genome is also increasing, particularly in organisms which are being sequenced as part of genome mapping projects. Combined, the techniques and sequences provide a very wide range of possible approaches which can be brought to bear on questions concerning population structure, breeding systems and taxonomy.

Unfortunately, many researchers as yet lack the experience and knowledge to take full advantage of what is available. Scientists often behave like sheep. When funding is tight there is pressure to apply a technique which is known to work rather than an untried alternative which may be more appropriate. Sadly, neither the sequencing of a single small piece of the mitochondrial genome, nor the typing of a sample set for a number of microsatellites, will produce answers to all the questions which need to be addressed. It is vital, if the latent power of the technology is to be used to best advantage, that the techniques are matched carefully to the questions which need to be answered.

In the case of taxonomic questions, DNA sequencing looks set to become more and more effective. However, the choice of which sequence to use is critical. Ideally, the regions will be evolving at a rate such that several changes characterise each branch of the tree. In order to safeguard against evolutionary events which cause sudden changes in the rate of evolution, it is always best to sequence several completely different regions independently and look for congruence. In the past, this might have been prohibitive in terms of time and money. These days, direct sequencing and PCR mean that there is little excuse not to follow this line.

To measure gene flow and population substructure there are a number of options. Mitochondrial DNA analysis is perhaps the best all-rounder, since this molecule offers regions covering a wide range of rates of evolution, from the rapidly changing D-loop to the more sluggist genic regions. However, a picture which relies entirely on mtDNA may be misleading due to this sequence's maternal inheritance. A full study should therefore use both mitochondrial and nuclear markers, and make good use of molecular sexing methods. Microsatellites have the potential to be excellent markers, showing high levels of variability and being selectively neutral. However, their full potential seems unlikely to be realised until we understand better the underpinning mutation processes which affect them.

At the finest level of resolution are studies requiring parentage testing and the determination of close family relationships. For these there now exist a number of possible options, among which again, any particular study will be more suited to some rather than others. If there is little information on one or both parents and large numbers of samples must be typed, the best option is undoubtedly to use microsatellite polymorphisms, even if this requires a year of cloning and primer derivation. With small sample sizes, and if each test case can be fitted on a single gel, DNA fingerprinting may be a better option, particularly if a colleague can be persuaded to let one plug into a system which is up and running. If time is short, one might resort to RAPDs, although this may be a sign of total desperation!

Summary

Ecologists, population biologists and everyone interested in the study of evolution are now faced with a powerful range of genetic techniques. With care and forethought, this power can contribute greatly to our understanding of the natural world and how it evolved. Equally, however, these techniques are open to abuse. Just as the wrong-sized screwdriver will often both fail to undo the screw and destroy the screw head, so poor matching of methodology to problem can result in misleading conclusions and wasted experimental effort.

Suggested reading

BRUFORD, M. W. and WAYNE R. K. (1993). Microsatellites and their application to population genetic studies. *Current Opinion in Genetics and Development*, **3**, 939–943.
A recent review of the application of microsatellite analysis to questions in behavioural ecology.

BURKE, T. (1989). DNA fingerprinting and other methods for the study of mating success. *Trends in Ecology and Evolution*, **4**, 139–144.
A good if now rather dated review of DNA fingerprinting.

HADRYS, H., BALICK, M. and SCHIERWATER, B. (1992). Applications of random amplified polymorphic DNA (RAPD) in molecular ecology. *Molecular Ecology*, **1**, 55–63.
A nice review of how RAPDs can be applied to population genetics.

HAGELBERG, E. (1994). Ancient DNA studies. *Evolutionary Anthropology*, **2(6)**, 199–207.
An excellent review of the exciting field of ancient DNA including both the amazing potential and the pitfalls.

MORITZ, C. (1994). Applications of mitochondrial DNA analysis in conservation: a critical review. *Molecular Ecology*, **3**, 401–411.
A well-balanced account which summarises the field nicely.

PACKER, C. A., GILBERT, D. A., PUSEY, A. E. and O'BRIEN, S. J. (1991). A molecular genetic analysis of kinship and cooperation in African lions. *Nature*, **351**, 562–565.
One of the classic studies in which DNA fingerprinting was used to reveal the evolutionary rationale for complex and apparently cruel behaviour in a well-known but poorly understood mammal.

Chapter 10

The genesis of species

Charles Darwin's *The Origin of Species* gave rise to the modern science of evolutionary biology. In this, Darwin describes meticulously how organisms will change with time, how evolution could strengthen a bird's bill to cope with tougher seeds or lengthen the giraffe's neck to reach ever higher foliage. *The Origin* is largely concerned with how natural selection brought about change in the heritable characteristics of different lineages over time (**anagenesis**), and explains why species differ from one another. However, the diversity of modern life shows clearly that this is only half the story. Given there was probably only one ancestral species, and that extinctions inevitably occur, there must be ways in which the number of species can increase, necessarily involving instances where lineages split into two separately evolving lines (**cladogenesis**). This process of splitting, or speciation, is the focus of much modern debate and is the primary focus of this chapter.

The biological species concept

Before considering the various models of how speciation can occur, we must first confront one of the most difficult biological problems, how to define a species. In 1942, Ernst Mayr produced the **biological species concept**, stating that species were 'groups of actually or potentially interbreeding populations which are reproductively isolated from other such groups' (Mayr, 1942). However, the concept and usefulness of Mayr's definition of the species has not gone unchallenged (see Templeton (1989) for a review). To begin with, the criterion of potential interbreeding often cannot be tested. Second, and importantly, Mayr's definition of a species is one which is very 'animal centred'. Although many groups of animals are 'good species' in Mayr's sense, many plants do not fit so easily into this classification. Amongst plants, hybridisation events are not uncommon, leading to the introgression of genes between groups of organisms that are quite widely diverged. Third, Mayr did not even attempt to include asexual species within his definition. Consequently, most micro-organisms, many invertebrates and all vegetatively reproducing plants are excluded from his species concept.

Some argue that the term 'species' is not worth defining. Logically, what we see are many organisms, all of which are more or less interconnected by gene flow. Groups of organisms which interbreed freely amongst themselves, yet which do not interbreed with other such groups, are traditional 'good species'. Less clear-cut instances, groups of organisms between which gene flow may occur, could potentially be described by their degree of isolation. This would negate the need to define an arbitrary dividing line between species and non-species. It would also remove the problem of trying to name geographically isolated groups which cannot interbreed, not because of their biology, but because of some impassable barrier. For example, most British mammals, and even some insects such as the swallowtail butterfly, are now cut off from their continental counterparts. With an alteration in geography or biology, gene flow might become re-established. As Kondrashov (1992) puts it:

To argue about whether similar but separated populations are already different species is as futile as to discuss how many grains constitute a heap. . . . we cannot discover new species like new islands but, instead, face continuous diversity and have to describe it.

With time, continued isolation will lead to genetic divergence and speciation proper. However, it is unclear how much time this takes. There is always the possibility that a change in geography will allow reversion to a single state. For example, a succession of ice ages have first created and then removed significant barriers to the movement of both aquatic and terrestrial organisms, and even changed patterns of gene flow among many plant populations. On a much smaller scale, a tree might fall and come to lie across a river, providing a potential crossing for small mammals and non-flying invertebrates. A topical and controversial anthropogenic example is provided by the opening of the Channel Tunnel, which now reunites Britain with the rest of mainland Europe!

The alternative viewpoint is that the species, however difficult it is to define, remains a valuable concept. For the most part the term works at least adequately and the alternative, describing all life in terms of gene flow between groups, however desirable, is clearly impractical. As yet, we have enough trouble determining simple parameters such as breeding systems, let alone measuring accurately the degree of interchange between all combinations of similar organisms. In practice, therefore, the species taxon must be considered useful and should be retained. Whether this will always be so is open to debate. As ever more sophisticated molecular tools are developed with which to probe patterns of genetic mixing, we may well develop a better understanding of what proportion of the species behave as Mayr would like, and this may alter our views.

Before leaving the species definition, it is perhaps fun to consider a recent example of how molecular techniques have uncovered an unexpected problem. The red wolf is endangered. During the early part of this century, it was hunted to virtual extinction for its reddish fur, and to reduce its perceived threat to livestock. With our current interest in conservation, attempts have been made to save this beautiful canid. A captive colony has been set up and a programme of genetic research has attempted to identify how red wolves relate to their nearest relatives, the smaller jackal and the grey wolf. In addition, genetic typing was also carried out using DNA extracted from the preserved skins of dead animals. The results surprised everyone (Wayne and Jenks, 1991). It turns out that the red wolf is a fully fertile hybrid between a female jackal and a male grey wolf! What were considered on the basis of morphology to be three species now appear (in the terms of the biological species concept) to be just one, with three populations.

Speciation

Consider how a species can divide to give two separately evolving lineages. By definition, this process must begin with some form of heterogeneity, which then becomes exaggerated and, eventually, consolidated into two (or more) distinct groups. When thinking about how this initial heterogeneity can come about, the most obvious scenario is spatial separation through some form of geographic barrier. However, there are other possibilities. Separation may occur because two groups have different feeding preferences or different behavioural patterns. In the latter context, we can generate two gene pools if one group has diurnal activity patterns, and the other nocturnal.

Swayed by the fact that geographic separation is by far the most obvious mechanism, historical treatments of speciation have used this as the basis for nomenclature, dividing the process into cases where speciation occurs in the same region (**sympatric speciation**) and those where the groups are at least initially separated (**allopatric speciation**). A further form, **parapatric speciation**, where the groups exist in adjacent regions with a zone of contact has also been defined, but the value of this extra definition is unclear: either there is no mixing, in which case the situation is functionally allopatric, or there is mixing, in which case the situation is effectively sympatric.

Whatever the geographic relationships of the populations concerned, in order to understand any given speciation event, we have to get to grips with three main processes: what was the origin of the initial heterogeneity, how did this heterogeneity become strengthened and, finally, what are the origins of the barriers which prevent interbreeding today? Having said this, for the purposes of this chapter, we will illustrate these processes with respect to allopatry and sympatry.

Allopatric speciation

In allopatric speciation events the initial heterogeneity involves creating two more or less isolated populations from a single progenitor. Two main mechanisms are considered: either a new population may arise *de novo* through colonisation of a previously empty habitat, or a single initial population may fragment into two or more smaller units (Fig. 10.1).

(i) Colonisation of new, empty habitat

(ii) Fragmentation of habitat at edge of species range

(iii) Division of habitat into two by (e.g.) movement of ice sheet

Fig. 10.1 Three mechanisms capable of producing isolated populations.

Island colonisation

During the 1960s, the map of the world changed. A volcanic eruption off the coast of Iceland created a new island, the island of Surtsey. Once cool, this island presented passing organisms with a virgin habitat which could be colonised. To begin with, progress was slow. A few marine algae began to grow around the shores and spores of lichens added the first vegetation. Presumably, seabirds inspected it but found little of interest beyond a possible nesting site. The first terrestrial animal was a spider. With time, islands such as Surtsey will become filled with a wide diversity of life, each species to some extent isolated from its mother population. The situation appears ripe for speciation to occur.

The end products of events such as the Surtsey eruption can be appreciated best by examining volcanic islands which came into being long enough ago for more extensive colonisation to have taken place. One such example is the St Kilda archipelago off north-west Scotland. On these islands we find a flourishing community of plants which is ostensibly similar to equivalent mainland sites, but which tends to be impoverished, having one representative of a genus where there would be two or three on the mainland. With respect to vertebrates, the situation is more extreme. Only two terrestrial mammals live there, one of which, the house mouse, is recognised as a subspecies; the other, the primitive Soay sheep, was put there by humans. The only other vertebrates are birds, which generally come and go as they please, with the fascinating exception of the St Kilda wren, which, like the house mouse, is now recognised as a subspecies.

St Kilda illustrates two important points about island speciation. To begin with, the probability of a coloniser arriving must lie between certain limits. Too low, and a population will never be established. For example, it is difficult to imagine how any terrestrial mammals would arrive without help from humans. In addition, sexual species have the added requirement that colonisation is by either a gravid female, or a pair who arrive together. The converse case is where colonisation occurs too frequently. Seabirds regularly fly hundreds or even thousands of miles. Therefore, any one breeding colony can breed freely with almost any other, such that a new island presents no barrier to gene flow and will merely become incorporated into that species' range. For speciation to occur, the arrival of colonisers must be possible, but rare, such that the new population experiences significant isolation from its closest neighbours.

The second point illustrated by St Kilda is one of resource availability. There appears to be no good reason why time and again only one of two closely related species of plant should have made the crossing. Much more likely is that most islands offer reduced habitat diversity and resources, such that, should both species manage to get a foothold, one or other species will eventually be eliminated by interspecific competition. Indeed this concept forms the basis of a general observation about islands; they tend to carry fewer representatives of any given taxon than equivalent mainland populations.

Before leaving islands, it should be remembered that islands are not just areas of land separated by water. Lakes and rivers are habitat islands which are separated from each other by land. Individual trees or groups of trees may be considered islands by lichens, epiphytes and insects. Ponds which periodically dry up can be thought of as islands in time. For life adapted to high altitude, mountain tops may be islands. On a much smaller scale, most bacteria face a patchwork of island habitats, defined by humidity, levels of oxygen, acidity and other chemical parameters. Any, or all, of these can qualify as islands, depending on the dispersal power and biology of the organism concerned.

A good example of an unusual island form is provided by the arctic/alpine species of brown butterflies of the genus *Erebia*. Many of these species are confined to restricted areas which vary greatly in size. Some occur spread over several mountain ranges; others are endemic to a single range. Some species are even confined to a single mountain. More interesting still are those species which are confined to a specific altitude band on a mountain. All these species are highly adapted, both to particular altitudes and/or temperatures, and to regions which harbour their various food plants. The result is a plethora of species, isolated from each other by intervening lowlands in any of a wide diversity of ways.

Population division

The second way by which two isolated populations can be created is by fragmentation, such as that which is often found at the edge of a species' range. Anyone who enjoys hill walking will be used to hardier species such as the mountain ash hiding in river gullies well above the tree line, taking advantage of extra moisture, soil and shelter from the high winds. Such instances illustrate how, as conditions become progressively less and less favourable, a species may become restricted more and more to outpost pockets where the habitat is friendlier.

The extent to which edge-of-range effects are likely to promote speciation depends on many factors, but the criteria are generally the same as for island colonisation events. Indeed, each habitat pocket can be thought of as an island which is prone to colonisation. After all, one way of viewing an island is as a discontinuous extension to a species' range. The main difference is that whereas islands tend to be difficult to get to but friendly once there, edge habitats tend to be easy to get to but offer little support to their denizens. Consequently, many edge habitats see repeated cycles of establishment, brief persistence and then demise. Of course, this is a generalisation and there is much variation. Larger pockets which lie further from the main population may well show considerable genetic isolation coupled with the potential for long-term persistence.

An alternative way in which populations can become fragmented is when some event creates a barrier within a previously continuous range. Examples of this cover a vast range of scale. At one extreme, the movement of tectonic plates, ice sheets and changes in sea level create the biggest effect. When Australia moved away from the rest of Asia, it isolated and protected a diversity of marsupials away from the speciation explosion which followed the evolution of placentals. More recently, the closure of the Panama isthmus created a barrier between the Pacific and Atlantic oceans. Some of the most wide-ranging effects have been felt as a result of climatic changes. For example, a succession of ice ages has had profound influence on the distributions of both flora and fauna. At the other extreme, human activities are creating barriers everywhere we look. Dams divide rivers, roads and cities divide the land and habitat destruction such as logging turns once continuous forests into series of isolated patches.

Of course, barriers do not have to be obvious physical obstructions, but instead may act indirectly. Parasites and pathogens will experience subdivision by anything which restricts the movements of their hosts. Climatic or other changes which alter the distribution of insects will have knock-on effects on the dispersal of plant species which rely on these species for pollen dispersal. There are many possibilities. However, when all is said and done, population fragmentation is very similar to island colonisation, with the crucial difference that the new region is precolonised.

Maintenance of isolates

If a new, isolated population is going to stand any chance of giving rise to a new species, it has to remain isolated long enough to diverge from other groups. Island colonisation events face particular hazards associated with the fact that they usually involve few individuals, or even a single family. To give a sense of scale, Mayr (1963) estimated that for each isolate that perpetuates, between 50 and 100 would have been formed but then gone extinct.

The dangers faced by prospective colonisers are diverse. Where the number of progeny produced per female is small, there is a finite probability that the first family is all the same sex, as exemplified by Adam and Eve's failure to produce a daughter. At worst, this would be terminal, at best this would compound the second danger, inbreeding depression. Lowered fertility through inbreeding depression is likely to affect almost all colonisers, given that most will be forced into sib–sib or parent–offspring matings. Its effect will be strongest over the first few generations but will weaken progressively as time goes by. This is because any problems, such as the expression of recessive lethals, effectively select for their own removal. If the first few generations are negotiated, then genetic problems become less important, being replaced by the inherent hazards of small population size. All populations tend to fluctuate in number, depending on good and bad years, on the weather, disease, predator numbers and other factors, and small populations run the continual risk of what might be termed stochastic extinction. All these problems are discussed at length in treatments on conservation biology.

The case of a subdivided population is rather different. Here, one assumes that the initial problems associated with extremely low numbers do not exist. The question is now whether the smaller populations are viable. For a number of species, resource exploitation or patterns of movement mean that subdivision automatically implies extinction. For example, fish such as salmon, which migrate between the source of a river and the sea, may be eliminated by a dam: half a river is no good. In a parallel but different way, the huge nutritional requirements of elephants dictate a minimum habitat size that will support a population. Regions smaller than this quickly become degraded, leading to starvation and eventual local extinction.

Summarising, geographic heterogeneity comes in many forms, and includes islands, barriers and general fragmentation. However, whereas these categories help us to discuss different scenarios, it is important to remember that there is much functional overlap between categories. From the point of view of speciation, it is much better to treat each case individually in terms of the properties which actually affect its evolutionary fate: how much gene flow links it to other populations, how big is it, how long is it likely to survive and what are the likely selective forces which will promote change?

Genetic divergence

What prevents populations which have started down the path of species divergence from reuniting into a single species? Of course, if members from the two populations never meet, this problem never arises. However, in many cases the geographical barriers which prevented members of each population from mixing break down. The populations come into what is termed secondary contact. Unless other biological barriers have formed which prevent gene flow, the two populations will mix and reform a single species. We call factors which prevent gene flow **reproductive isolating mechanisms**. Isolating mechanisms fall into two classes, defined by whether they reduce the likelihood of formation of a hybrid zygote (**pre-zygotic**) or whether they reduce the ability of a hybrid individual that has formed to survive and reproduce (**post-zygotic**) (Mayr, 1963). Pre-zygotic mechanisms involve those characters which influence whether or not a fertilised egg is produced and cover a broad range of phenomena. At one extreme lie traits, such as habitat usage and courtship behaviours, which may determine that members of different populations do not meet or do not choose each other as mates. At the other extreme, mating occurs but some combination of physical and/or physiological incompatibilities prevents effective fertilisation. Post-zygotic isolation mechanisms assume that although mating and fertilisation occur, progeny have reduced or zero viability/fertility. Such mechanisms include developmental problems and incompatibilities between mother and offspring. The essence of post-zygotic isolation is that even when hybridisation between the two populations is 'successful', gene flow still does not occur.

Pre-zygotic isolation

Populations that are isolated may diverge in many aspects of their biology. Divergence in any facet of the process of locating and obtaining a mate is likely to be reflected in a degree of pre-zygotic isolation on subsequent contact of the populations. For instance, if selection in a plant has acted on flowering time, such that individuals in two populations produce and are receptive to pollen at different times, this may limit inter-populational gene flow. Beyond such adaptations to different environments, plants may have a reduced probability of hybridisation if they have coevolved with different pollinators whilst in allopatry (Baker, 1959). However, the most potent forces creating pre-zygotic isolation probably derive from within the species, from sexual selection.

Consider the case of sperm competition. The interests of a male lie in ensuring that a female which multiply mates uses his sperm, rather than the sperm of other males. As reviewed in Chapter 6, a variety of male behaviours for ensuring paternity have evolved. Males block females up with plugs to prevent other males from inseminating them, they scrape previous sperm out of the female, and even place toxins in the accessory fluid to kill other sperm. Many of these male behaviours are deleterious to the female. Selection, unsurprisingly, favours females which avoid/temper the deleterious male tactics. The result can be a morphological arms race. On the one hand, the male penis evolves a morphology which places the sperm in the exact best place to be at the front of the queue. On the other hand, the female genital tract evolves to reduce the effectiveness of these tactics. We expect female and male genital morphology to evolve almost continuously. Under these circumstances, males and females from diverged populations are unlikely to 'fit together' properly for sex. Thus, even if they locate each other and copulate, successful transfer of sperm may be prevented.

Interestingly, entomologists have long known that genital morphology is one of the best ways to tell insect species apart. It has also been observed that the male penis and female tract often fit together so precisely that interspecies mating is prohibited, since the female acts as a lock to which there is only a single male key. Sperm competition driving a genital 'arms' race provides both a plausible explanation for why this should be so, and a potent mechanism for the evolution of **reproductive isolation** (Eberhard, 1985).

Selection may also produce divergence in the mating preferences of females in different populations (Lande, 1981). In many species, females exercise a choice over which male they mate with. Female mating preferences can evolve at any time and for a variety of reasons (see Chapter 6). Male traits evolve in concert with them. In a population of choosy females, a male either has the preferred character, or loses an opportunity for mating. The male trait thus evolves in response to the female preference. Under these circumstances, males may evolve exaggerated versions of any trait that females prefer, becoming lumbered with extraordinary ornaments such as the peacock's tail, the remarkable plumage of the birds of paradise, and the conspicuous wing patterns of birdwing butterflies. Preferred male traits are not restricted to physical ornamentation, but may also include sounds, smells and behavioural displays.

If different female mating preferences evolve in the two populations, then males from one population will be unattractive to females from the other. Even should they come back into sympatry, gene flow will be restricted by the preference to mate with a male from the same population.

Post-zygotic isolation

Post-zygotic reproductive isolation occurs if the hybrid zygotes have either reduced viability or fertility. One of the major questions in the study of speciation concerns the genetic basis of post-zygotic isolation and its evolutionary origins. Another goal of workers in this field is to provide a genetic explanation for an observation made by J. B. S. Haldane (1922). Haldane noted that inviability and sterility were most profound in the heterogametic sex. In *Drosophila* and mammals, where males are heterogametic, male hybrids are less fit than females, but in Lepidoptera and birds, where females are heterogametic, the opposite is the case. We consider four different causes for hybrid breakdown.

Unselected genomic incompatibilities   Populations that are isolated will inevitably diverge through the processes of genetic drift and natural selection. New mutations enter the population and may spread. However, they will spread only if they are compatible with the genes at other loci in the individual in which they are found. Mutations which disrupt function will be selected out of the population. When genetic combinations within an individual are compatible, the genome is said to be coadapted. However, as Dobzhansky (1937) and Muller (1942) noted, genes that are coadapted within their own population will not necessarily be compatible with the genetic background of any other population (Fig. 10.2). In the hybrid context, they may cause disruption of development or gametogenesis even if they were selectively favourable on their own genetic background.

At first sight, hybrid disruption through the breakdown of coadaptation seems unlikely. Why should genes interact to cause breakdown? Most changes should not be expected to produce incompatibility in the hybrid context. Two factors explain the high incidence of hybrid breakdown. The first of these is that development and gametogenesis are very delicate processes that involve many genes which interact, and whose doses and activities must correspond. The second factor is that divergence is occurring at many loci simultaneously. Thus despite the fact that individual changes are unlikely to be incompatible, some inevitably will be.

Empirical studies of hybrid breakdown in the genus *Drosophila* provide evidence that the mechanism envisaged by Dobzhansky and Muller does operate. Genes in species within this genus can be mapped with relative ease. Studies of the interactions between genes within and between drosophilid species provide not one, but many examples of loci that, although compatible within a population, interact and cause inviability or sterility in hybrids (Coyne, 1992; Wu and Palopoli, 1994).

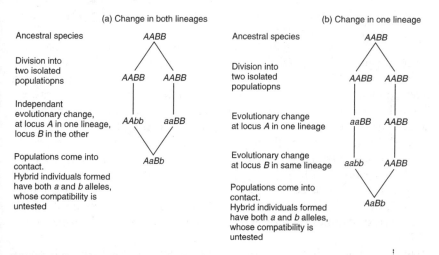

Fig. 10.2 Coadaptation and genetic divergence. Consider evolution at two loci in two sister lineages. A mutant arises and is fixed at one locus in one population in the first case, and at the other locus in the second. These two new alleles are potentially a source of hybrid breakdown, for they have never been on the same background. In the second case, the same situation of untested combinations arises by change in one lineage only.

Divergent selection and reduced hybrid fitness   When a species is separated into two populations, these populations will change in relation to their different environments. Even if the environments are similar, variations in coevolutionary interactions within the two populations are likely to lead to divergence of the two. In a similar way, coevolution may occur within the genome, to repress selfish elements.

When hybrids are formed between two diverged populations, they will be intermediate in form, a mix between the two populations. These intermediate hybrids may be less fit than either parental type because the integrity of both parental morphologies has been compromised. A good case is provided by the butterfly *Papilio dardanus*, which was discussed in Chapter 4. This butterfly is mimetic on more than one model, and within a population the mimetic forms segregate as if they were controlled by a single multiallelic locus with a hierarchy of full dominance relationships between the alleles. When butterflies from different regions are crossed, both the dominance and the pattern of segregation break down. The forms produced are variable, intermediate and lack the precise resemblance to models shown by their parental races. This breakdown of the perfection of mimicry means that such hybrids are at a great selective disadvantage.

It is to be expected that whenever two populations are adapted to different environments, then the hybrids are likely to suffer reduced fitness because they are adapted to neither. A good example involves the Tatra Mountain ibex, *Capra ibex ibex*. During conservation efforts, a Czechoslovakian population was augmented by individuals from a second population in Sinai and Turkey (Greig, 1979). The resulting crosses were fertile, but the reproductive cycle became displaced, causing kids to be born in February when it was too cold. Consequently the kids died, as ultimately did the population.

Differences between populations in their patterns of coevolution may also produce disruption in the hybrid context. This is well illustrated by observations made by Lee Ehrman and co-workers of the causes of sterility between different populations in the *Drosophila paulistorum* complex. Crosses between individuals from different populations produce viable offspring, but the hybrid males are sterile. However, if this cross is performed after the female parent has been treated with antibiotics, the male hybrid subsequently formed is fertile. In this case the cause of the sterility is an inherited bacterium (a streptococcus) which has coevolved with its host in each of the populations. Presence of the bacterium is thought to be beneficial to the host. However, the coadaptation between host and bacterium that exists within a species breaks down in the hybrid context, causing sterility (Kernaghan and Ehrman, 1970; Somerson *et al.*, 1984).

Hybrid breakdown because of the loss of coadaptation between coevolving partners extends beyond interspecific interactions. As discussed in Chapter 8, genetic elements such as the P-transposable elements may increase in copy number despite causing progeny to be sterile in certain circumstances. The invasion of these rogue elements is followed shortly by the spread of genes which repress them. A population that has been invaded by P-elements generally also possesses repressors. However, when a hybrid is formed, the genetic background that creates the repression is frequently lost. The sterility effect returns. The importance of coevolution with parasitic elements within the genome to the creation of hybrid sterility and inviability has been the subject of much recent speculation (Frank, 1991; Hurst and Pomiankowski, 1991). However, further data are needed before a full assessment of its importance can be made.

Chromosomal effects   As discussed in Chapter 2, the constitution of a species with respect to chromosome number and content is subject to evolution. Divergence in karyotype (chromosome number and structure) between populations may cause dysfunction at meiosis and result in sterility. Differences in chromosome number have been identified as the basis of reproductive isolation between populations of mice of the genus *Mus*, and also between populations of dik-dik antelope (*Madoqua* species). In each case, hybrids from different populations are inviable or sterile because individuals in the two populations have different numbers of chromosomes (Ryder *et al.*, 1989).

Chromosome divergence may induce hybrid sterility even when the two sets of chromosomes appear to match, form bivalents and separate. This is because, although much homology exists, the distributions of genes on chromosomes may have altered as a result of inversions or translocations. If say a translocation becomes fixed within one of the two populations, it is possible for pairing to occur satisfactorily, but the random segregation of the chromosome sets results in the production of gametes which bear some genes in excess, and others not at all (see Chapter 2). Such duplications or deletions may result in progeny which are inviable, maladapted or sterile.

The unusual case of *Wolbachia*   In all the above cases, it is divergence in the nuclear genome of two populations that causes hybrid breakdown. Recent studies of insects suggest that hybrid inviability can be produced without any change in the genome of the host. The change is instead the result of invasion of the population by the cytoplasmically inherited bacterium, *Wolbachia*. This bacterium was first discovered in mosquitoes when attempts to cross individuals from different populations persistently failed (Yen and Barr, 1973). Additional crosses succeeded, but only when males were first given antibiotics. The males in each population bear different strains of *Wolbachia*, and this bacterium places a chemical in sperm which renders inviable those zygotes that do not bear it. There is said to be cytoplasmic incompatibility. Within a population all individuals (male and female) bear the same strain of bacterium, because females without the bacterium produce no viable progeny when they mate with infected males. Within-population matings are therefore fertile. However, different populations have different strains of the bacterium with the result that individuals from one population are incompatible with those of another.

Wolbachia has so far been observed only in arthropods, but the indications are that it is quite widespread in both incidence and effect (O'Neill *et al.*, 1992; Rousset *et al.*, 1992). Potentially it is an important source of reproductive isolation in insects.

Secondary contact between populations

Allopatry offers the possibility of reduced gene flow, allowing genetic divergence. Given an initial divergence, there are a number of possible pathways that will lead to complete reproductive isolation, and hence full speciation. Simplest is if the two populations never encounter each other again. Alternatively, populations may come into contact but gene flow between them is prevented because hybrid matings produce either sterile or inviable offspring. The two species will be maintained so long as they are sufficiently ecologically different that they do not exclude each other. Differences between the two may involve the range of the two species (they may then be in contact, but not overlap), or they may occupy different niches within sympatry. If they are maintained, natural selection will rapidly promote any trait – behavioural, physiological or other – that reduces the likelihood of hybrid matings, a process termed reproductive character displacement (Butlin, 1987).

The situation for two populations that contact but are not completely isolated is more complicated. When these populations touch, hybrids will at first be formed. If reproductive isolation is weak, gene flow will occur and the two populations will tend to merge back into one. However, if hybrid fitness is significantly lowered then a hybrid zone may form. There will be a continual diffusion of individuals into the zone from either side. In opposition to this movement, there will be progressively stronger selection against hybrids as individuals increasingly invade the space occupied by the other population, because the frequency of hybrid matings increases. The two populations can thus be maintained integrally, with a stable hybrid zone between them.

When two somewhat diverged populations come to coexist, and the formation of hybrids occurs continually, selection may favour genes producing female behaviours which avoid the deleterious effects of hybrid matings. This controversial hypothesis, first conceived by Dobzhansky (1937), is termed **reinforcement** (Blair, 1955). The thesis is that mating preferences (usually expressed by females) evolve to reinforce partial reproductive isolation, leading to complete isolation and speciation.

Cementing the partition: reinforcement and decreased hybrid fitness

Despite the fact that the notion of reinforcement was first conceived over 50 years ago, there is still no consensus on its importance in speciation. The primary problem is that when hybrid fitness is reduced but non-zero, by definition gene flow will continue between the two populations. However, most genes which favour assortative mating will be effective only when they are restricted to one or other population. If there is gene flow, a trait arising in one population will soon spread to the other, so nullifying its effect. Critics argue that this fundamental problem means that the conditions under which reinforcement can occur are too narrow. In most cases the likely outcomes are that a hybrid zone without mating preferences is maintained, or that one or other population goes extinct, or that significant introgression results in rapid and complete mixing.

The true importance of reinforcement in speciation is currently unresolved. Arguments for and against are complicated by the number of parameters they must take into account. These include the level of hybrid disadvantage, the rate of gene flow between the populations, the number of traits that would be reliable indicators of population origin, the genetics of mating preferences and their preferred traits and whether there is a cost to mate choice. As a general rule, reinforcement is favoured by low hybrid fitness, the presence of reliable population indicators, low gene flow between populations, low costs to mate choice and mate preferences being controlled by one or at most two loci rather than **polygenically.**

Empirical evidence for reinforcement is sparse, but does exist. Investigating isolating features in *Drosophila*, Jerry Coyne and H. Allen Orr examined the degree of pre-zygotic and post-zygotic isolation between pairs of species of *Drosophila*, and also used genetic distance to estimate how recently they diverged (Coyne and Orr, 1989). They observed that pre-zygotic isolation was found to occur more strongly among sympatric species pairs than between allopatric pairs. Since pre-zygotic isolation is the expected outcome of reinforcement, this result is consistent with, but not proof of, the possibility that reinforcement can play a role in sympatric speciation. More recent work by Mohamed Noor on the sibling species *Drosophila persimilis* and *D. pseudoobscura* has strengthened this contention. Noor (1995) showed that the discrimination by females of both species against non-conspecific males was greater when the females were drawn from areas where the two species occurred sympatrically, rather than from allopatric populations.

However, many still argue that reinforcement, if it occurs, will be rare. They cite two points of evidence against its role. First, laboratory simulations designed to generate reinforcement have been unsuccessful and empirical studies have given mixed results (Butlin, 1989; Rice and Hostert, 1993). Second, they cite the results of theoretical models of the process (Butlin, 1987). These tend to predict that reinforcement will be rare, and that usually there will be either complete introgression or one or other population will be driven to extinction. However, these models can be criticised for the necessary simplicity of their underlying assumptions. For example, many models assume that mating occurs randomly, yet most organisms have evolved highly structured mating systems which strike a sensitive balance between the need to avoid inbreeding and the optimal strategies of each sex. Such structure may well have a profound effect, potentially shielding some parts of the population from the effects of introgression and hence buying valuable time for the evolution of mechanisms promoting full reproductive isolation.

Sympatric speciation

Sympatric speciation is the process by which two populations that coexist in the same geographic area manage to diverge. It is widely regarded as the most difficult type of speciation to visualise and to explain. However, before considering how sympatric speciation may occur, it is worth re-emphasising the conceptual ambiguity which is inherent in this terminology. Functionally, it is the presence of linking gene flow which makes speciation difficult to explain, not the positions of individual organisms. Thus, although gene flow and geographic separation are often correlated, this does not have to be the case.

By acknowledging this distinction, some of the primary conceptual problems can be bypassed. There is no longer any need to erect special categories for rare cases such as parapatry, and the necessity to treat poorly isolated island populations separately is circumvented. Instead, we are free to concentrate on attacking the true heart of the problem: what factors cause divergence and how do these overcome and/or modify levels of gene flow? With this in mind, we can now progress to consider examples of speciation that have previously been considered either as sympatric or parapatric, where gene flow is restricted but not negligible.

Dobzhansky, Endler and others have pointed out that evidence of speciation is often found in habitats that show great heterogeneity, where large changes occur over a small range. Such habitats are common. As every wine buff will know, conditions for vine growth vary with every slight change in altitude, aspect to the sun, soil type and drainage, a variation which is transmitted with surprising fidelity to the resulting wine. From the point of view of an organism, rather than of an interested consumer, the differences in selection pressure can be equally great, and wherever gene flow is restricted, local adaptation can result. A nice case in point is that of alpine plants, where differences in altitude can result in appreciable differences in the timing of the growth period and hence flowering. In turn, this may result in a degree of genetic isolation.

Once local adaptation has occurred, natural selection will promote further genetic isolation. This is because matings between individuals from different regions or habitats will tend to produce maladapted offspring. Thus, any traits which promote mating within, rather than between, groups, be they behavioural, physiological or of some other form, will be selected. The most obvious trait would be a simple reduction in dispersal. Individuals who stay at home will be more likely to mate with an individual carrying locally adapted gene combinations. Any genes which reduce mobility or the tendency to move will, therefore, be favoured and will spread, albeit slowly. As a result, gene flow will become further reduced, and the opportunity for even greater local adaptation will be realised (Fig. 10.3). Although not usually considered as such, these processes differ little from those involved in reinforcement, as discussed on pages 230–1.

Toxic, heavy-metal-rich spoil heaps are a by-product of mining operations in some parts of Wales, and provide the setting for a classic case study. The story concerns two common grasses, *Agrostis tenuis* and *Anthoxanthum odoratum*, living on and near the spoil heaps. Only plants showing tolerance to heavy metals can survive. It was noticed that invasion of these heaps was slow, and that the plants looked sickly. Under experimental conditions it was shown that the progeny of plants collected from spoil heaps were adapted to them, in that they showed much higher tolerance to heavy metals than the progeny of plants collected elsewhere. However, heavy metal tolerance has a cost. Plants with high levels of tolerance have decreased competitive ability in unpolluted soils (McNeilly, 1968; Cook *et al.*, 1972).

The conditions for speciation appear ripe. Plants growing on heaps will produce less fit progeny when they cross with non-tolerant plants and vice versa. Selection would thus be predicted to favour any mechanisms which reduce the frequency of between-habitat crosses. This is exactly what was found. Plants growing in and around mines tend to mate assortatively. At Trelogan and Drws y Coed, McNeilly and Antonovics (1968) found plants on and off the mine to have different flowering times. Plants on the mine flower earlier than those off the mine, a characteristic which has since been shown to be heritable (Nicholls, 1979). This effect is particularly pronounced at the edge of the mine where one might expect gene flow to be greatest.

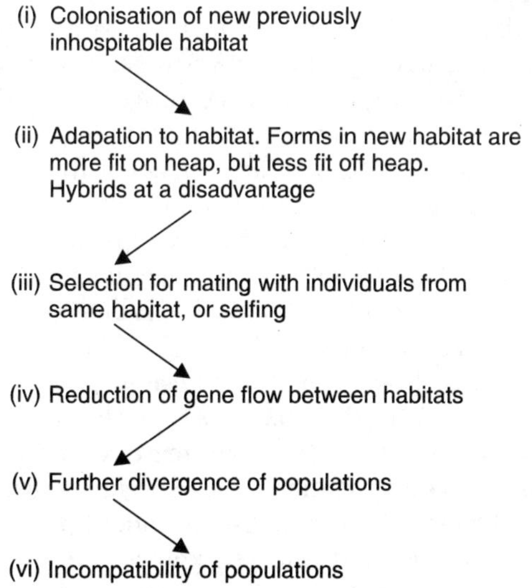

(i) Colonisation of new previously
   inhospitable habitat

(ii) Adapation to habitat. Forms in new habitat are
    more fit on heap, but less fit off heap.
    Hybrids at a disadvantage

(iii) Selection for mating with individuals from
     same habitat, or selfing

(iv) Reduction of gene flow between habitats

(v) Further divergence of populations

(vi) Incompatibility of populations

Fig. 10.3 Population division in a stepped cline: heavy metal tolerance in plants.

In this case, history appears to provide a vindication of Endler's (1977) predictions about how habitat heterogeneity can promote speciation. However, there is still room for ambiguity. The earlier flowering time of the mine population could be a direct consequence of heavy metal poisoning and have nothing to do with selection for reduced gene flow. Fortunately, the rigour of the study suggests this alternative may be rejected. Both the increasing strength of the trend towards the mine edge and the observation that plants on the mine show increased self-compatibility (Antonovics, 1968) lend support to Endler's model. In less thorough studies, the potential room for doubt is clear.

It could be argued that the case of plants invading a hostile environment is special, and has more in common with colonisation of an island habitat than with true sympatric speciation. Perhaps so, although one could counter by pointing out that all species at the edge of their range face a similar situation, even though the abruptness of habitat changes may appear less. In addition, sharp habitat boundaries are much commoner than one might expect. For example, many phytophagous insects feed on two or more plants. The same can be said for many parasite–host and predator–prey interactions. All such cases provide a habitat separation with the potential to drive speciation.

Undoubtedly the best example of a species' niche being divided by food preference has been revealed by Guy Bush and his co-workers during their classic study of the haw fly, *Rhagoletis pomonella* (Bush, 1969; Feder *et al.*, 1990). This species is native on hawthorn in the United States and the female lays eggs inside fallen haws, where the larvae subsequently develop. In the late eighteenth century, apple trees were introduced into the United States. Today the haw fly may be found living and breeding successfully on both hawthorn and on apple.

Allozyme studies show that the populations living on hawthorn and apple are genetically distinct. Population differentiation is marked, with homogeneity between flies living on hawthorn, homogeneity between flies living on apple, but heterogeneity between the two. In addition, the progeny of individuals from hawthorn and apple have preference for their natal host plant, and appear to be adapted to it. A single species of apple maggot fly has become split into two sympatric forms, with the strong likelihood that full speciation is not far away. The key element here is the fact that courtship, mating and egg-laying all take place on the food plant. In this way both habitat separation and a mechanism promoting assortative mating are linked, allowing specialisation and eventual speciation along classical lines (Feder *et al.*, 1994).

The two examples given so far, heavy metal tolerance in plants and food plant preferences, both involve effectively binary habitat states. Such cases appear to be relatively common, but it is also interesting to examine whether more gradual clines can also initiate speciation. The expectation here is that any discontinuity in a natural cline will be emphasised. Indeed, Endler (1977) argues that such discontinuities exist and are well known; they are in fact hybrid zones. Traditionally, hybrid zones have been thought of primarily as sites where two populations have come back into contact having previously diverged in allopatry. However, there seems no good reason why some might not be. Rather, as Endler says, they may represent regions of tension where sympatric heterogeneity is in the process of forcing a break.

Whether Endler is correct remains unclear. The principal counter argument runs that gradual clines are likely to involve adaptive trends affecting many loci, each with its own particular, different cline. Consequently, the point along the cline at which assortative mating is favoured will vary from locus to locus, making any potential break point highly diffuse and reducing greatly the ability of selection to promote divergence. This argument is strong, but does not hold for all conditions. If the loci involved are linked, or if fitness is the product of epistatic interactions between loci, changes in gene frequency could be correlated between loci. Under these circumstances a natural break point could form, a point on the cline either side of which assortative mating would be promoted by selection. Whether such genetic conditions exist in nature or merely provide a theoretical possibility has yet to be determined.

Speciation in a homogeneous population

Speciation in a homogeneous population is an extreme form of sympatric speciation where there is no help offered by habitat differentiation. Consequently, this is the most controversial scenario of all. Just how barriers to gene flow could arise while individuals continue to interbreed freely is not immediately obvious. It is, therefore, ironic that a speciation event of this kind is one of the best recorded.

Problems in the way chromosomes pair and segregate during meiosis present a major barrier to fertility. Usually, this is caused by differences between members of a homologous pair. A simple way to remove many of these problems is to double the chromosome number. By so doing, the members of any troublesome chromosome pairs will immediately gain identical partners. As noted in Chapter 2, polyploidy is rare in animals. In contrast, amongst plants polyploidy (particularly allopolyploidy) is common, and represents a crucial means by which hybrid infertility can be overcome. At the same time, the act of doubling the chromosome number, in one fell swoop, produces complete genetic isolation and hence 'speciation'.

This process was first observed in Kew Gardens, London, in 1899. In these gardens, a new species of *Primula* was found, and was named *Primula kewensis*. Subsequent experiments showed the new species to be a hybrid between two related species that already grew in the gardens, *Primula floribunda* and *Primula verticillata*. Although sterile, because the chromosomes were not homologous and did not pair properly at meiosis, the new plant did propagate vegetatively (Digby, 1912; Newton and Pellew, 1929). Later, three instances of this plant setting good seed were observed, a feat shown to be associated with the doubling of its chromosome complement. A new species had arisen. *Primula kewensis* had become fertile within itself. At the same time it achieved reproductive isolation from its diploid parental species, since cross-pollination between it and its parental species resulted in sterile triploids.

Speciation through polyploidy can be inferred from the karyotypes of a great many groups of plants, including wheats (Triticacae) (see Chapter 2) and members of the onion family (*Allium*). It is also recorded in algae and bryophytes. Polyploidy is accepted as one mechanism of sympatric speciation, but is it the only one? Many believe so, notably Ernst Mayr (1963). Some accept that other routes are theoretically possible, but view them as unimportant. This leaves the final camp containing those who believe sympatric speciation without polyploidy plays an important role in the evolution of some groups (Tauber and Tauber, 1989; Bush, 1994).

Adaptive radiations as case studies in speciation

Darwin's observations of the Galapagos finches were of great importance to his thesis that species are not static, but change. The **adaptive radiations** found on islands like the Galapagos have since played an important role in our understanding of the process of speciation. The most widely accepted hypothesis for speciation in the Galapagos finches follows the work of David Lack (1947) (see Grant (1984) for a review).

The Galapagos island group consists of a series of small islands in the Pacific Ocean, off the coast of South America. Thirteen species of finch, each with specific morphologies and ecologies, inhabit the Galapagos archipelago. Evidence suggests these all derive from one founder species which first colonised the island within the last two million years. Lack imagined the speciation process to have both an allopatric and a sympatric phase. Initially, the islands were colonised by one species of finch. As population size increased, birds dispersed to other islands in the complex, forming several geographically isolated populations. These populations would have probably diverged genetically, particularly in respect of resource use. Following an allopatric period, recolonisation of the island from the parental population may have occurred. Being now, at least in part, reproductively isolated, and utilising different resources, the two species could coexist. Selection would promote further divergence in resource usage (character displacement) and aspects of courtship behaviour that prevented hybridisation.

The repetition of this process would generate more isolated populations, and so more new species. On the Galapagos, therefore, it appears to be the presence of many islands which allows new populations to form and diverge. The variety of habitat on the large islands has allowed coexistence, and Grant (1984) speculates that the thirteen species found on the islands today reflect what is probably a packed habitat. The different species of finches now present are specialists, with some feeding upon cacti, others on seeds of particular sizes, and yet others insects. Differentiation in resources within the island appears to be limited, and it is unlikely that additional species can be supported.

The conditions that foster radiations have also been examined in Hawaiian fruit-flies, which exhibit one of the most dramatic explosions of speciation. The islands of Hawaii are volcanic and are themselves changing geologically. Every now and again new eruptions add another island to the easterly end of the chain. In Hawaii, *Drosophila* have apparently speciated again and again. Over 450 species have been described, and many more await description.

What conditions on the islands of Hawaii make speciation so common? When first formed, each island presented an empty environment, free from competition. Between islands, movement of the weak-flying *Drosophila* must be rare, allowing colonisation without extensive linking gene flow. It is therefore possible to get several integral populations which may then diverge. In addition, the topography of each island creates an extreme diversity of habitats. For example, on Maui, sites only eleven kilometres apart, one at the top of a mountain and one at its base, experience annual rainfalls of 1000 and 30 centimetres respectively. Finally, occasional lava flows act to divide populations. All these factors probably played a part in producing the species diversity we see today. The island habitat produces continual division and isolation for a weak-flying insect. New populations form and diverge rapidly. To these must be added a role for sexual selection which has had such a profound affect on the morphology of Hawaiian *Drosophila*. Sexual selection appears to have had an unusually important role in speciation in this context. Closely related species which appear to have only weak post-zygotic isolation exhibit amost complete pre-zygotic isolation (see Kaneshiro and Boake (1987) for a review). The picture of diversification and speciation is highly complex, involving repeated colonisation of new habitats, division of pre-existing populations, and return to sympatry. The presence of marked habitat diversity allows many species to coexist, and the role of sexual selection in speciation adds a further dimension that may not as yet be fully appreciated.

Radiations appear to occur when there is both a capacity for isolation of populations and habitat diversity sufficient to support large numbers of closely related species. Islands such as the Galapagos and Hawaii, and perhaps a few other archipelagos such as the Canary Islands, provide the perfect situation. When the finches arrived on Galapagos, there were many empty niches. Each new isolated lineage could survive. However, it should be realised that most islands are not like this. Many islands are not sufficiently heterogeneous to support many species. The degree of radiation that occurs depends crucially on the diversity of habitat. Islands like St Kilda, where there is lower species diversity than on the mainland, and as yet no truly endemic species, might prove a more representative model for islands.

Discussion and summary

Speciation is one of the most fundamental subjects in the science of evolutionary biology. Speciation occurs in three stages. First, populations become separated, and gene flow is restricted. Second, these populations diverge. This divergence may produce reproductive isolation. Third, if contact is restored before full isolation, selection may favour traits and behaviours which avoid hybrid matings. Modes of speciation have often been classified in terms of how the original separation was brought about. Speciation is allopatric if the populations are geographically separated, or sympatric if they are not. In reality, what we have is a continuum of levels of gene flow. In pure allopatric speciation, the gene flow is very low. In sympatric speciation, it may be higher. The major difference between the models is that sympatric models require change within the initial population to restrict gene flow within it. The initial reduction of gene flow in allopatric models does not involve genetic changes within the species, rather it arises from external influences.

Accepting this clarification, the problem of speciation becomes rather more straightforward. Where gene flow is low, the two populations are free to diverge either passively and slowly through genetic drift, or more actively and faster through natural selection acting to improve the fit of each group to its local niche. Eventually, as long as isolation continues, speciation will occur.

Within sympatric speciation, therefore, it is the mechanisms by which gene flow can be reduced and eventually eliminated that form the focus of most debate. Hypotheses concerning the mechanisms of speciation in these cases cluster around scenarios in which habitat differences and assortative mating can be linked. There are many possibilities, two classic examples being via food choice, as in the haw fly, and via physiological changes such as the earlier flowering of grasses growing on spoil heaps. Molecular techniques offer much to this field. They allow us to quantify gene flow between populations and to dissect in detail the mating systems within a species. With these tools, we can look again at old problems and measure the important parameters.

Following separation into isolated lineages, the two populations will continue as separate lineages if members of the populations do not make contact, or if they make contact but do not mate with each other, or if divergence renders hybrid progeny incompatible when they do so. A large body of workers are currently interested in assessing what type of genetic and evolutionary changes produce incompatibility. Currently, the genomic locations of some genes causing incompatibility are being located, pinpointing the genes responsible. This allows us to look for patterns in the chromosomal locations of genes which may give insights into the rationale for Haldane's rule. Soon, it is to be hoped that not just the location, but the nature of the genes involved will become clear.

In the future, the increasing power of molecular genetic analysis may offer novel approaches. The importance of sympatric speciation, for instance, has had to be reassesed in the light of work on the phylogeny of cichlid fishes. In a recent piece of work, Schliewen *et al.* (1994) used mtDNA sequence analysis to investigate the origins of the species flocks in two African lakes. They found both to be monophyletic, indicating that the species in each lake were derived from one colonist. The nature of these lakes (they are described as ecologically monotonous) argues for a role sympatric speciation in these radiations.

Suggested Reading

BRIGGS, D. and WALTERS, S. M. (1984). *Plant variation and evolution*. Cambridge University Press.
Note particularly Chapters 9–11.

BUSH, G. L. (1994). Sympatric speciation in animals: new wine in old bottles. *Trends in Ecology and Evolution*, **9**, 285–288.

BUTLIN, R. (1987). Speciation by re-inforcement. *Trends in Ecology and Evolution*, **2**, 8–13.

COYNE, J. (1992). Genetics and speciation. *Nature*, **355**, 511–515.

OTTE, D. and ENDLER, J. (eds) (1989). *Speciation and its consequences*. Sinauer, Massachusetts.

Chapter 11

Phylogenetic reconstruction and molecular clocks

A fundamental concept which runs through much of evolutionary thought is that of genetic distance: how individuals, populations, species and higher taxonomic groups are related to each other. Often, the relevant empirical data involve multiple comparisons between many different genetic elements. Although such data could be presented in a table, they become far more readily accessible when converted into a pictorial format in which all distances are summarised in a single branching, tree-like diagram. Representations of this nature are called phylogenetic trees and are used extremely widely. This chapter examines what trees can be used for, how they are constructed and the extent to which their inherent attractiveness may mask important weaknesses.

Uses of trees

First and foremost, phylogenetic trees are used as a way of reconstructing historical events to give a better idea of how organisms are related to one another. Examples cover the whole range of evolutionary distances, from questions about the origins of life itself, through sorting out how the major metazoan radiations relate to one another to, perhaps, discovering which reptilian lineages gave rise to birds and mammals. On a finer scale, there are many extant species whose relationships are unclear, particularly those with no obvious close relatives, such as the guinea pig (Thomas, 1994). Phylogenetic trees can, in principle, be used to establish their closest affinities. Below the level of the species, trees offer a powerful means to study how a species' component populations came into being.

Once phylogenetic relationships have been established, a wide range of evolutionary questions can be answered. At the broadest level we can begin to understand major steps which led to the evolution of the principal kingdoms we see today. By relating phylogeny to patterns of development we can find out about how characters such as segmentation and limbs evolved. In many cases good phylogenies can provide information about whether life history traits such as sex and parthenogenesis, eusociality or paternal care have evolved once within a group, or many times (see, for example, Stern, 1994). Findings of this nature can have a profound influence on our understanding of how these traits evolve.

Phylogenetic trees can also be used to attach a time-scale to evolution. Many mutations can be thought of as occurring randomly, hitting the DNA like raindrops in a never-ending shower, their rate not varying over time. Under these circumstances, the degree of genetic divergence between two lineages will provide a measure of evolutionary time since they last shared a common ancestor. This possibility was first recognised through the observation that the rate of change of mitochondrial DNA was similar in diverse groups of organisms, and was quickly christened the molecular clock hypothesis.

One of the best known examples of where DNA sequences were used to date an historical event involved using mitochondrial DNA analysis to study human origins (Cann *et al.*, 1987; Vigilant *et al.*, 1991). From the deepest branch in their tree, the authors claim support for an African origin of mankind and use the convergence point of the tree to propose a hypothetical basal human female lineage, a 'mitochondrial Eve', who lived an estimated 200 000 years ago. This view has since come under attack. However, whether correct or not, there can be no doubt that this paper opened up a stimulating debate in which established ideas held by both anthropologists and molecular biologists have been challenged.

A third, and very recent use of trees is as a means to deduce something about a population's history, an approach pioneered by Paul Harvey and colleagues in Oxford (Harvey *et al.*, 1994; Holmes and Garnett, 1994). As we saw in Chapter 3, the amount of genetic variation supported by a population depends on its size, large populations carrying more variability than smaller ones. Consequently, when a population or phylogenetic group is large, its tree is relatively bushy, and when it is small, the tree is rather spindly. Wherever one can construct a reasonable tree, its structure will therefore say something about what has happened in the past: increases and decreases in bushiness will tend to reflect growths and declines. Thus, phylogenetic trees constructed from DNA sequences of the AIDS virus reveal a pattern of exponential growth consistent with what we know about its epidemiology (Nee *et al.*, 1995). By contrast, the same analysis applied to Dengue fever reveals a pattern of greater than exponential growth (Nee *et al.*, 1995), but this is again in line with expectations because Dengue fever epidemics have been increasing in frequency and regions of hyperendemism have been expanding (Monath, 1994).

Tree controversies

With such immense potential for helping us to understand the course of evolution, it is vital that the strengths and weaknesses of tree building are appreciated fully. Unfortunately, there is much greater scope for mistakes than is implied by the trees one sees in published papers. As an example, consider a recent revised phylogeny of whales. In their paper, Michel Milinkovitch and colleagues have taken a fresh look at cetacean phylogeny using DNA sequences from the mitochondrial 12S rRNA gene, concluding that, contrary to widespread belief, the sperm whale is in fact most closely related to baleen whales (Milinkovitch et al., 1993). Such a finding caused quite a stir. Very soon afterwards a second group headed by Úlfur Árnason published an alternative phylogeny based on mitochondrial cytochrome b sequences which supported the traditional view that sperm whales belong with the other toothed whales (Árnason and Gullberg, 1994).

This example shows how two contradictory phylogenies, both based on mitochondrial DNA sequences, can both get published in *Nature*, despite the fact that at least one of them must be wrong. The important point is that there is enough doubt and uncertainty in the tree-building process to allow two groups of respected molecular biologists to produce acceptable phylogenies, either of which would stand as acceptable and convincing in the absence of the other. In fact, the matter remains unresolved. Both groups have subsequently produced further evidence to reinforce their respective viewpoints and others have reanalysed the original data (see, for example, Adachi and Hasegawa (1995), Milinkovitch et al. (1995) and references therein).

Given such disparate views about the phylogeny of whales, it is hardly surprising that mitochondrial Eve should find herself under sustained fire (Hedges et al., 1991; Templeton, 1991). A number of weaknesses in the original analysis have been identified and the tree has even been reconstructed by others using 'better' methods. When this is done both the dating of Eve's existence and the African origin of humans come into question. Interestingly, the latest evidence to impinge on this controversy bolsters the original story. A recent study of variability on the human Y chromosome reveals so little (zero!) variation worldwide that the authors conclude that modern humans probably shared a single common ancestor, plausibly around the date given to Eve (Dorit et al., 1995).

The conclusion from these two examples is that trees which go out on a limb, so to speak, will receive much more scrutiny than those that do not. And yet, as long as we accept that trees are used to further our understanding rather than to confirm that which is already known, such a situation must inevitably bias our view of evolution against new ideas. Trees which support a particular current view are not questioned seriously and are usually considered to add good support. Those which appear to show something importantly different are examined critically and, worryingly, tend to reveal a range of potential and actual shortcomings. This chapter takes a critical look at the process of tree building in order to find out where the problems lie.

Trees based on amino acid sequences

Early phylogenetic trees were constructed using the sequences of amino acids in proteins, such as haemoglobin, myoglobin and cytochrome b. These first trees were not constructed to answer particularly challenging questions, tending to concentrate on confirmation that fish are fish and mammals are mammals, and that members of the same order group together. In a way, they were tentative, their aim being more to confirm the theoretical possibility of using sequence data to reconstruct phylogenies, rather than to elucidate new relationships.

By comparing the same protein in different species it soon became apparent that different proteins evolve at very different rates. Some, such as the histones, change little over aeons, whilst others, such as fibrinogen (one of the blood-clotting proteins) change extremely rapidly. Quite reasonably, it was suggested that the rate of evolution depends on the degree to which each protein is constrained by its function. Mutations which cause changes in the active site of an enzyme often result in inactivity and, therefore, tend to be deleterious and eliminated by natural selection. Conversely, mutations which fall in regions of a protein where the identity of each amino acid is unimportant, for example regions which act as spacers linking two active domains, may be neutral with respect to selection (see Chapter 4).

Changes in the amino acid sequence, particularly at constrained sites, will tend to occur infrequently and reflect functional improvements to the protein which have become fixed by natural selection. By contrast, neutral changes will occur more frequently but, by definition, will have little or no effect on the function of the protein. Much debate has focused upon whether the small number of selected changes or the large number of neutral changes has had a greater effect on the course of evolution. Although interesting, this debate is of little relevance to the construction of trees. The important point is that sites which are constrained by function behave less predictably than neutral sites, and hence add an unnecessary and unwelcome complication to the process of tree building.

Nucleotides are better than amino acids

Although useful, trees constructed using the amino acid sequences of proteins have their limitations. To begin with, protein sequencing is difficult and time-consuming. More importantly, the information gained is biased because it is restricted to protein-coding genes, is of poor quality because proteins are more likely to be selected than non-coding regions, and is wasteful because the genetic code uses 64 possible nucleotide combinations to specify just 20 amino acids. By contrast, DNA sequencing is rapid, technically easier and can make use of the entire genome, not being restricted to the small fraction which is transcribed into proteins. This latter point is brought home by the fact that protein-coding genes make up at most 5 per cent of the human genome, and probably comprise only 1–2 per cent (Jones, 1995).

Nucleotide sequences, as with amino acid sequences, evolve at widely different evolutionary rates. At the base of the tree of life, a handful of genes have been retained by all cellular life forms. Even between man and fungi, the giant redwood tree and bacteria, the same basic sequence can be recognised, allowing the various kingdoms to be related to one another. The gene used most widely for this broad-scale analysis is the small subunit rRNA gene (see p. 254), one of the genes that codes for the three RNA molecules which underpin ribosome function. It is perhaps not surprising that, were any sequence to be preserved over four billion years or so of evolution, it would be one which codes for a key element in the process of translating the genetic code of DNA into proteins.

Such fundamental genes have remained recognisable because they continue to carry out the same basic function, and this sets constraints on the changes that can occur within them. For the ribosomal RNA genes, the 'evolutionary strait-jacket' has been tied particularly tightly. In other genes, it is only key elements which have been preserved through time, elements such as the domains which bind to DNA or to a variety of co-factors. At the other end of the scale of conservation, many genes have regions which effectively lack function, for example stretches of amino acids which act as passive spacers. Finally, outside coding regions there is the rest of the genome, much of which appears to lack function, allowing it to change as fast as the mutation pressure dictates. Effectively a spectrum exists, from the highly conserved rDNA genes down to regions which appear to lack any form of constraint.

However, non-constrained regions do not represent the most extreme state. Some DNA sequences are prone to processes which actively promote change, either because they have an unstable structure or because they interact with proteins which promote rearrangement. For example, minisatellites (see Chapter 2) contain a short, conserved motif which bears strong homology to an ancient bacterial recombination signal (Jeffreys *et al.*, 1985) and which appears to be recognised by a protein which catalyses recombination (Collick and Jeffreys, 1990; Collick *et al.*, 1991) and results in great mutability. Individual arrays mutate as often as once every six generations (Wong *et al.*, 1986). It is ironic that this extreme instability appears to result from a motif which has remained the functional target for its protein partner over billions of years!

The construction of good phylogenetic trees requires a rate of evolution which is appropriate to the taxonomic questions being asked. On the one hand, highly conserved sequences like the rDNA genes are no good for distinguishing subspecies because the number of differences will be too few to make meaningful comparisons. On the other hand, functionless regions are poor at resolving differences over large periods of evolutionary time because any useful information becomes swamped by back mutations. Ideally, a sequence should be sought in which the branches of the tree differ by, say, 5–10 per cent of their nucleotides.

Which bits of DNA should be used for trees?

When scientists first began to construct trees using DNA sequences, there was a tendency to concentrate on genes which coded for the proteins which had been used for protein trees. For example, amino acid sequences of haemoglobin and myoglobin were converted directly to their genic equivalents. However, genes which code for proteins are not necessarily the best sequences to use for DNA-based trees. Indeed, there are strong arguments against using coding sequences. The most potent of these concern functional constraint. For most proteins there are regions which, through selection for preserved or improved function, tend to be more similar or more different than would be expected on the basis of chance. Such regions will confuse the process of tree building by distorting the linear correlation of genetic change with time.

A second problem arises from the redundancy of the genetic code. Most amino acids can be specified by two or more different codons, such that some mutations result in a change in the amino acid sequence and some do not. Mutations which cause a change at the level of the DNA, but not in the protein, are called silent substitutions. As a general rule, mutations in the third position of a codon often specify the same amino acid, while mutations at the first and second positions usually cause a change in the amino acid sequence of the protein. Therefore, it is not surprising that third-position nucleotides are much more polymorphic than either first- or second-position nucleotides.

It is now becoming apparent that equivalent codons are not used randomly (Gojobori *et al.*, 1982). Often, one or two of several possible codons are used to the virtual exclusion of all others. For example, in bacterial genes there is a general bias in favour of Gs and Cs in the third position over As and Ts. Interestingly, this bias is much less pronounced over the first 10–20 bases (Eyre-Walker and Bulmer, 1993). Such complexities are interesting to evolutionary biologists, but for tree building they must be viewed as another unwelcome level of complexity. Phylogenetic analysis would be simplified considerably if it could safely be assumed that all nucleotides are used equally.

Slowly, as we learn more about how DNA sequences evolve, a consensus is being reached on which genes make the best trees. For large taxonomic distances, such as those between kingdoms, phyla, and major branches within phyla, most studies concentrate on the rDNA coding regions. There are in fact three ribosomal RNA genes, each identified by its size. The smallest of these genes is the 5S rRNA, just over 100 bases long, and this has been used widely in phylogenetic studies (Hendriks *et al.*, 1986). However, a recent analysis suggests that it is too small to support meaningful inferences (Halanych, 1991). More informative trees have been constructed using the two longer genes. The nomenclature of these genes is slightly confusing because it is based on size, as measured by the sedimentation coefficient S, and the genes tend to be shorter in lower organisms. Thus, the 12S, 16S and 18S genes are all homologous to each other, while the other gene includes incarnations as 22S, 26S and 28S. Perhaps the best answer is to refer to the former, which is used most often for phylogenetic analysis, as the small subunit rRNA, referring to its position in the ribosome.

The small subunit rRNA gene has been used to elucidate relationships ranging from the major kingdoms (Gray *et al.*, 1984; Woese *et al.*, 1990) down to families of nematodes (Fitch *et al.*, 1995). However, for shorter evolutionary distances, the rRNA genes show progressively fewer differences and hence lose resolution. Consequently, for phylogenies involving comparisons within a class, attention turns to faster-evolving genes such as the globins (Bailey *et al.*, 1992) and various mitochondrial genes. Of these, the nuclear studies make up only a small proportion. The large majority of studies, at least among higher organisms, use mitochondrial DNA sequences.

In a recent review, David Rand (1994) suggested that, if one attempted to design the perfect sequence for phylogenetic analysis, one would probably reinvent the mitochondrial genome. The reasons for this are discussed in greater depth in Chapters 2 and 9, but include rapid evolution, a predominance of single base-pair substitutions, lack of recombination, multiple copies in each cell and probable selective neutrality. As with the nuclear genome, the mitochondrial genome contains a mixture of coding and non-coding regions which vary in their rate of change and so in their uses. The fastest evolving region is the D-loop or control region, a non-coding region which evolves rapidly enough to be suitable for intra-population analysis (Edwards, 1993). Among the genes, one of those sequenced most frequently is that coding for the enzyme cytochrome b (Árnason and Gullberg, 1994), which evolves of the order of five times slower than the D-loop.

Building a good tree

Tree building is used widely by evolutionary biologists, yet it is one of the most difficult exercises to perform well. The process involves a number of discrete stages, each of which is subject to its own errors and opportunities for mis-interpretation. Despite this, it is common to find confidence limits placed on branching patterns. The best way to appreciate how misleading such limits can be is to examine the various stages in tree construction and their pitfalls.

Generating raw sequence data

Having selected a particular section of the genome with which to build one's tree, that section has to be sequenced. In the final tree, the species in question will usually be represented by a single sequence. Which individual and which population is this taken from? Is it representative? Is it accurate?

As the technique of PCR comes to dominate molecular biology, the process of cloning is becoming increasingly rare, and more and more sequences are being determined using direct sequencing methods, going more or less directly from PCR product to DNA sequence. Unfortunately, PCR itself can introduce errors. Not only does it have its own inbuilt error rate, causing mutation during the amplification process, but it will also, on occasion, amplify sequences other than the intended targets. This may happen either through contamination from related research or when a related sequence exists else-where in the genome. It is the latter which is particularly difficult to control for.

Occasionally genes become duplicated. When this happens, the two copies will probably diverge, either because one is hijacked for another function, or because one copy 'dies', picking up mutations which make it a non-functional pseudogene. As long as the change is not too great, there is every chance that the PCR primers will inadvertently amplify the 'wrong' sequence, potentially making nonsense of the whole phylogeny. Problems of this sort have been encountered both with mitochondrial sequences which have become trans-ferred to the nuclear genome (Smith *et al.*, 1992), and with a duplication of the 16S ribosomal RNA gene in a halophilic bacterium, *Haloarcula marismor-tui* (Mylvaganam and Dennis, 1992).

Finally, even when the target sequence has been correctly amplified, there is still the question of whether it is representative of the sequence in that parti-cular species. Michael Lynch and Paul Jarrell (1993) have recently looked at methods for calibrating molecular clocks in 12 sets of data for different regions of the mitochondrial genome. They found that, although there was often good agreement about the rate of divergence, in no case did the slope of change against time intercept at 0,0. Instead, at zero time, their extrapolations indi-cated some divergence. Retrospectively this makes perfect sense, because spe-cies do not diverge from a single lineage but from populations. The divergence at zero time is therefore merely an indication of the polymorphism which exists in all populations. By implication, simple molecular clocks will appear to start off fast and then slow down, often making them inaccurate over short evolu-tionary periods.

Alignment, event assessment and weighting

In order to compare two sequences, it is first necessary to align them, such that at each position homologous bases are compared. Unfortunately, alignment is not as easy as it may sound, particularly when the sequences being compared differ in length. Inevitably there will be ambiguities in interpretation, some of which are illustrated in Fig. 11.1. In practice, alignment is performed by any one of a number of sophisticated computer programs which search, base by base, for regions of strongest similarity and then work to produce the best fit, inserting gaps and extra bases where required. The critical parameter is how reluctant the program is to insert gaps. Increasing the number of gaps allows a greater proportion of bases to be aligned but at the same time becomes pro-gressively less biologically realistic. Ultimately, the final alignment must reflect a compromise between an attempt to minimise the number of mismatched bases and our current view of how often the study sequence suffers insertion or deletion events.

One of the key advantages of the mitochondrial genome is that it is itself constrained by size. Insertions and deletions are relatively rare. For this reason,

Fig. 11.1 Examples of ambiguities which may arise during a sequence alignment. Consider the following short sequences:

1	2	3	4	5	6	7	8	9	0	1	
G	A	T	A	T	A	C	G	A	A	C	Upper strand
G	A	T	A		G	C	G	A	C	T	Lower strand

This alignment is not bad: there is one gap and 7 of 11 bases match. Let us adopt the nomenclature that GL6 means the G in the lower strand position 6. AU0 thus refers to the A in the upper strand at position 10.

Positions 5 and 6 are ambiguous. Should the G go opposite the A (making a transition) or opposite the T (making a transversion)? A single gap is the simplest solution, but one could argue that the TA in positions U3 and U4 is most likely to have originated through slippage. In this case, the G at L6 must be an insertion. Unless, that is, the gap at L5 should extend to include L6, allowing AU0 and CU1 to pair with AL9 and CL0 to pair at the right-hand end. Clearly, even in a short sequence such as this there is no one right answer; any alignment program can only take a best guess.

alignment of mtDNA sequences tends to be easier than for many other sequences. Coding sequences also share this advantage, since any changes in length of necessity involve multiples of three bases. However, there always remains an element of circularity. Our view of the best alignment depends on our current view of evolution. This in turn depends on information gleaned from previous alignments. In other words, if we believe sequence x is prone to insertions and deletions then we will allow the alignment program greater latitude and it will respond by producing a 'gappy' alignment which will only reinforce our preconception.

In order to use sequence data to best effect for constructing phylogenies it is important to weight characters according to their informativeness. For example, if one saw two people with mid-brown hair standing at an English bus stop one could conclude little about their relatedness. Mid-brown hair is common and therefore relatively uninformative. By contrast, if these two people were instead both true albinos, it would be reasonable to assume that they were relatives. Albinism is rare and is therefore a powerful indicator of genetic affiliation. In just the same way, differences between DNA sequences can also be ranked according to phylogenetic significance.

Early tree-building algorithms used a simplistic model in which all possible base changes occur with equal probability (the Jukes–Cantor model; Jukes and Cantor, 1969). However, we have already seen in Chapter 2 how far this can be from the truth. For mutations affecting coding regions, silent changes are much more likely than those which result in a change to the final protein. In almost all DNA sequences, transitions outnumber transversions and deletions and insertions tend to be rarer than single base substitutions. On a broader scale, any particular change may be influenced by its position both within a gene

(Eyre-Walker and Bulmer, 1993) and within a chromosome (see isochores in Chapter 2). By incorporating appropriate weightings for some or all of these factors, the process of tree building can be improved.

A second factor which should be considered is the extent to which each event is independent. All other things being equal, a pair of species which differ by one mutation will be, on average, twice as closely related as a pair which differ by two mutations. However, this only holds if all mutations occur independently of one another. Again as discussed in Chapter 2, wherever there is functional secondary structure, such as the complex stem–loop organisation of the ribosomal RNA molecules (Hancock et al., 1988; Hancock and Dover 1990), there exists the possibility of compensatory mutations. A change at one site will either be reversed, or it will trigger a change at another site.

The known complexities of the genome present prospective tree-builders with a wide range of problems, and probably many others which remain to be uncovered. A number of these are discussed in the preceding paragraph, sufficient to give a flavour of how difficult event assessment can be. There is, however, one final factor to consider, one which is in some ways the inverse of all the others. In any particular comparison between homologous DNA sequences, the important measurement is less the absolute number of differences, but more the number of differences per unit length. Consequently, it is important to consider not only biases affecting mutations which do occur, but also factors which prevent change. Whenever a sequence gains, loses or changes function its mutation rate is likely to change, potentially disrupting a molecular phylogeny.

Tree building itself

Having aligned a number of sequences and assessed the events which generated any differences, the next step is to feed the data into a computer program which will build a tree. A number of these are available, and include names such as PAUP, PHYLIP, CLUSTAL, and others. The approaches used by these programs fall into two main categories: distance methods, which attempt to construct the tree stepwise from its constituent parts; and whole-tree methods, which compare complete trees for their fit to the data.

There are two primary whole-tree methods, **maximum likelihood** and **parsimony**. In maximum likelihood, alternative trees are assessed for their fit to the data by calculating the actual probability of observing each alternative tree based on the composite probabilities of observing each of the events that has occurred during the tree's evolution, including both the order of branching and the lengths of all the individual branches. Maximum likelihood is arguably the most sophisticated method, but is also the most expensive in terms of computer time (Yang et al., 1994). Parsimony, on the other hand, is more concerned with finding the tree that requires the minimum number of independent mutation events to explain it. Parsimony is sometimes referred to by the descriptive name of minimum evolution.

Whole-tree programs work best when all possible trees are examined. Unfortunately, the computer time required to do this increases exponentially with the number of elements in the tree. As a general rule, if there are more than 15 elements, the task will take prohibitively long. Consequently, most real exercises have to be satisfied with only a proportion of all trees being examined, often only a small proportion. Various algorithms have been devised to attempt to maximise the chance that the subsample of trees which are actually looked at has the best chance of including the best-fit tree.

Among a number of pairwise methods we will look at two which are among those used most widely. The simplest method, UPGMA, works by adding one sequence at a time, beginning with the two most closely related. At each step, the growing tree is considered to be a single sequence equivalent to the average of its parts (Fig. 11.2). This method suffers from the drawback that the topology of the final tree depends greatly on the accuracy by which the first few branches are defined, and is particularly sensitive to differences in evolutionary rate between branches. A second method is the neighbour-joining method of Saitou (Saitou and Nei, 1987). This method works almost in reverse, starting with a completely unstructured tree, a simple star phylogeny where every sequence is equidistant from every other, and then adding internal branches to improve the fit to the data. Neighbour joining copes better with evolutionary rate variation and appears to be an efficient way to arrive at an acceptable topology.

Whatever tree-building method used, it is often helpful to provide a fixed starting point in the form of an 'outgroup', a sequence which is known to lie outside the rest of the data set. An example of an outgroup would be the inclusion of a chimpanzee sequence when considering the phylogeny of human racial groups, or the inclusion of a crustacean when considering radiations within the insects. Outgroups help to pinpoint the 'root' of the tree, something which is particularly important for methods such as neighbour joining, which do not themselves identify a root.

How accurate is the clock?

Any piece of DNA that accumulates mutations at a constant rate can, in principle, be used as a molecular clock by linking the number of changes to the time of divergence. Different sequences accumulate mutations at different rates and will therefore tick either faster or slower. In order to find out what that rate is, it is necessary to link the degree of genetic divergence to some evolutionary event with a known date. Useful events may be geological, such as an ice age or a change in sea level, or a defined taxonomic branch-point which can be identified accurately in the fossil record. A further method by which to calibrate a molecular clock is to make use of highly specific associations

Fig. 11.2 The unweighted pair group method of averages (UPGMA) method of phylogenetic tree construction. This method is arguably the least sophisticated method, but is also the easiest to understand. Consider the following table of genetic distances between four species, A, B, C and D.

	B	C	D
A	5	2	11
B		7	13
C			12

The shortest distance is A–C, which forms the start of the tree, each branch having half the distance A–C, in other words, 1. AC are now considered as a single fixed unit, and the table is recalculated such that distance X–AC is the average of X–A and X–C:

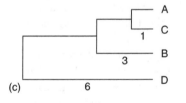

	B	D
AC	6	11.5
B		13

(a)

Now the next branch can be added, to form the ACB cluster. The total distance is six, therefore each branch must be three units in length.

(b)

This leaves only the final branch, D, whose length is again half the average of all its components (six) and yielding the final tree:

(c)

between two different organisms. For example, the symbiotic bacterium *Buchnera* lives inside the cells of aphids and therefore must speciate when they do. Moran *et al.* (1993) have confirmed that the evolutionary paths of these two organisms are comfortingly coincident, allowing the clock of the bacterium to be set against the phylogeny of its aphid host. Whatever the method used, each clock is best calibrated against as many events as possible, not only to derive a more robust estimate of its rate of ticking, but also to check for irregularities.

To what extent can time-scales produced by molecular clocks be trusted? This question remains the subject of heated debate. In some parlances the mitochondrial clock is accepted to the point that sequence differences are converted directly into time using a universal conversion factor of 2 per cent divergence per million years (Wilson *et al.*, 1985). Others are more sceptical, pointing out differences between warm-blooded and cold-blooded animals (Martin *et al.*, 1992; Martin and Palumbi, 1993; Rand, 1994), evidence for the action of natural selection (Ballard and Kreitman, 1994), the occurrence of hypervariable mtDNA haplotypes (Avise *et al.*, 1989) and the fact that data do not fit the expected Poisson distribution (Takahata, 1991). However, recent treatments suggest that the majority of problems can be allowed for mathematically (Lynch and Jarrell, 1993). A balanced view would probably conclude that molecular clocks appear to work reasonably well when applied to DNA sequences for which large amounts of sequence data have been collected. Under these circumstances, most important problems are likely to be identified and allowed for. However, there are bound to be exceptions, and the less well a sequence is understood, the greater will be the chance that timings based upon it will go awry.

How good is a tree?

A great weakness of tree building is the fact that there is no good way to find out how good the final tree is. The reason for this is fairly obvious. Tree building does not estimate a parameter, the true phylogeny, it searches for a tree which fits the available data best. Errors thus stem from several unquantifiable sources: the efficiency of the computer algorithm in finding the 'best' tree, the extent to which the 'best' tree according to the model of evolution used is in fact an approximation to the true phylogeny, the precise relationship between the quality of the 'best' tree and the 'next best' tree and others. However, this has not stopped people trying. The most commonly encountered method is called bootstrapping.

Bootstrapping works by resampling, asking the question, 'if some variable sites are ignored and other sites are counted twice or three times, how often do I get the same tree?' (the whole procedure is outlined in Box 11.1). Effectively, the procedure examines whether or not a small number of sites play a disproportionate role in determining the tree's form. To perform a bootstrap analysis, many replicate data sets are produced by random resampling of the original data, and each one is used to construct a new tree. Branch points or nodes of the 'best' tree which are supported by only one or two sites will appear rarely in bootstrap replicates. By contrast, nodes which are supported by many sites will appear in almost all bootstrap replicates. Thus, bootstrap values, the percentage of replicates which support each node, can be used as indicators of the stability of the shape of the final tree.

Bootstrapping is undoubtedly useful in terms of providing some idea of how well the data support a particular tree shape. However, it must be emphasised that the method does not provide formal confidence limits. True nodes may receive low bootstrap values and high values do not guarantee accuracy. More

Box 11.1: The bootstrap procedure

Bootstrapping is the name given to a statistical method for placing some form of confidence limits on a set of observations without making too many assumptions. It is thus well-suited to the analysis of phylogenetic trees, a fact pointed out by Joe Felsenstein (1985a). The principle of bootstrapping is simple: one asks how often random samples from the original data set give concordant answers. In the case of a phylogenetic tree, each data point is taken as a nucleotide base which differs between the species being studied.

Consider a phylogenetic tree based on a sequence with ten variable sites. Different sites will provide evidence of a link between different species. Overall, however, a tree-building program will arrive at a single 'best' topology. To bootstrap this tree one would generate a new data set by making ten random selections from the original data, such that some sites are omitted while others appear twice or more. From this new data set a new tree is constructed and the process repeated 100 times. For each branch in the original tree, we can now assign a bootstrap value, the number of times out of 100 that the same branch appeared.

It is clear that high bootstrap values will only result when many sites support a particular branch. Conversely, branch points that are supported by only one or a few sites will yield low values. By extension, bootstrapping will tend to work best when all sites are equally informative and worst when a few sites are disproportionately important. For example, a bootstrap analysis of morphological traits trying to place whales among the vertebrates might well show strong support for them being fish, due to the large number of adaptations to the aquatic environment.

importantly, bootstrapping only considers the accuracy of a given program analysing a given data set that has been aligned in a given way. All of the many possible errors that may have occurred on the way to the final program output running are ignored. The effect of this omission is difficult to evaluate, but one can easily imagine circumstances in which a relatively minor modification to, for example the alignment, could result in a much greater change in the shape of the final 'best' tree.

Three examples of tree building

The process of tree building and the problems it faces are perhaps best illustrated by examples. The first of these attempts to reconstruct the relationships between the kingdoms, using rDNA sequences; the second looks at very recent, rapid evolution, relating different genotypes associated with different infections of the AIDS virus; and the third is a pioneering study which for the first time attempts to verify a tree by using genetic engineering to reverse evolution.

The ribosomal RNA genes and the kingdoms

Five principal types of genome have so far been identified: archaebacterial, eubacterial, nuclear, mitochondrial and **plastid**. The only genes which are recognisably common between all these groups are those coding for transfer and ribosomal RNAs. Of these, the transfer RNAs are too aberrant in mammalian mitochondria to allow meaningful sequence alignment, and the 5S rRNA is absent from most mitochondria. The molecule chosen by Gray *et al.* (1984) is the small subunit rRNA gene (see p. 244).

The comparative analysis of this subunit across the kingdoms is not straightforward. To begin with, the basic sequence alignment is difficult. Although undoubtedly homologous, the sequences differ in length by a factor of two, from 953 bp in human mitochondria to 1955 bp in wheat mitochondria. In order to make meaningful comparisons the authors take advantage of the underlying structure of the molecule. Over its length, not all regions are equivalent. The molecule is a patchwork of regions that show almost complete conservation, those that show sequence variability and those that vary in both length and sequence. In this study, alignment was achieved by using the conserved regions to act as landmarks by which to navigate. Additional anchor points were obtained by using single invariant and semi-invariant residues. The complexity of the problem is illustrated in Fig. 11.3, where the extent to which the single thread folds back on itself to create a three-dimensional molecule is shown in two dimensions.

Following alignment, the second major problem to be faced was that the number of sequences to be analysed (16) was too great for the available computational power. This problem was circumvented in several different ways. First, those sequences with 'known' affinities, for example all the eukaryotes, were pre-grouped. Second, small trees were constructed from subsets of the data. Thus, the relationships amongst mitochondria were examined by building a tree from all mitochondrial sequences plus a couple of outgroups. The point of this exercise is to identify, early on, which sequences are behaving well and which are not. In the current example, both animal and fungal mitochondrial sequences are found to cluster together, but the wheat mitochondrial sequence appears more distant, suggesting that two invasion events of eubacteria into archaebacteria have produced mitochondrial-like structures. The final step was to use the information from each of the subtrees to place restrictions on the grouping in the final tree. The result (Fig. 11.4) shows that nuclear sequences are closest to the archaebacteria, that mitochondria may be **diphyletic**, that

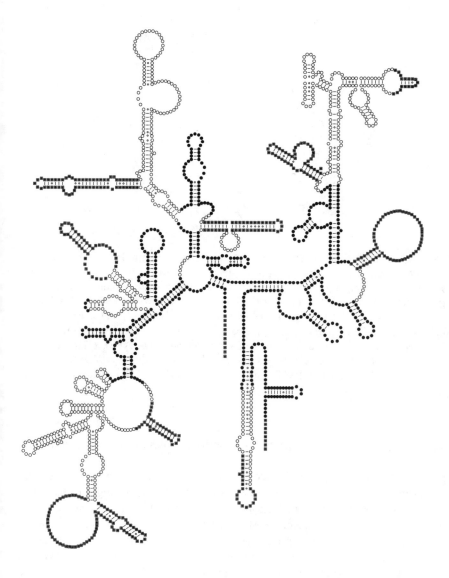

Fig. 11.3 Complexity of folding of small subunit ribosomal RNA. This figure is based on a consensus structure derived from the study of rRNA molecules from many organisms and illustrates how the single linear molecule can fold back on itself to form a maze of stems and loops. Features universal to all organisms are indicated by filled circles, features present in many but not all organisms by open circles and regions which vary between organisms by square symbols. (After Gray *et al.* 1984.)

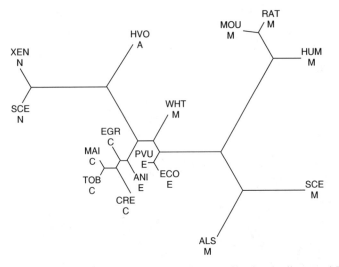

Fig. 11.4 An early tree of life, based on small subunit ribosomal RNA sequences. Five different major subclasses of sequence were analysed: N = nuclear, M = mitochondrial, A = archaebacterial, C = chloroplast, E = eubacterial. Species represented are: human (*Homo sapiens*, HUM), mouse (*Mus musculus,* MOU), rat (*Rattus norvegicus*, RAT), frog (*Xenopus laevis*, XEN), wheat (*Triticum aestivum*, WHT), maize (*Zea mays*, MAI), tobacco (*Nicotiana tobaccum*, TOB), yeast (*Saccharomyces cerevisiae*, SCE), fungus (*Aspergillus nidulans*, ALS), *Chlamydomonas reinhardtii* (CRE), *Euglena gracilis* (EGR), *Halobacterium volcani* (HVO), *Escherichia coli* (ECO), *Proteus vulgaris* (PVU), and *Anacystic nidulans* (ANI). (After Gray *et al.*, 1984.)

plastids had a **monophyletic** origin, and that organelles originated as endosymbiotic eubacteria.

Is this rRNA tree a good one? Possibly, but there are reasons to suspect not. To begin with, its branch lengths are almost certainly misleading. For example, we already know that even mitochondrial DNA sequences, which some would say offer the best molecular clock, do not keep anywhere near perfect time, as discussed above and in Chapter 2. A second assumption of the tree-building model is also contravened. It is known that the base composition of the rDNA sequences can vary greatly between taxa (Hancock and Dover, 1988; Hasegawa and Hashimoto, 1993; Steel *et al.*, 1993). It follows that different lineages must have experienced different mutational biases during their evolution, and yet these are ignored. There is also the pertinent question of what has been the driving force behind change. The ribosomal RNA molecule has been around for billions of years. It seems inconceivable that it has yet to arrive at its optimum. Instead, we have to consider two alternative models.

In the first model, the rRNA molecule evolved rapidly towards its optimum, which was reached more than a billion years ago, before most of the major extant radiations had come into existence. Since then, changes have been restricted to minor back and forth shuffling among those sites which can tolerate change. This scenario makes the testable prediction that an rRNA sequence from one species should be able to function if genetically engineered into another. The second model is one in which the rRNA molecule continues to evolve. In this case, we must envisage a dynamic evolutionary equilibrium between the rRNA molecules, the ribosomal proteins and the cellular milieu. This model makes the prediction that the rRNA will have gone through periods of stasis interrupted by bouts of sudden change associated with major evolutionary events, such as the evolution of photosynthesis, the transition to multicellularity, the initial mitochondrial symbiosis or the evolution of warm-bloodedness.

Until we find out which of these two models is closer to the truth, it is difficult to assess how accurate the estimates of branch length really are. In this light it is interesting to consider other, more recent, studies which look at the same problem. Such trees appear widely different, particularly with respect to their branch lengths, even if the same basic elements tend to be retained (Woese *et al.*, 1990; Cammarano *et al.*, 1992). At the same time, more recent treatments both solve some problems (Hasegawa and Hashimoto, 1993; Steel *et al.*, 1993) and identify others (Forterre *et al.*, 1993; Sogin *et al.*, 1993). As elsewhere, it seems as though we will have to understand rather more about the evolution of the rDNA before firmer conclusions can be drawn.

The AIDS virus and dentistry

Human immunodeficiency viruses (HIV) are retroviruses with RNA genomes that are apparently responsible for AIDS (acquired immune deficiency syndrome). A number of instances have been reported of health workers contracting AIDS from their patients. Recently, an instance of the reverse path was identified when a patient was found who was thought to have contracted AIDS from a Florida dentist (Ou *et al.*, 1992). When alerted, the dentist duly contacted all his other patients, recommending that they be tested for HIV. A further nine cases of HIV-positive patients were identified. How many, if any, of these were infected from the dentist?

One way to tackle this question is to construct a phylogenetic tree linking each of the viral isolates. In theory, infections from a single source should be more similar to each other than to viruses from other sources. The task is made both easier and more difficult by the extreme rate of evolution of HIV, which is perhaps a million times as fast as eukaryotic nuclear genomes. For the Florida dentist case, a single gene, the *gp120* gene, was sequenced several times from each of the ten patients, and from a number of control individuals who had AIDS, but who were not on this particular dentist's list.

The reconstruction of phylogenies based on viral sequences faces a different set of problems to those faced by the rRNA tree of life. First, the mutation rate is so high that the virus undergoes considerable evolution during its existence in each infected patient. It is important, therefore, to collect samples as soon after the estimated date of infection as possible, and to sequence the gene several times from each patient. The latter allows one to look at intra-individual viral variability. In the current case, the trees that were constructed were based on the two most extreme sequences found from each patient. A second problem relates to mutational bias. Most DNA sequences show a bias, often a pronounced bias, in favour of transitions rather than transversions. For HIV sequences, the picture is much more complex. Each possible change appears to occur with its own characteristic probability, and these vary considerably. For example, A→G occurs twice as frequently as its converse, G→A, and 40 times as often as C→G.

In the original study, the trees that were built appeared to make sense. Five patients (later six) fell onto a single branch which included the dentist himself. Four other patients could be placed elsewhere in the tree. When the background of each patient was examined, it was found that none of the patients who clustered with the dentist, the so-called dental clade, were at risk from other sources, whereas all the rest were considered to be at high risk for behavioural reasons (homosexuals, drug users, promiscuous heterosexuals and the like). This is what would be expected if some of these people had contracted AIDS from the dentist and some from other sources. Interestingly, several different tree-building programs all identified the dental clade with high probability.

In this case, tree building appears to have worked. And it probably did. However, a number of factors were ignored which might have affected the outcome. The HIV virus is under intense selection from the host immune system, which it evades by mutating constantly. People may well differ in how good their bodies are at fighting AIDS, and hence how much selection pressure they exert on the virus to change. In the Florida dentist saga, no account was taken of any therapy that the patients were given, which might have either slowed or accelerated the rate of change. Also, there is a clear pattern to the way in which an HIV infection progresses (Ou *et al.*, 1992), yet the stage of the disease (at sequencing) was not modelled. Leaving the special circumstances that apply mainly, or only, to AIDS, the tree-building programs used were not able to allow for non-independent mutation events, even though this is a real possibility. In many ways, it is surprising that the dental clade ever fell out so nicely. These arguments are covered well in a review article about the accuracy of molecular phylogenies (Hillis *et al.*, 1994).

The story of the Florida dentist and his hapless patients has a sting in its tail. In a later paper, the sequence data are augmented and re-analysed using different tree-building programs (DeBry *et al.*, 1993). With extra controls, the re-analysis was able to show that the dental clade was not as pure as first thought, and often includes control isolates. Of course, since no one knows what happened during the evolution of the virus in these individuals, nor the precise pattern of transmission, we can never know the true picture. This highlights the ultimate problem with trees: in the absence of other evidence, it is impossible to tell whether the computer programs have been successful in their task, or have, so to speak, barked up the wrong tree.

Artiodactyl ribonucleases

Finally, a look towards the future. With most trees, there is no good test of whether the branches are correct. We cannot go back in time to find and test a common ancestor. However, a recent study concerning the evolution of ribonuclease enzymes points the way to an alternative test. At every stage during evolution, a given gene must have been functional. By using genetic engineering, the predicted sequence of these ancestral proteins can be constructed and then expressed in bacteria. Such proteins can then be isolated and their functionality tested using standard biochemical techniques. This pioneering approach has recently been used by Thomas Jermann and colleagues (1995) to examine the evolution of the enzyme RNase A.

The RNase A superfamily is a fascinating group of proteins which have been associated with a number of diverse functions, including antitumour activity, immunosuppression and cytostatic activity. Many of these functions appear to have been adopted since the mammals split from the reptiles and involve several gene duplication events. Jermann and colleagues selected the enzyme digestive RNase A, which catalyses the breakdown of RNA, and compared sequences, catalytic activities and thermal stability among both modern artiodactyls and predicted ancestral sequences based on their molecular phylogeny. They were able to show that reconstructed proteins were both functional and retained similar thermal stabilities. More telling, however, was the finding that proteins reconstructed from the deepest branch points actually showed a fivefold *increase* in activity against double-stranded RNA. This finding is precisely in line with expectations based on changes that are likely to have occurred during the evolution of ruminant foregut digestion.

This study can be, and has been, criticised on a number of grounds. Just as with the other two examples, the tree has yet to be fully substantiated. However, the RNase A phylogeny is still a milestone paper because of the principle it employs to test phylogenies. Genetic engineering has now reached a level where it is no longer particularly difficult to construct an expressible gene according to a predicted sequence. Ideally, as in the current instance, there should be some firm prediction about the protein's properties which may be tested. However, and particularly for greater levels of divergence where the predicted sequence differs significantly from any known real sequence, possession of the correct function is a good start.

Can trees be trusted and will they get better?

Thus far, we have examined tree building from a generally negative viewpoint, pointing out a diverse range of pitfalls that undermine the validity of many trees. We believe this scepticism is necessary because, far too often, tree building is performed on autopilot, data being fed blindly into the nearest or most widely available computer program. But are all trees rotten? Almost certainly not. It is an inescapable fact that DNA sequence similarity almost invariably indicates taxonomic proximity, and that the sequences one derives hold within them much of the information which is needed to reconstruct the past. At the same time, it is equally true to say that some information is missing, and that no program will ever be able to construct the perfect tree.

Three factors give cause for optimism. First, the rate at which sequence data are being collected means that many genes and genomic regions are becoming ever better characterised. In terms of tree building, this additional knowledge will enable us to improve our understanding of what mutations occur in which lineages, where and how often. In addition we will be able to test more carefully for the presence of natural selection. Second, our capacity to implement theory is also improving. Computers are becoming faster and more powerful, and the models by which trees are constructed are becoming more sophisticated, incorporating ever more data about the nature and frequency of mutational events. Finally, and not least, with the advent of PCR and developments which allow the genetic typing of large numbers of individual sperm, routes are opening up by which we may measure mutation rates directly. This last option provides an exciting opportunity to test the molecular clock directly, by comparing mutation rates in different taxa.

However good our database of sequences becomes, there always remains the problem of circularity. If one branch appears longer than any other, is this because that lineage is older, or because its mutation rate is greater? We cannot tell. Unfortunately, a sophisticated program might well 'decide' that the mutation rate is greater and then introduce a compensation factor to allow for this. In other words, the more sophisticated the program, the more it will be possible to compensate for differences in mutation rate and mutation bias between lineages. At the same time, such programs will inevitably run the risk of ironing out genuine differences which reflect actual past events. Once again, there is no perfect solution. If evolution has misbehaved we can only guess at how.

Possibly the most worrying aspect of tree construction is that it encourages poor science. At almost every stage in the process, the scientist is faced with subjective decisions: which alignment to choose, which sequence is representative, which tree is best supported, how many times should the computer run the program and so on. When the end result is a tree which the experimenter sees is clearly at odds with what is already believed about the evolutionary relationships of the material being compared, i.e. from morphological, palaeontological and other evidence, the program is often rerun and the system 'tweaked' to encourage other, more plausible outcomes. In some cases, 'obvious' errors, such as the separation of two accepted sister species, are counteracted by putting constraints on the program, fixing a link, and letting the rest of the tree form around it. This often causes a sudden flip in topology, perhaps giving more acceptable tree shapes.

The bottom line is that tree building has dangerous levels of 'slop'. Trees that clash too strongly with morphological data will not be published. Trees which agree with everything that has gone before do not make headlines. There is thus strong selection in the literature for trees which differ from expectations just enough to raise the odd eyebrow and provoke interesting discussion while remaining plausible. Unfortunately, for every tree that gets published, probably hundreds are drawn, each one either as a replicate or using slightly different parameter settings. Consequently, it is up to individuals to pick the one topology that will represent their data. If the first tree which emerges agrees with expectations, does everyone really try as hard to create alternative topologies as if the first tree makes no sense at all? Finally, among trees submitted for publication, to what extent do referees and journal editors block radical and confirmatory trees from being published.

Summary

Two things are certain. First, phylogenetic proximity is related strongly to DNA sequence similarity. More closely related species will be more similar than more distantly related ones. Second, the perfect phylogenetic tree will never be constructed. We can never know the precise train of evolutionary events that led to the present. We can only infer a reasonable approximation. The reality of tree building lies somewhere in between.

At present, our ability to construct accurate trees is adequate, but probably not as good as most people think. There are just too many unknowns. What affects the relative numbers of transitions and transversions? Which sequences are constrained by function and which are not? What is the role played by higher orders of genome structure? What are the most important mutational mechanisms affecting each specific sequence? These are just some of the current crop of unanswered questions.

As more sequences are analysed, we can expect that the room available for unexpected deviations from our simple models of evolution will decrease. Our ability to build trees based on well-known sequences must improve. The down side is a certain circularity. Unexpectedly long branches are usually interpreted as a speeding up of the clock in the lineage concerned, and may well be compensated for algebraically. However, the more we refine our models in this way, the more we will tend to throw out potentially useful information. Inevitably, the price of increased uniformity and precision is an overall reduction in the resolution of the trees.

The conclusion is simple. Trees are extremely useful, but imperfect. It is probably true to say that 95 per cent of the branches on 95 per cent of trees are probably placed in the correct position and approximate to a meaningful length. The problem is that as yet we have no good way to identify which are the problem branches. To avoid placing too much faith in any particular feature, we must remain aware of all the sources of error which are inherent in tree building. This involves erring on the side of scepticism, wherever possible using several different sequences and several different tree-building programs and trying to use trees as substantive evidence alongside as many other character comparisons as can be assembled.

Suggested reading

HALANYCH, K. M. (1991). 5S ribosomal RNA sequences inappropriate for phylogenetic reconstruction. *Molecular Biology and Evolution*, **8**, 249–253.
A well-argued attack on the use of short sequences in phylogenetic analysis.

HASEGAWA, M. and HASHIMOTO, T. (1993). Ribosomal RNA trees misleading. *Nature*, **361**, 23.
Another good paper pointing out a range of tree-building pitfalls.

HILLIS, D. M., HUELSENBECK, J. P. and CUNNINGHAM, C. W. (1994). Application and accuracy of molecular phylogenies. *Science*, **264**, 671–676.
An excellent appraisal of many aspects of tree building.

HOLMES, E. C. and GARNETT, G. P. (1994). Genes, trees and infections: molecular evidence in epidemiology. *Trends in Ecology and Evolution*, **9**, 256–260.
New uses for trees. Forefront applications of phylogenetic reconstruction.

MILINKOVITCH, M. C. (1995). Molecular phylogeny of cetaceans prompts revision of morphological transformations. *Trends in Genetics*, **10**, 328–334.
An object lesson in combining solid tree-building methods with an awareness of morphology and more traditional approaches.

WOESE, C. R., KANDLER, O. and WHEELIS, M. L. (1990). Towards a natural system of organisms: proposal for the domains Archaea, Bacteria and Eucarya. *Proceedings of the National Academy of Science USA*, **87**, 4576–4579.
A slightly more modern tree of life. The arguments continue, but this is one of the best attempts.

Chapter 12

The design of genetic systems

The quest of the evolutionary biologist, as envisaged in this book, is to understand the way organisms are, how they came to be that way, and why they differ. The examination of design involves combining the probability of survival and reproduction of individuals which have a particular genetic make-up with the rules that govern the transmission of genes to form new individuals. These transmission rules may vary between species. Some species are asexual, individuals showing fidelity of transmission of genes to their progeny, whereas other species are sexual. The way in which genes combine to make an individual also varies. Sexually reproducing species exhibit an alternation of haploid and diploid phases. The predominant phase in lower plants (e.g. algae, mosses and ferns) is haploid, whereas in flowering plants and most animals the diploid dominates. Some asexual species show the same type of cycle. In this chapter we discuss what might be considered the evolution of genetics: why these rules have evolved.

The evolution of eukaryotes

The genetic differences between eukaryotes and **prokaryotes** are of fundamental importance to how they evolve. The genetic material of the prokaryote is maintained within one circular chromosome in contrast to the genetic material of eukaryotes, which is dispersed between linear chromosomes, which usually exist in pairs and undergo recombination. Thus, whereas the genetic material of a prokaryote is within one tight linkage group, with all loci in linkage disequilibrium, the genetic material of eukaryotes is in many linkage groups with much lower levels of disequilibrium. Even within a eukaryote linkage group, disequilibrium is liable to be lost through recombination.

In evolutionary terms, prokaryotes behave as asexual haploids, whereas eukaryotes have many, independently segregating loci. This has several consequences. First, in haploids, the action of a gene cannot be masked, because there are no dominants or recessives. Favourable mutations therefore spread more rapidly. Second, because prokaryotes are frequently asexual, mutants which are advantageous in isolation do not necessarily spread. In an asexual prokaryote, the effect of selection will depend on the fitness of the genome containing the mutant. Further to this, eukaryotes, with a few notable exceptions, bear extra-nuclear DNA within organelles. This DNA has subtly different transmission genetics from nuclear DNA, leading to the interesting evolutionary conflicts described in Chapter 8.

The question must be raised as to how eukaryotes evolved and why they became successful. During the 1970s, the view held by the majority of biologists derived from Lynne Margulis' illuminating study of the similarities and differences between the DNA and protein synthesis machinery of the eukaryote extra-nuclear genome (plastids, mitochondria), the eukaryote nuclear genome, and the prokaryote genome. Margulis noted that a significant number of the features of the extra-nuclear and prokaryotic genomes were similar. Both mitochondria and prokaryotes possess circular DNA molecules. The ribosomes found in mitochondria and prokaryotes are also similar, and different from the eukaryotic ribosomes. These and other similarities suggested to her that the evolution of eukaryotes was based on the formation of symbioses between prokaryotes following ingestion of one prokaryote by another. The ingesting prokaryote formed a nuclear membrane to protect its DNA, and the ingested bacteria evolved to become the extra-nuclear genome. To Margulis, the key feature in eukaryote evolution is that of ingestion via endocytosis, a behaviour not observed in prokaryotes (Margulis, 1970).

The endosymbiotic theory of the origin of the extra-nuclear genome has become accepted in the scientific community. In addition to the weight of evidence provided by Margulis, phylogenetic analysis of the rRNA genes of mitochondria and chloroplasts shows them to be a derivative of the purple (proteus-like) bacteria, in contrast to eukaryote nuclear DNA, which appears to be derived from the archaebacteria (Gray et al., 1984). However, the view that the initial event in the evolution of eukaryotes was the formation of endosymbioses is seriously in question. Tom Cavalier-Smith (1987) has argued persuasively that the feature that defines eukaryotes is not the extra-nuclear genome, but the presence of a nucleus, a cytoskeleton and an endomembrane system. Evidence supporting this view comes from study of ancient groups of protists such as the microsporidia, a group of intracellular parasites of animals. Microsporidia have no extra-nuclear genome. Is this because they have lost an extra-nuclear genome that they once possessed, or is it because, as Cavalier-Smith maintains, the original eukaryote had no extra-nuclear genome, and microsporidia are an early branch of this radiation?

Examination of the cell biology and genetics of microsporidia suggests strongly that microsporidia are in fact ancient protists, which evolved before the extra-nuclear genome. Their protein synthesis machinery is more resemblant of prokaryotes than eukaryotes, their ribosomes being small (70S) compared to those found in higher eukaryotes (80S). Furthermore, when Charles Vossbrink and Carl Woese (1986) studied the ribosomal DNA of one microsporidian, *Varimorpha necatrix*, they found that this species lacked 5.8S rRNA, a size of rRNA not found in prokaryotes but which was previously considered ubiquitous in eukaryotes. Further, they found that the sequence of the small subunit rRNA was intermediate between those of other eukaryotes and those of prokaryotes (Vossbrink *et al.*, 1987). That microsporidia are derived from one of the earliest radiations of the eukaryotes lends credence to the notion that the early eukaryotes did not have mitochondria, rather than the view that the microsporidia are a eukaryotic group which has lost the extra-nuclear genome.

The evolution of sex

The mixing of genes as a result of recombination, random assortment, and gamete fusion produces immense differences in the way sexual and asexual organisms respond to selection. Eukaryotic sexual reproduction involves the formation of a new individual by the fusion of gametes from two individuals (syngamy). Each gamete is formed by meiosis, involving recombination between homologous chromosomes followed by their random assortment.

From an evolutionary viewpoint, the commonness of sex presents some-things of a quandary, for sexual reproduction is an inefficient method of pro-pagation compared to asexual reproduction. In an asexual population, every individual makes some energetic investment in raising progeny, be this provi-sioning the new individual at the point of formation, or through care there-after. In a sexual population, the males generally place little or no investment in the care of the next generation. In terms of their contribution to reproductive rate, they are a waste. If we examine the result of competition between sexual and asexual types, it is clear that the asexual type will hold sway in the short term. Consider an environment colonised by an inseminated female insect and an asexual contemporary. The growth in number of the asexual type is far more rapid than the sexual (Fig. 12.1). The lack of investment by males in progeny constitutes what is frequently referred to as the **twofold cost of sex**. The question the twofold cost poses may be simply stated: why do females halve their genetic contribution to progeny by allowing a male (who does not help) to contribute to them?

Sex does not always involve such a large amount of waste. Sex in isogamous organisms (ones where the fusing gametes are of similar size) involves equal parental investment by the two parties. Furthermore, male parental care is seen in some anisogamous species. However, there are other ubiquitous costs asso-ciated with the act of sex itself. These costs include both those associated with finding a mate and those associated with the sexual act itself. In the latter context, both the physical act of mating and the possibility of contracting disease during sex may impose costs.

Sexual reproduction is thus costly compared to asexual reproduction. Despite this, sex is very common among the eukaryotes. Explaining the com-monness of sexual reproduction is a problem that has become something of a preoccupation for evolutionary biologists. The literature on this subject is

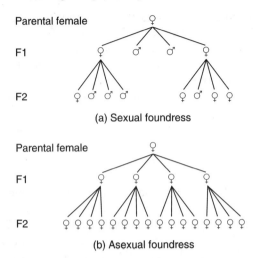

Parental female

F1

F2

(a) Sexual foundress

Parental female

F1

F2

(b) Asexual foundress

Fig. 12.1 The twofold cost of sex. Comparison of multiplication in (a) a sexual species with a 1:1 sex ratio and where the male does not give parental care, and (b) an asexual species. Each female (sexual or asexual) produces four progeny in this example.

immense. Sex, according to Bell (1982), is 'the Queen of all evolutionary problems'. And it is indeed an old one:

If we bear in mind that in sexual propagation twice as many individuals are required in order to produce any number of descendants and if we further remember the important morphological differentiations which must take place in order to render sexual propagation possible, we are led to the conviction that sexual propagation must confer immense benefit on organic life (Weissmann, 1887).

As Felsenstein (1974, 1985b) notes, hypotheses to account for the near ubiquity of sex fall into two major categories. In one category are hypotheses where sex is considered to confer an immediate benefit to the individual (immediate benefit hypotheses). The other category contains hypotheses where the maintenance of sex is favoured because of benefits associated with the genetic processes of sex, random assortment and recombination (variation and selection hypotheses).

Immediate benefits hypotheses

Kondrashov (1993) considers there to be three types of immediate benefit that may accrue from sexual reproduction. First, sex may increase fitness because it allows the production of fitter offspring by selection of mates. Second, sex may be beneficial because it aids the repair of deleterious mutations. Third, sex may be favoured because progeny with two parents have a higher fitness than progeny with one.

The first of these hypotheses proposes an immediate benefit to sex from the fact that sex incorporates choice. As discussed in Chapter 6, two types of choice exist. There is the choice of males by females (or vice versa), and the choice of gametes from a given male. Empirical studies have shown that females can and do choose partners with 'good genes'. Choice may also be extended to gametes. In either case, the opportunity exists for sex to confer an immediate benefit by improving the genetic quality of the offspring. On the other hand, these benefits must be weighed against potential costs. Mate choice and gamete choice are both energetically wasteful. In isogamous species some gametes are wasted whereas in polygamous species some males are excluded from mating.

The second type of immediate benefit to sex may accrue from the repair of mutations. During the course of meiosis, homologous chromosomes become paired, a stage known as synapsis. Once aligned, it appears that cellular enzymes recognise differences which they then repair, converting the sequence of one homologue to that of the other. If this process results in the elimination of deletions and deleterious mutations, then it could confer a benefit to sexuality. However, whilst chromosome conjugation and repair during meiosis are now well-documented, there is no evidence that sequence conversions regularly favour the 'beneficial' copy. Indeed, our current understanding of meiosis does not yield a plausible mechanism by which the deleterious copy could be identified.

The third immediate benefit that has been proposed is the advantage of having two parents instead of one. Biparental care, present in some anisogamous species, could be construed as giving an advantage to sex. Here, the presence of two parents provisioning offspring could act as an insurance policy against the chance death of one parent. Clearly, this advantage will be restricted to species where there is biparental care, excluding all plants and many animals.

Variation and selection hypotheses

Sexual reproduction has two main effects on the distribution of genetic variability, one acting at the level of the family, the second at the level of the population. First, whereas offspring produced by an asexual individual are essentially genetically identical, chromosomal reassortment ensures that those of a sexual pair differ from one another. Several models have been advanced for how offspring diversity can yield long-term benefits (see below). Second, sex, through reassortment and recombination, can break down associations between loci. By so doing an advantage may accrue, either because deleterious combinations have been eliminated or because beneficial combinations have been created (see Chapter 3) (Box 12.1). Such options are not open to an asexual organism.

The advantage of variation among progeny: reduced sibling competition   The environments which many species inhabit vary in time and space. This variation, it was argued by Williams (1975), may promote the production of

---

Box 12.1: Sex and linkage disequilibrium

Consider two loci, A and B, each with two alleles, A and a, B and b. Each of these alleles is at frequency 0.5 in the population. During the haploid phase there are four possible genotypes, AB, Ab, aB and ab. If there is no association between the loci, then each of the four genotypes should be at frequency 0.25. If there is deviation from this value, we say the loci are in disequilibrium.

The circumstance of maximum disequilibrium would be when half the individuals were of genotype AB and the other half of genotype ab. In an asexual population, this disequilibrium would not change between generations. No individuals of type Ab or aB would be created without mutation. However, sex acts to push the population toward random association. In the above case, the two types of gamete would fuse. One-quarter of the individuals would be $AABB$, one-quarter $aabb$ and one-half $AaBb$. If the two loci are unlinked, then this last class of individual will segregate to create all four genotypes in equal proportion.

After sex, therefore, the gamete genotype frequencies are AB 0.375: Ab 0.125: aB 0.125: ab 0.375. Each generation, sex acts to decrease disequilibrium and push the loci toward more random association. The amount by which it does so depends upon whether or not the loci are linked, and if they are linked, how tightly they are linked. For fuller details of the causes and dynamics of disequilibrium between loci, see Chapter 3.

---

offspring sexually rather than asexually. This can be seen by consideration of a species which inhabits a coarse-grained environment. A coarse-grained environment is one in which the species is found in many different types of patch, and an individual must attempt to grow and mature in the patch in which it lands in competition with the other types in that patch. An individual once mature produces offspring which themselves disperse. Williams argues that sex will be favoured in such an environment if siblings tend to disperse together and enter the same patch, and if there is strong competition between individuals within a patch.

The advantage to sex in these circumstances arises because sexual offspring which disperse together are likely to be genetically different. When they land in a patch, because they are variable, it is likely that one of them is fitted to the patch and can thrive in the strong competition. However, it is unlikely that two of them are fitted to the same patch. In contrast, when asexual offspring land in a patch together, they are genetically identical. It is less likely that any of them are fitted to the patch in which they land. On the rare occasion when a fit does exist thereby allowing them to thrive, all individuals are fit, but intense sib-sib competition will curtail the benefits.

This argument likens sex to a lottery. A sexual organism selects tickets (offspring) bearing a variety of numbers. When these land in a patch, it is likely that one of them will be the most adapted and win the prize of successful maturation. An asexual organism selects tickets which bear the same number. When these land in a given patch, there is a relatively low chance that any of them win the prize. Most importantly, when they do win the prize, it is still only one of them that progresses to claim it, because of the severity of competition within a patch. When there are two winners, each does not receive the full prize. Rather, they fight for it. The asexual individual thus wins fewer prizes than a sexual one. Mathematical analysis suggests that under these circumstances, sex can gain the necessary twofold advantage (Maynard Smith, 1976a).

Sibling competition may produce a benefit to sex under other ecological circumstance. In a hypothesis he entitled 'the **tangled bank**' theory, Graham Bell (1982) postulated that sex would be favoured if offspring were placed into the same local environment and individuals that were genetically different exploited slightly different niches within this environment. A sexual parent would benefit because its progeny would suffer less mutual interference. This theory differs subtly from that of Williams, in that it does not rely on having large numbers of siblings being dispersed *en masse* by the parent. Rather, the variation in the siblings results in their dispersion.

In conclusion, sex makes the progeny of a given individual more diverse. This variation can provide an advantage to sex if siblings co-exist and mutually interfere. However, the sibling competition theories, although possibly providing an advantage to sex in some cases, cannot be the sole explanation of the commonness of sex. Many organisms do not have competition between siblings, but do have sex. Many marine invertebrates are a case in point, being sexual but with widely dispersed planktonic larvae.

Long-term advantages associated with destroying disequilibrium In a perfect world, it might be argued, the environment would be constant, and mutations would not exist. In such a world, sex would be unnecessary. An organism which is perfectly adapted to its environment is best off having perfect fidelity of transmission of genes between generations. However, the world is not like this. Organisms live in a constantly changing environment to which they must adapt. Further, they are constantly being disrupted in their function by mutations. Even were the environment constant, the presence of deleterious mutation would disrupt adaptation. Many of the hypotheses concerning the maintenance of sex postulate that the main benefit of sex is that it maintains a better environment–genotype match at the population level. Hypotheses can be split into two types. The first of these focuses on how sex can speed the spread of advantageous alleles through the population. The contention is that sex speeds adaptation by breaking down linkage disequilibrium. The second type of hypothesis centres on the advantage of sex in maintaining a population which is relatively undisrupted by the recurrent problem of deleterious mutation.

Beneficial mutations and the advantage of sex

In an environment that is constant in space and time, the individuals that would evolve are those with the perfect make-up for that environment. In a 'frozen niche', the best individual is an asexual one with a perfectly coadapted gene complex. Sex would not be favoured in a frozen niche, because sex would disrupt adaptation. If the fittest form was heterozygous at one locus, then asexual populations could maintain perfect adaptation, whereas sexual ones could not. However, environments are rarely constant. Rather, they are heterogeneous in time and space. Sex can speed the process of adaptation if change is occurring at two loci simultaneously. There are two reasons for this. The first of these was recognised in the 1930s by Fisher and Muller, and concerns the speed with which the optimal genotype is created in sexual and asexual populations. Their argument runs thus. Consider a population of haploids where, initially, the best genotype at two loci is AB. Now, suppose the environment changes in such a way that two new alleles at these loci, a and b, were favoured simultaneously. In an asexual population of medium size, mutation would rapidly generate the favoured Ab and aB clones. However, the genotype ab would be unlikely to arise in the short term as it would require two mutations in one lineage in a very short time. It is only likely to arise following the spread of either Ab or aB clones. In contrast, ab genotypes would arise through sex and recombination in a sexual population, once Ab and aB individuals were present (Fisher, 1930; Muller, 1932) (Fig. 12.2). Where a novel advantageous form must arise by two independent mutations (which will be the case in all but very large populations), sex is beneficial in that it decreases disequilibrium and allows the rapid formation of the favoured type.

The second way in which sex can speed the process of adaptation is more difficult to understand. The advantage of sex in this case is that it speeds the spread of two or more advantageous alleles in a population which initially contains all of the possible combinations. In the terms of the previous example, the most advantaged form, ab, already exists in both asexual and sexual populations. An advantage to sex will accrue when the fitness effects of the two loci are not independent, but instead, the fitness of an ab individual is less than would be expected from simply compounding the fitness effects of each allele (diminishing epistasis) (Eshel and Feldman, 1970).

The argument proposes that under these conditions of selection, the most advantaged type ab is not rapidly favoured over types Ab and aB. Sex mixes the types, such that selection favouring types Ab and aB can contribute to the creation of the most favoured type ab. With diminishing epistasis, sex creates the optimal type ab more often than it breaks it down. Under these conditions

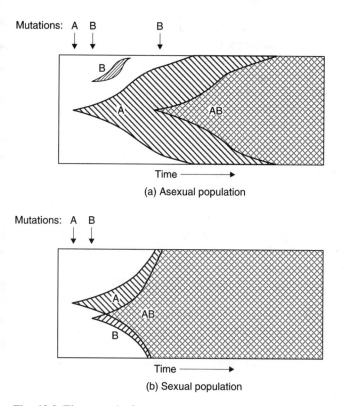

Mutations: A B   B

B

A

AB

Time ———▶

(a) Asexual population

Mutations: A B

A

AB

B

Time ———▶

(b) Sexual population

Fig. 12.2 The spread of two novel advantageous mutants in a sexual and an asexual population of medium size with a reasonably high mutation rate. Because of sex and recombination, both mutations are found in all individuals in a sexual population more rapidly than in an asexual one. (After Fisher, 1930, and Muller, 1932.)

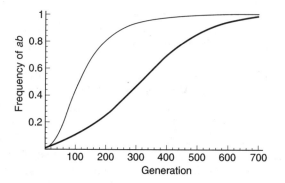

Fig. 12.3 The spread of two beneficial mutations through a sexual and an asexual haploid population with diminishing epistasis. In this case we examine the spread of two alleles in a population originally at equilibrium, with 1 per cent of individuals being of type ab. The fitnesses of the genotypes are $AB = 1.0$, $Ab = 1.04$, $aB = 1.04$ and $ab = 1.05$. The fine line on the graph represents the change in the frequency of ab in a sexual population. The bold line indicates the change in frequency of ab in an asexual population.

of selection (which Maynard Smith (1978) considers to occur commonly) sex accelerates the process of adaptation and improves the genotype–environment match (Fig. 12.3).

It is thus clear that if environmental change occurs such that more than one mutation at a time is favoured, sex may increase the mean fitness of members of the population. Beneficial mutations in two different asexual lineages can never come together. Furthermore, even if a double mutation in one lineage does result in their co-occurrence, it is likely that selection will act more effectively on the sexual population. The evolution of sex is thus likely to be promoted in changing environments. We might say that these theories of sex are *carpe diem* theories. Sex produces the best individuals most rapidly, and sexual populations adapt more quickly. Sex produces populations of individuals that can 'seize the day'.

The advantage to sexual reproduction will depend on the rapidity of change. It might be argued that stochastic environmental change is unlikely to occur on a regular enough basis to prevent periodic invasion by asexual types once the optimal genotype has been reached. However, if the selective pressure fluctuates in direction, such that alleles do not have the time to be fixed and disequilibria need constantly to be broken down, then this invasion by asexual types may not occur. As discussed in Chapter 7, the interaction between individuals of different species may provide a regular source of variation in the optimal type. This is particularly the case in host–parasite systems. Parasites and their hosts are expected to enter into coevolutionary cycles, with frequent changes in the direction of selection.

The cycles produced by biotic interactions are thought to produce one further benefit to sex. Hamilton *et al.* (1990) have argued strongly that if cycles occur very rapidly, then sex will be favoured because it buffers the population against extreme changes that could result in the loss (or reduction to very low levels) of alleles and genotypes which, although deleterious during one part of a host–parasite coevolutionary cycle, are likely to be advantageous in another part of the cycle. Hamilton states that sex may be favoured because it allows the storage of 'temporarily bad alleles' in a population where selection fluctuates rapidly in direction.

To illustrate his point, Hamilton suggests consideration of selection acting upon one locus with two alleles (see Kondrashov, 1993). In this population, heterozygotes are completely inviable in odd generations, and homozygotes are completely inviable in even generations. He then asks us to consider the fate of three populations. The first is an asexual clone which is heterozygous. This goes extinct after the first generation. The second is an asexual clone which is homozygous. This does very well in the first generation, but goes extinct after the second generation. The third is a sexual population. During the first generation, heterozygotes die, but the homozygotes survive and mate. These homozygotes produce all three genotypes. In the second generation, the heterozygotes survive, mate, and again produce all three genotypes. Under this radically altering selective regime, sex is clearly favoured, because it allows the reconstitution of temporarily bad allele combinations. Although one of the asexual clones is better under each selective regime, sex is better overall.

The model given above is clearly an oversimplification of the cycles that occur in nature. The models of Hamilton and co-workers show that the principle of 'storing temporarily bad alleles' can work under genetic conditions of host–parasite interaction that are more reasonable.

Parasitism and sex: empirical evidence

The importance of parasitism in providing an advantage to sex is not easy to ascertain. One line of research that has been followed is to make field observations of the incidence of sex and parasitism. The corollary of the view that parasitism promotes sex is that asexuality is likely to occur in habitats where parasitism is uncommon. Parasitism is expected to be uncommon in highly disturbed habitats where the coexistence of host and parasite is rare. This creates certain predictions as to species in which asexuality remains, and, where a species is present in a variety of habitats, which populations of a particular species would be expected to have retained (or re-evolved) asexuality.

This correlation has been examined for both animals and plants (Levin, 1975; Glesener and Tilman, 1978). Sex is most common in the stable tropics, where biotic interactions predominate. This correlation certainly does suggest that it is heterogeneity over time that matters, and that it is biological rather than ecological heterogeneity that counts. Peripheral habitats, which are exposed to greater variation in terms of abiotic environment, are those with the highest count of asexual species. Abiotic disturbance does not correlate well with the incidence of sex, whereas biotic disturbances appear to.

Other evidence for the importance of parasites in producing a benefit to sex derives from studies of natural intraspecific variation. Lively (1987), for instance, studied the freshwater snail *Potamopyrgus antipodarum*. This snail is susceptible to infection by a trematode, *Microphallus*, and he observed a correlation between the amount of sex (measured here by the proportion of males in a habitat) and the level of parasitism observed. This he held to be evidence that the existence of parasites promotes sex. However, as Kondrashov (1993) notes, it could equally be true that the correlation is produced because sex promotes parasites (parasites could become more prevalent in sexual populations if they transmit between hosts during copulation). Alternatively, parasites and sex could each have been promoted by an independent factor.

Testing Hamilton's argument by examining the incidences of sex and parasites is further complicated by the fact that a correlation of sex with high infection rate by parasites may be interpreted as evidence for parasitism favouring sex, but a correlation of sex with a low infection rate by parasites may be interpreted as an example of the efficacy of sex in decreasing parasite load. Both low and high parasite loads of sexual species have been interpreted as evidence for Hamilton's thesis. This is a situation where science begins to flounder. All observations are consistent with the parasitism-drives-sex hypothesis, so in reality neither observation provides any evidence. This is not to say that the hypothesis is false. Rather, we must be careful to delineate the exact predictions of our hypotheses, and seek observations that are less ambiguous.

Observations of the incidence of parasitism and sex have thus so far failed to provide unequivocal evidence for the parasitism-drives-sex view. The alternative approach to the problem of assessing the importance of this view is to test whether the assumptions underlying the models are met. It appears that the conditions of parasite–host interaction underlying Hamilton's models (that parasites are narrowly specialised to one genotype of host and vice versa) are not a universal feature of host–parasite interactions. Although selection experiments have given evidence for genetic specialisation of parasites on their hosts (Jaenike, 1993), Matthew Parker (1994) has questioned how frequently plant hosts are specialised to particular parasite types. He contends that plants may be genetic 'jacks of all trades', resistant to many types at once.

The question has been raised as to the reality of coevolutionary cycles between parasites and hosts. Barrett (1988), for instance, comments that the best evidence of parasite–host cycles derives from agricultural data sets, where the behaviour of man in choosing cultivars resistant to the prevailing parasite genotypes has driven cyclical change in parasite gene frequencies. The relationship between parasite and host may here be an artefact resulting from the planting of masses of fields of genetically identical individuals. We should seek data sets on parasite–host coevolution from natural populations.

Deleterious mutations and the evolution of sex

It is arguable that the environment–genotype match can never be maintained. Even if the environment is constant and there are no coevolving enemies, mutation still destroys adaptation. Asexual populations may, even under constant conditions, have lower fitness than sexual ones because they accumulate mutations. Although mutations are essential to generate variation for adaptation, a greater proportion of mutations are deleterious than are advantageous. This places a cost upon being asexual. To envisage this cost, consider an asexual lineage that is perfectly adapted. In each generation, some slightly deleterious mutations will arise. These mutations will occur randomly among the individuals in the population. In each generation, selection will promote those individuals with fewest mutations. However, in an asexual organism, if we ignore the extremely unlikely event of precise back mutation, the minimum number of mutations can only go up, it cannot go down. The process of mutation accumulation was first recognised by Muller (1964), in what has been termed '**Muller's ratchet**'. A useful analogy is that of Ridley (1993). Ridley imagines asexual reproduction to be like a photocopying process. Photocopiers produce very good but not perfect copies of the original. The quality of the best copy decreases between generations, just as the 'quality' of the best asexually produced individual decreases.

Sex, it is argued, may be favoured because by combining parts from various photocopies, it will sometimes retrieve a copy as good as the original which will then be favoured by selection. The reassortment associated with sex destroys disequilibria between mutations and can restore the perfect genotype, defeating Muller's ratchet.

The importance of Muller's ratchet in decreasing the fitness of asexual populations is controversial. The effect of sex on mutation accumulation has been the subject of much theoretical study. The 'problem' posed by the ratchet increases with both decreasing population size (more chance of losing the optimal genotype) and increasing size of the functional genome (greater mutation rate).

Deleterious mutations may lower mean population fitness even without the threat of random loss of the best type through the ratchet mechanism. Sex may, under certain conditions of selection, lower the equilibrium level of genome contamination. As Kondrashov (1988) points out, sex will decrease the contamination of the genome with deleterious mutations when each additional mutation decreases fitness by more than the one before (positive epistasis). When the effect of mutation on fitness is non-linear in this way, sex is advantageous, because sex increases the variance of the population for deleterious mutations (some have very few, some very many).

This effect can be seen most easily if we consider the extreme situation, where selection removes (say) the 20 per cent of individuals with the most mutations (truncation selection). Let us start with an asexual and a sexual population with the same mean and variance in per genome mutation load, and let the between-genome variation in load be small. After a round of reproduction, the variance for mutation number in an asexual population is still low, increased only slightly by the process of mutation. Then, the 20 per cent of individuals with the most mutations are killed. These will bear only a few more mutations than the mean. However, in the sexual population, the next generation has increased between-genome variation in contamination. The 20 per cent of individuals with the most mutations in this population bear many more mutations than the mean. Thus, selection purges more mutations per round in a sexual population than an asexual one. Given that the sexual and the asexual population will have a similar mutation rate, the greater removal of mutations per round of selection will render sexual populations less contaminated by deleterious mutation than asexual ones.

Can the advantage of a decreased load of deleterious mutation maintain sex? One test is the per genome mutation rate. Kondrashov (1982) suggests that for the deleterious mutation with positive epistasis to produce the necessary advantage to sex, on average one mutation per genome per generation is required. However, it has recently been argued by Rosemary Redfield that sex is at least partly the cause of deleterious mutations (Redfield, 1994). Redfield notes that the mutation rate in the male is higher than that in the female (see Chapter 2). In the few cases where it has been measured, the mutation rate in the male is higher by a factor of between two and eight. She reworks Kondrashov's models to incorporate this fact. With significant male-biased mutation, the conditions for deleterious mutations to provide a significant advantage to sex are more strict than envisaged by Kondrashov. Around two mutations per genome per generation are required.

Discussion: the benefits of sex

The evolution of sex remains something of an enigma. At present, there is perhaps a slightly disturbing degree of polarity in the debate. There is a tendency to consider either parasites, or deleterious mutations, or environmental variation as important in the maintenance of sex. Just as most evolutionary biologists would now argue that both neutral drift and natural selection are evolutionary processes, the simple answer may be that there is actually no single truth to the maintenance of sex. The existence of counter-arguments to all theories suggests that all the forces discussed above may be important in combination in dictating the incidence of sexual reproduction. Recently, a model incorporating both mutation- and parasitism-based advantages to sex has been published (Howard and Lively, 1994). Such models promise to provide a fruitful new avenue.

What data would allow us to evaluate the hypotheses fully? This is not straightforward. The mutation hypothesis may be open to test because technological innovations should enable the per genome rates of mutation in males and females to be measured in a wide range of species. However, there is still the problem of quantifying the fitness effect of the mutations if a measure of the deleterious mutation rate is to be obtained. On the other hand, environmental hypotheses are likely to be difficult to test, for these require observation of selection over considerable periods of time. That said, at least the genetic assumption underlying Hamilton's models should be amenable to analysis.

There is one further complication to the study of the evolution of sex. This is that it is not necessarily easy to evolve an asexual type from a sexual one. In mammals, for instance, an asexual mutant is doomed to failure because of genomic imprinting (see later in this chapter). Examination of animals which have managed to evolve asexuality show many of them to have what Maynard Smith (1986) referred to as 'sexual hang-ups'. Asexual bark beetles, for instance, sometimes need to be mated and for sperm to fuse with their egg before the egg will develop. The reason is simply that the development of a sexually produced zygote is initiated by a signal of fertilisation. The first asexual dispensed with the DNA in the sperm but still required the signal the sperm contained. The evolution of fully liberated asexuality requires two evolutionary events: the ability to produce a zygote without halving the number of chromosomes via normal meiosis, and the ability to trigger zygote development without the act of copulation. As it is unlikely that a single muta-tional event could confer both these abilities, secondary asexuality may be rare simply because the required elements are unlikely to arise together. It is perhaps not surprising that parthenogenesis is commoner among the haplo-diploid Hymenoptera, in which unfertilised female gametes in sexually reproducing species do develop (into males), than in any other group of higher animal.

The reverse transition, from asexuality to sexuality, is also likely to be diffi-cult. Meiotic sex is a delicate process that (usually) results in an accurate partitioning and recombination of chromosomes to produce four gametes containing the correct number of chromosomes without excessive loss of genes from inaccurate recombination. This complexity of sexuality makes the initial evolution of sex an even more difficult problem than its maintenance. The above analyses assume meiotic sex to be perfect. However, during the initial evolution of sex, this cannot be assumed. Sex involves more than fusion, it also involves an accurate reduction division. Unsurprisingly, the question of the pathway leading to accurate meiotic sex remains unresolved.

The evolution of haploid, diploid and haplo-diplontic life histories

The effect of selection depends on ploidy. Advantageous alleles spread more slowly in diploids, because the effect of an advantageous gene on phenotype is masked initially if it is recessive, and the selective advantage is sometimes diluted if it is only partially dominant. One of the great triumphs of nine-teenth-century cytology was to reveal the regular pattern of ploidy of sexual organisms. Sexual organisms show alternation of ploidy, with a haploid stage following meiosis (or gametogenesis), and a diploid stage following syngamy (fertilisation) (Fig. 12.4).

In most animals and higher plants the life cycle is a classic diploid one. Here, the haploid stage (the gamete or gametophyte) is brief, does not have an independent existence, and involves few or no mitoses. Gametes fuse in the process of fertilisation to form a diploid stage (or sporophyte) which exists independently of the haploid, undergoes mitoses, and only after a period of time produces the haploid stage via meiosis. By contrast, most green algae and all bryophytes (mosses and liverworts) show a haploid life cycle. In these groups, mitoses occur in the haploid (gametophyte) stage, and the diploid (sporophyte) stage is transitory, involving either few or no mitoses. In mosses and liverworts, the diploid stage depends on the haploid stage for nutrition. Members of the red algae, the Foraminifera and the Pteridophyta (ferns, horse-tails and club mosses) do not show a predominance of either the haploid or the diploid phase, and are said to have a **haplo-diplontic** life cycle. Both haploid and diploid stage are free living and mitoses are seen in both phases. Alternation of ploidy is also seen in some asexual protists. Here, the organism usually possesses either two or many copies of the same genome, but this is

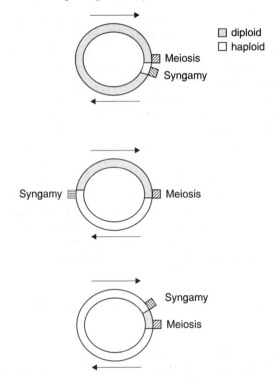

Fig. 12.4 The different life cycles of animals and plants.
(a) The diploid life cycle.
(b) The haplo-diplont life cycle.
(c) The haploid life cycle.

periodically reduced to one by undergoing mitoses without DNA replication, from which the original number is regained by DNA replication without cell division (Kondrashov, 1994).

Why should this diversity of lifestyles exist? Bell (1982) commented that the paucity of attempts to explain the diversity of systems constituted a major scandal. Since this time, the subject has received more consideration, but, as yet, no single truths can be said to have attained acceptance.

Factors influencing the relative importance of haploid and diploid phases

The diploid lifestyle is considered by many to be favoured because of intrinsic genetic advantages that the diploid state offers. For instance, in the diploid state, mutations are not necessarily expressed. The effect of deleterious somatic mutations is lessened in the diploid phase so long as they are partially recessive (Orr, 1995). In addition, inherited deleterious recessive mutations may be masked, producing a benefit to diploidy either in the short term (Crow and Kimura, 1965), or, under certain conditions, in the long term (Kondrashov and Crow, 1991; Perrot *et al.*, 1991). Diploidy also permits the faster acquisition of new advantageous genes, because the mutation rate is increased (Lewis, 1985). Where **overdominance** (heterozygote advantage) exists, diploidy permits the creation of fitter individuals than haploidy would allow (Crow and Kimura, 1965).

An alternative view to such 'genetic advantage' hypotheses is that the extension of the diploid phase is an adaptation against intracellular parasites. Sex sometimes involves the mixing of cytoplasms, and thus the mixing of cytoplasmic symbionts. Cell division may cause the segregation of these symbionts, and so lower parasite load per individual cell. As mitosis is a faster form of division than meiosis, an extended diploid phase would be selectively advantageous (Hurst, 1990). This may explain the initial evolution of a long diploid phase. However, in species where there is uniparental inheritance of cytoplasm, the maintenance of such a phase cannot be explained by this type of advantage. Against this background of potential advantages to diploidy, we have also to explain the continuation of haploidy in many groups. What factors could compensate for the genetic cost of haploidy? The small genome size (and therefore cell size) associated with haploidy is thought to produce certain ecological advantages. Haploid cells, with less DNA to replicate, may achieve a higher rate of division than comparable diploids (Cavalier-Smith, 1978). The lower quantity of DNA may also lower nutrient requirements for division and so speed the cell cycle (Lewis, 1985). However, how far these factors affect multicellular haploids has yet to be ascertained.

The factors responsible for the evolution and maintenance of haploid and diploid phases of equal length are perhaps even more obscure. It has been argued that the equality of the haploid and diploid stages may be a function of the adaptation of the haploid stage to one season, and the diploid to another, by virtue of the effects of ploidy on their size and nutrient requirement (Stebbins and Hill, 1980). However, this is not a general advantage, because in some haplo-diplontic cycles, morphology and physiology do not vary with ploidy. One recent view is that this system is favoured in some taxa because it halves the cost of sex. If diploid and haploid phases are constrained in length (for instance, an organism is constrained to being less than 65 cells in size), then having both diploid and haploid phases of maximum length may be favoured, for then syngamy happens half as frequently as when either the diploid or haploid phase predominates. Syngamy being wasteful (two individuals fuse to create one), the haplo-diplontic lifestyle may be favoured (Richerd *et al.*, 1993).

The evolution of sexes

In most animals and some plants there are two sexes, male and female. The male produces sperm or pollen which are small and motile, and, on fusion with the gamete of the female, contribute little cytoplasm to the resulting zygote. This difference in size of the gametes is termed **anisogamy**.

Anisogamy, and competition between gametes

The classical theory that attempts to explain the evolution of anisogamy was derived by Geoff Parker and co-workers (1972), and is based upon gamete competition. Anisogamy, it is argued, is the result of an arms race within a species. Before anisogamy evolved, the population would have consisted of individuals which all produced similarly sized gametes. If a mutation results in one individual producing a slightly smaller gamete, then this individual will be able to produce more gametes. It is clear that, whilst all others are producing large gametes, the reproductive success of this individual will increase. However, as the gene for producing smaller gametes spreads, so the probability of two small gametes fusing increases. This may result in the production of an individual with insufficient soma to develop successfully. This inviability produces selection for small gametes which fertilise only large gametes, so forming the basis of anisogamy. Furthermore, as competition between individuals that produce small gametes intensifies, selection will favour even smaller gametes (to increase both the number of sperm produced and possibly their mobility). A similar pattern will affect those individuals producing large gametes. Selection will favour increased gamete size and reduced motility of the ovum. Thus, Parker shows that an arms race within a species to increase the number of fertilisations may account for anisogamy, and the evolution of males and females.

Sexes and the inheritance of cytoplasm

Although Parker's theory was widely accepted, more recent theories (Hurst, 1990; Hastings, 1992; Hurst and Hamilton, 1992; Law and Hutson, 1992) have suggested that the benefit to anisogamy might lie in its association with uniparental inheritance of cytoplasmic genes. One set of these hypotheses suggests that the exclusion of cytoplasmic parasites is the benefit of anisogamy. Consider a species in which cytoplasmic parasites are common; let us say they infect one-quarter of all individuals. If an individual derives its cytoplasmic constituents from both parents, as might be observed in an isogamous mating, the probability that it will inherit a cytoplasmic parasite is $7/16$ ($1-0.75^2$). However, if cytoplasmic constituents are only derived from one parent, as is the case in anisogamous species, then the probability of inheriting a parasite is reduced to $1/4$. Thus, an advantage to anisogamy is that it reduces cytoplasmic parasite load.

The advantage of uniparental inheritance may go beyond reducing rates of parasitism. Hastings (1992) and Hurst and Hamilton (1992) suggest that **bipar-**

ental inheritance of cytoplasms may be deleterious because it allows paternal and maternal organelles to mix. They imagine that if mixing occurs, conflict between maternal and paternal organelles would be likely. There would be competition between organelles of different origin for transmission to the next generation a conflict which could be deleterious to the organism. Uniparental inheritance might thus have evolved because it reduces this conflict within the cytoplasm.

Both these theories of the evolution of anisogamy are based upon the potential advantage of uniparental inheritance. Uniparental inheritance of cytoplasmic contents is not restricted to anisogamous organisms, however. In many isogamous Protozoa, there are two mating types. Gametes can be sorted into two groups which will not fuse with each other, but will fuse with the opposite type. A given organelle is always derived from the gametes of one of these pools (the pool of donor individuals). Although the diploid stages are not differentiated, the gamete stages are functionally differentiated with respect to cytoplasmic inheritance in the same manner as in anisogamous species.

Hurst and Hamilton (1992) suggest that the exclusion of one set of cytoplasmic genes in isogamous species may be analogous to anisogamy in evolutionary logic. They conjecture that the driving force behind the evolution of uniparental inheritance (and hence sexes) is the need to minimise conflict between cytoplasmic genomes. Anisogamy and **isogamy** represent different solutions to the same basic evolutionary problem.

The data with which to judge the 'cytoplasmic conflict' theories of sexes and uniparental inheritance are not yet available. The basic requirement is that organelles have the potential to be in conflict. The mechanism of ensuring uniparental inheritance is suggestive of such a process in some species. For example, the observation in mice and bees that paternal mitochondria which enter the zygote are not transmitted suggests that the separation of paternal and maternal cytoplasmic constituents is of importance in evolution (Meusel and Moritz, 1993; Kaneda et al., 1995). However, the importance of cytoplasmic parasites in the evolution of anisogamy is not yet known, and it would be premature to discount ecological theories which explain this phenomenon. Furthermore, it is worth stressing that more than one motive force may have played a part in the evolution of sexes, and that the relative importance of different factors in individual evolutionary events need not have been the same.

The evolution of sex chromosomes

Mendel's first law recognises that genes occur in pairs, and that in a diploid, each gene and chromosome has a homologue. Homologous chromosomes pair at meiosis, and may form chiasmata and recombine. However, it is not true that all genes have a pair. In a great many species of animal and some plants, the chromosomes on which the sex determination genes are located are heteromorphic (i.e. they are different in form, and are not fully homologous). The human male bears two sex chromosomes, X and Y, which are morphologically and genetically very different. The X chromosome bears many functional genes, but the Y chromosome is, in large part, genetically redundant. Furthermore, as well as being different in genetic constitution, the different sex chromosomes do not recombine over much of their length. A small region exists, termed the **pseudoautosomal region**, where they can pair and do recombine, but apart from this they are separate.

The absence of recombination between non-homologous regions of the X and Y means that the evolutionary biology of genes on sex chromosomes is subtly different from that of genes on autosomes. Genes on the X chromosome will, in general, respond more rapidly to selection than those on autosomes. This is because recessive alleles on the X will be expressed in males: their phenotypic effects cannot be masked by dominant alleles. In addition, the population size of the X chromosome is three times that of the Y, making the Y chromosome more susceptible to drift effects than the X. Furthermore, the absence of recombination between non-pairing regions of the X and Y (or between Y chromosomes) means that the genes on these chromosomes, with the exception of the pseudoautosomal region, have separate evolutionary fates. New genes may be formed on the X with no homologue on the Y.

Why are heteromorphic sex chromosomes so common? The evolution of sexes associated with anisogamy requires some genetic control over sex. In many cases it also promotes selection to produce equal numbers of male and female offspring (see Chapter 4), and, therefore, selection for an even segregation of sex-determining alleles. The initial method of producing even segregation is likely to have been for sex to be determined at one locus with two alleles, with one sex being heterozygous, the other homozygous. This is a system of genic sex determination. In order to understand the evolution of sex chromosomes, we must understand why the chromosomes associated with the sex-determination genes stopped recombining and why the genes on one of the chromosomes degraded.

The motive force for the evolution of sex chromosomes is thought to be the presence of sexually antagonistic genes. Sexually antagonistic genes are those which have a positive effect when in one sex (here, through intra-sexual competition in males), but a substantial cost in the other. In addition to selection for the sexes to develop correctly, there is also often selection for males and females to be different. Males, for instance, are often selected for size and weaponry to maximise the number of matings they achieve during their lifetime, whereas females are selected to maximise their lifetime reproductive success through apportioning energy to survival and care of offspring.

Sexually antagonistic genes may spread easily if they become linked to the sex-determining locus. A mutation which produces antlers and is tightly linked to the male sex-determining gene will clearly spread, because it is rarely expressed in the female. However, any recombination in the region bearing the sex-determining locus would be deleterious. Recombination would bring the sexually antagonistic gene into the wrong context with respect to the sex-determining gene. Selection will therefore favour any factor that prevents recombination between the regions of chromosomes bearing different sex-determination alleles. This, Rice (1987) suggests, is the evolutionary force that accounts for the absence of recombination associated with sex chromosomes.

The absence of recombination between newly evolved sex chromosomes provides a scenario that explains the degeneration of the Y chromosome (Charlesworth, 1978, 1991). Consider the effect of mutation on the sex chromosomes in a species where the male is heterogametic. If the Y chromosome never recombines, it is an asexual lineage, and will be subject to the accumulation of mutations through Muller's ratchet (see p. 274). Charlesworth argues that this mutational load will lead to the evolution of decreased expression of the Y and increased expression of the X chromosome. Further, once expression of the Y chromosome has ceased, the genes on it will be expected to degenerate further; there will be no selection for their maintenance, and they will gradually be lost.

The basic components of the above model are given in Fig. 12.5. Simulations by William Rice, using populations of the fruit-fly, *Drosophila melanogaster*, as a model system have also shown that they have some validity. In two elegant experiments, Rice (1992, 1994) showed first that sexually antagonistic genes do become associated with sex-determining loci, and second, that loss of recombination does result in a remarkably rapid build up of mutation. Further to these experiments, observations in a second species of fruit-fly, *Drosophila miranda*, suggested a molecular mechanism for degeneration. *Drosophila miranda* has a Y chromosome composed of two parts, a classical Y section, and a part which has been translocated to the Y from chromosome 3. This latter part is now in the process of degeneration, due largely to transposable element activity (Steinemann and Steinemann, 1992; Steinemann *et al.*, 1993).

Fig. 12.5 The current view of the evolution of heteromorphic sex chromosomes.
(a) Sexually antagonistic gene (beneficial in male) spreads if genetically linked to male sex-determining allele.

(b) Occasional recombination produces inappropriate conjunction of sexually antagonistic gene and the female-determining allele.

(c) Selection for reduced recombination between chromosomes bearing different sex-determining alleles.

(d) Reduced recombination leads to formation of a male-determining chromosome which never recombines, begins to accumulate mutations, and degenerates.

(e) Selection for non-expression of the degenerated genes on the male-determining chromosome. Male- and female-determining chromosomes diverge over much of their length. They become heteromorphic.

Transposition into functional genes causes loss of function, and there is no recombination to allow the retention of fully functional chromosomes.

Evolution and gene expression

Gene expression varies. The pattern of gene expression of all species varies to an extent with differing environments. Beyond this, some alleles are dominant over others. In some, gene expression relates to parental origin, a phenomenon termed genomic imprinting. The observation is that only one copy of a gene is expressed, and that this copy comes regularly from either mother or father. We here consider the relationship between genotype and phenotype, how dominance can evolve, and why genomic imprinting may have arisen.

Phenotypic plasticity

When light-skinned people sunbathe, they tan. Tanning is a result of the increased production of melanins in the skin. The expression of genes involved with melanin production varies according to environmental factors. Tanning might be said to be an example of phenotypic plasticity, the alteration of gene expression according to environmental conditions.

All organisms show a degree of phenotypic plasticity because environments fluctuate. This is seen most spectacularly when the same species of plant is grown in different light environments. Ones kept in dim light grow in height very rapidly compared with those grown in sunshine. The evolutionary rationale is clear. A plant which remains in dim light will die. Growing rapidly increases the probability of survival because it might then overgrow any source of shading. Plants in general show considerable phenotypic plasticity because they do not move. Because they do not move, they must be able to react to any environment in which they find themselves. On the other hand, an animal often has two alternative strategies. It can either alter phenotype, or it can alter location.

Phenotypic plasticity involves the modulation of the expression of one or more genes by external influences. Gene expression can be controlled before the transcription of the gene (control over the quantity of transcript). Expression can also be controlled after transcription, by affecting the rate of splicing of the mRNA molecules. Each gene has what is termed a norm of reaction, which may be envisaged graphically as a plot of expression against a given environmental variable.

The norm of reaction can be considered a product of selection. When environments are variable, and a good phenotype–environment match is required for survival/reproduction, selection will favour a wide norm of reaction. To go back to the tanning analogy, there is a very narrow norm of reaction of pigment with varying sunlight amongst people of Afro-Caribbean descent, because the environment in this region is largely predictable. A high density of melanins in skin is beneficial at all times. In higher latitudes, the intensity of sunlight is far more variable. In winter it would be wasteful to produce melanins. However, in summer a lack of melanins equates with increased skin cancer. A wide norm of reaction has therefore evolved. Variation in norm of reaction probably derives from mutation in the area of DNA upstream from the gene, for factors which bind here affect transcriptional activity. Mutation in the sequence of this region may alter the propensity of different factors to bind, and thus affect gene expression.

The evolution of dominance

Genes occur in pairs, and the relationship between genotype and phenotype is crucial in determining the rate of spread of novel mutants. Dominant alleles invade more rapidly but become fixed more slowly than recessives (see p. 61). Dominance of a wild type over mutants is probably often an intrinsic property of the alleles involved. As Muller (1933) and Wright (1934) have noted, mutants will often be deletions, with the result that they produce less or no product. In such cases, dominance arises naturally if the wild-type allele produces enough product for correct function. Given that wild-type genes are selected to be expressed such that they withstand perturbations in the level of expression, they should therefore be expected to be 'naturally' dominant over deleted mutants.

However, dominance is not necessarily an intrinsic property of an allele. Fisher (1928, 1930) theorised that dominance might be an evolved trait, being a selective consequence of the inferiority of heterozygotes. He considered that if there were a wild-type gene in a population which was being disrupted in its function by recurrent mutation, then selection would favour modification of its expression. Modifier genes would spread if they altered gene expression such that the wild-type allele was dominant over mutants.

There is little support for Fisher's theory of dominance as stated above. Brian Charlesworth (1979) notes that the expectation from Fisher's theory is that mildly deleterious mutations should be expected to be more completely recessive than fully lethal ones. This is because mildly deleterious alleles will be maintained in a population for a longer time, and thus produce stronger selection for dominance against them. However, observation suggests that mildly deleterious alleles are in fact less likely to be recessive than lethals. Evolved dominance of a wild type over mutant alleles is not a frequent occurrence. The data are more consistent with Muller's and Wright's view that dominance most frequently occurs when the mutant produces no product. The occurrence of dominance over a mutant which still produces a functional product, as envisaged in Fisher's theory, appears to be rare.

However, the evolution of dominance of a wild-type allele over a deleterious mutation is not the scenario in which Fisher thought dominance most likely to evolve, for as he noted, it requires continual challenge of a wild type by a mutant over a long period of time. He considered dominance to be more likely to evolve during the spread of a novel, advantageous allele. Indeed his notion that dominance might evolve was first conceived during his work on the evolution of polymorphic Batesian mimicry (see Turner (1985), and Chapter 4). During the early stages of spread of a recessive allele, there is a protracted period of opportunity for selection to increase dominance over the allele it is replacing. At low frequency, the proportion of novel alleles in heterozygotes rather than homozygotes is very high. Thus, if it is assumed that the new allele confers greatest advantage when fully expressed, selection will favour modifiers that increase the expression of the new allele in heterozygotes. Here it is dominance of an allele over one which it has replaced that evolves, rather than dominance over one which has been challenged by recurrent mutation.

We can extend this case to predict that dominance is most likely to evolve when there is a continual presence of disdavantaged heterozygotes. This will happen when, for instance, there is clinal habitat variation, such that different alleles are favoured in different geographic areas. If there is gene flow along the cline, heterozygotes will continually be produced. This continued production of heterozygotes will allow selection for modifiers that produce dominance.

Direct evidence for the evolution of dominance outlined above is hard to obtain. We need to observe selection producing dominance over a period of time. However, the theory does make testable predictions. Most pertinently, dominance of an allele should break down in hybrids. Observations of crosses of the polymorphic butterfly *Papilio dardanus* fulfil these conditions. When crosses were made between individuals from the same population, different wing patterns showed distinct dominance relationships. However, when crosses were made between individuals from different populations, no dominance was observed (Clarke and Sheppard, 1960c) (see Box 4.1).

In summary, dominance and recessiveness may be either intrinsic properties of the alleles, or they may evolve. Dominance of a wild type over mutant alleles appears to be most often explained as a natural occurrence. The mutant produces no product and does not affect phenotype. However, selection does appear to favour modifiers which produce dominance of an allele over one which it is in the process of replacing.

Genomic imprinting

The revolution in molecular biology has produced some remarkable insights into the pattern and mechanisms of heredity. Perhaps one of the most amazing observations is that the expression of a gene may correlate with its parental origin. This phenomenon, in which the parental genome has an effect on the expression of the genes in the progeny, is termed genomic imprinting and is exemplified by what Haig and Graham (1991) described as 'the strange case' of insulin-like growth factor two. Insulin-like growth factor II (IGF II) is expressed in the mouse foetus and the adult brain. Within the foetus, it promotes the movement of metabolites across the placenta, promoting the acquisition of resources from the mother. The paternal copy of the IGF II gene is fully expressed, whereas the copy inherited from the mother is not (De Chiara *et al.*, 1990, 1991). In contrast, the gene coding for a receptor for IGF II, the IGF II Type II receptor, shows a reversed pattern of expression. The maternal copy is expressed fully, whereas the paternal copy is not (Barlow *et al.*, 1991). The result of imprinting is that individuals bearing two copies of either maternal or paternal alleles are unbalanced in terms of gene expression. They have severely impaired survival. The existence of genomic imprinting prevents the evolution of parthenogenesis in mammals, because a foetus with only maternally imprinted genes will perish.

Why should genomic imprinting have evolved? Haig and Graham suggest that this may be a further case of intragenomic conflict (see Chapter 8). In this case the conflict is a result of the different interests of male and female parents. Haig and Graham argue that, if IGF II serves to acquire resources from the mother, it is in the interest of the father that his progeny get as much resources as possible. It is in the interest of the mother, however, to limit the amount of resources given to the developing foetus. In short, the mother has an interest in conserving resources for her future reproduction which the father of the foetus does not share. Envisage initially a gene in the male parent that produces increased expression of the paternally derived IGF II gene in the foetus. This is in conflict with the 'evolutionary interest' of the foetus's mother. Genes in the mother which imprint foetal IGF II genes in the opposite direction should then be favoured. This is exactly what we find: it appears that the IGF II Type II receptor, of which the maternal copy is expressed, is an inhibitor of the action of IGF II; that is to say it prevents the transfer of resources from mother to offspring.

Haig and Graham's hypotheses for evolution of imprinting is one of a number of hypotheses that have been formed. Other explanations centre on an advantage of imprinting to the individual. James Thomas (1995), for instance, considers genomic imprinting to be a mechanism of chromosome counting. Loci which are genomically imprinted in a foetus bearing the normal complement of chromosomes are expressed in just one copy. The lack of a chromosome in a foetus would then be indicated by the lack of a product, and the presence of an extra chromosome by double the amount of product. This difference in titre could be detectable by the mother and, it is argued, would be beneficial as it would allow early abortion of progeny with reduced viability. The reduced waste of early abortion would be the selective advantage driving the evolution of genomic imprinting.

Both of these hypotheses have logic to them. They also produce certain testable predictions. Importantly, Haig and Graham's hypothesis predicts that imprinting should not be observed in species where lifetime monogamy is the rule, since here there is no conflict between the evolutionary interests of the parents. Furthermore, Haig and Graham predict that imprinting should be associated primarily with genes which affect the passage of resources across the placenta. Thomas's hypothesis requires that abortions are associated with monosomy and trisomies, and that imprinted loci are the signal that elicit the abortions.

The evolutionary genetics of imprinting is in its infancy. New hypotheses to account for this curio of genetics are appearing regularly. It would be surprising (and perhaps disappointing) if they did not continue to do so, as more complete data on imprinted genes in humans, mice and other mammals are rapidly produced.

Discussion

The evolution of genetics is a subject which, because of its nature, will have many theories and predictions, not all of which will be amenable to empirical tests. The field is in a state of flux. Classical theories which account for the biology and genetics of organisms were based largely on either arguments associated with ecology or the problem of recurring mutation. The recent past has seen a rise in explanations invoking the importance of conflict as a motive force behind the design of genetics, of sex being a consequence of conflict between parasites and their hosts, of sexes evolving as a consequence of the need to prevent either the spread of cytoplasmic parasites, or of the need to prevent conflict within the cytoplasm, and of genomic imprinting being a manifestation of conflict between genes of different parental origin. The beauty of conflict arguments is that conflict provides recurring selection. Both genetic elements involved are capable of adaptation. Genetic conflict and the selfishness, even ultra-selfishness, of genetic elements, are currently in vogue because they add a new dimension to the way in which evolution is viewed. There is no denying that an understanding of the nature of conflicts between genetic elements may lead to significant advances in our understanding of the evolution of genetic systems. However, the current swell of interest in such ideas does not mean that they are necessarily true explanations in all cases, nor that all observations of genetics have a single fundamental explanation.

Summary

The genetic systems we observe are not static, but are subject to change. On a micro-evolutionary time-scale, the pattern of dominance and of expression of genes may evolve. Over longer time-scales, the number of chromosomes can change, as can the constitution of chromosomes associated with sex-determination loci. The time spent in the diploid phase may be an evolved response to the genetic advantages of diploidy and the ecological advantages of haploidy. The evolution of sexes may be the product of parasitism. Examining the rationales for these changes, it becomes apparent that mutation, conflict between individuals of the same and different species, and intragenomic conflict are powerful forces which may drive the evolution of genetic systems. Indeed, some features of genetic systems are difficult, if not impossible, to comprehend, except when viewed in the light of their selfish components.

Suggested reading

HURST, L. D. (1992). Intragenomic conflict as an evolutionary force. *Proceedings of the Royal Society of London B*, **248**, 135–140.

HURST, G. D. D., HURST, L. D. and JOHNSTONE, R. A. (1993). Intranuclear conflict and its role in evolution. *Trends in Ecology and Evolution*, **7**, 373–378.

Two pieces on how intragenomic conflict may be involved in the evolution of genetic systems. The former concentrates on intranuclear conflicts, the latter on conflicts between cytoplasmic and nuclear genes.

KONDRASHOV, A. S. (1993). Classification of hypotheses on the advantage of amphimixis. *Journal of Heredity*, **84**, 372–387.

LADLE, R. (1993). Parasites and sex: catching the red queen. *Trends in Ecology and Evolution*, **7**, 405–408.

Two pieces on the evolution of sex: the first is comprehensive, the latter a commentary on the parasite–host coevolution theory.

LUCCHESI, J. S. (1994). The evolution of heteromorphic sex chromosome. *Bioessays*, **16**, 81–83.

MOORE, T. and HAIG, D. (1991). Genomic imprinting in Mammalian development: a parental tug-of-war. *Trends Genetics*, **7**, 45–48.

A short review on the conflict theory of the evolution of genomic imprinting.

VALERO, M., RICHERD, S., PERROT, V. and DESTOMBE, C. (1992). Evolution of alternation of haploid and diploid phases in life cyles. *Trends in Ecology and Evolution*, **7**, 25–29.

Chapter 13

Conclusion

In the career spans of three generations of scientists, the study of evolution has come a long way. Darwin and Wallace's paper, 'On the Tendency of Species to form Varieties; and on the Perpetuation by Natural Means of Selection', was read to the Linnaean Society of London on 1st July 1858. *The Origin of Species* was published the following year. During the rest of the nineteenth century progress was slow.

Following the resurrection of Mendel's work in 1900, the rate of progress has been accelerating, at times gradually, at times in a more jerky fashion, with periods of consolidation being punctated by new insights, or the discovery of new techniques. The melding of the Mendelian laws of inheritance and Darwin's selection theory took about forty years, giving rise to the field of evolutionary genetics, and culminating in the formulation of the Neo-Darwinian synthesis. During these four decades, four scientists emerged, each of whom could rightly claim to have made a significant contribution to the foundations of this field. The works of Fisher, Haldane, Muller and Wright were, and still are, highly influential. This is not least because these four giants, who bestrode the field well into the second half of the century, often did not see eye to eye. Their disagreements were often fertile, raising alternatives that experimentalists, such as Ford and Dobzhansky, could aim to test.

In the 1940s, it became clear that nucleic acids are the principal chemicals that control the passage of biological characteristics down the generations. In 1953, Francis Crick and James Watson, after eighteen months collaboration, submitted a paper to *Nature* describing the structure of DNA. Now the pace really quickened. With the coming of an understanding of the genetic code came new theoretical insights, and the development of novel ways of looking at gene products, then genes, and finally the structure of genes and non-coding bits of DNA.

Genetic variation became amenable to study at more and more levels. The neutral mutation theory of molecular evolution was proposed and, eventually, embraced. The concept of a molecular clock was proposed, debated and, to a large extent, accepted. The theories of punctated equilibria and of concerted evolution were put forward, and are still argued over. And as the methods of examining and manipulating DNA become more and more varied and powerful, undoubtedly new findings will lead to new insights and greater understanding of the processes that shaped life on Earth.

In 1973, Jacob Bronowski wrote in *The Ascent of Man*, 'The theory of evolution is no longer a battleground.' This may be true if the question is, 'has evolution occurred?' But if the question is 'how has evolution occurred?', those who seek to investigate this question still fight in an extremely active and bloody field, but now with more and more powerful weapons.

Given the current rate of technological innovation it is perhaps foolhardy to speculate on the advances that may be made over the next decade or two. After all, who would have conceived twenty years ago that we would now have DNA fingerprinting, or the ability to amplify the DNA in a single sperm up to workable amounts. It is perhaps more pertinent to consider the areas of evolutionary research that may have become tractable as a result of recent technological advances, and are the most likely to be fruitful in providing novel insights and fuller understanding.

Most biological systems are too complex to reduce to a set of equations, even very long equations, without a host of simplifying assumptions. One of the most pervasive of these, and one which is seldom questioned, is that mating between individuals occurs randomly. Even the most astute and best-respected modellers make this assumption, particularly when treating complex subjects such as the outcome of sexual selection or the possibility of assortative mating preferences across a hybrid zone. More complex scenarios are not considered, first because they are probably too complex to approach deterministically and, second, because for most organisms we know effectively nothing about population substructure.

Here is a field in which the power of a molecular technique, DNA fingerprinting, has already been demonstrated. Although the number of studies that have used this technique to analyse mating strategies and population structure rigorously is still quite small, virtually every study emphasises that the assumption of random mating has nothing to do with reality.

An understanding of both mating strategies and population substructure is crucial to a range of evolutionary problems. As we have already seen, the evolution of conspicuous characteristics is considered to be primarily the province of selection except in small populations, when genetic drift may have a significant influence. If what is conceived as a large population in which drift is likely to have little significant effect is in reality subdivided into many small or very small subpopulations, drift may indeed play a significant role. The problem then becomes one of the amount of gene flow between the subpopulations, and so a knowledge of who mates with whom becomes crucial. If the amount of gene flow is greatly restricted by the movement of individuals between subpopulations, or other components of the mating system, considerations may move into the sphere of divergence of the subpopulations, isolation and speciation.

This is just one possible train of discovery that might surface as techniques and approaches allow us to become more rigorous. The same knowledge of population structure and mating strategies might equally shed light on Wright's shifting balance theory, or the chances that drift could aid the evolutionary passage between adaptive peaks. This type of information is also crucial in understanding the dynamics of ultra-selfish genes, the epidemiology of infectious diseases, and the role that group selection might play in the evolution of mutualism, or reduced virulence, or that of kin selection in the evolution of co-operation.

Other even newer techniques promise much for other areas. The development of automated sequencing technology is likely to allow us to collect sufficient data to assess the accuracy of evolutionary phylogenies and molecular clocks in the foreseeable future. Automated sequencing, in theory, does not allow us to do anything new. It just allows us to do it much quicker. When performed manually, DNA sequencing traditionally involved the use of radioactively labelled nucleotides, followed by autoradiography. The technology which allowed the process to be automated depends on substituting fluorescent dyes for radioactivity. The labelled DNA fragments can then be electrophoresed as normal, but are detected by a scanning laser as they fall off the bottom of the gel. Each band is then recorded as a spectral peak, each of the four bases being represented by a different colour. Sophisticated software can then interpret these peaks directly into a sequence, which is then downloaded into a database.

There are many advantages to the automated system and one significant disadvantage. Without going into detail, automated sequencing allows far more information to be obtained per man hour, perhaps by an order of magnitude. The down side is that at present sequencers of this type are neither cheap to buy nor to run. Cost, at present, imposes a severe limitation on the use of this technology in most phylogenetic studies. However, it is probable that the cost of the hardware and software will reduce and that systems will be developed that are both more user-friendly and more reliable, so reducing the man-hour expenses of maintaining and operating these machines. If so there are many areas of evolutionary biology that are likely to benefit from the wealth of data that could be generated, always assuming that the analytical software can cope.

Apart from allowing assessment of the accuracy of phylogenetic trees and molecular clocks by cross-referencing from different parts of the genome, such data should allow more rigorous investigation of which sequences are fully neutral and which are functionally constrained. The fate of, and changes in, specific genes in different lineages may be tracked, and as a result of advances in developmental genetics this information may shed light on evolutionary changes in gene function. Further, expansion of sequence data sets may provide new insights into the mechanisms and importance of genomic evolution and concerted evolution.

Complete genome mapping projects currently under way are also likely to yield huge amounts of data of use to evolutionary geneticists. The subjects chosen for complete genome mapping include the fruit-fly, *Drosophila melanogaster*, the nematode worm, *Caenorhabditis elegans*, the bacterium, *Escherichia coli*, and humans. While the map for *E. coli* is already near completion, the monumental task of reading all two billion bases in the human genome is going to take some time to complete. Again the spin-offs for evolutionary geneticists are likely to be in greater understanding of genomic evolution. What determines the movement of sequences of bases around the genome? How universal are isochores, and what is the significance of these huge sequence domains? And, again, which parts of the genome are most free to evolve and which are the most constrained?

One final technique deserves mention. The development of PCR has provided molecular geneticists with the power to amplify a product from as little as a single molecule. One of the applications of this new procedure is in mutation analysis. It is no longer essential to have access to large pedigrees. Far more efficient is to use blood (for somatic mutations) or sperm (for germ-line mutations), and to perform single molecule analysis. With the number of PCR reactions which can be set up per day increasing steadily, it is now becoming feasible to look for rare events. By analysing aliquots containing 100 sperm at a time, events as rare as one in a million can be quantified. For the first time we have a window through which to look at the process of mutation, and so at the roots of genetic variation, directly and rapidly.

Given the rate of innovation and discovery, it is already not possible for individual people to feel that they have a detailed knowledge of all aspects of biology that evolution impinges upon, for that is virtually all biology. However, cogent evolutionary biologists must recognise that even if they cannot be knowledgeable in all relevant disciplines, answers to all but the most trivial evolutionary questions are likely to employ techniques and be built on knowledge from a diverse range of fields. Theoretical population geneticists must have some notion of the natural history, the behaviour and the ecology of the organisms they are modelling if the models are to have any relevance. It is no good assuming random mating if the organisms do not mate randomly (and very few do!). Evolutionary ecologists must understand at least the uses and power of the molecular genetic technologies, even if they do not know how to set up a gel, if they are going to apply the most appropriate technique to their problem. Beginning to solve a problem without a knowledge of the tools that are available is usually an inefficient approach. To the same extent, the molecular biologist working away on their own pet gene or DNA sequence should have some notion of its position in the genome, its function, if indeed it has one, the way it interacts with other sequences, the sequences and genes that flank it, and preferably the biology of the whole organism that bears it. It is somewhat unhelpful to discover after investing significant effort, that a sequence that one has chosen to study because it is assumed to be neutral is sitting right next to a major gene that will have been subjected to intense selective pressures in the past.

Evolutionary biologists may be molecular geneticists or naturalists, developmental biologists or ethologists, mathematicians or ecological geneticists.

Whichever they are, there is no doubt that those with some appreciation of the other spheres of endeavour will be rewarded by greater understanding and insights into the captivating processes of evolution. Darwin described a wonder that is appropriate to all serious students of evolution in the final paragraph of *The Origin of Species*:

There is grandeur in this view of life, with its several powers, having been originally breathed into a few forms or into one; and that, whilst this planet has gone on cycling according to the fixed law of gravity, from so simple a beginning endless forms most beautiful and most wonderful have been, and are being, evolved.

Glossary

Acrocentric chromosome: A chromosome with a centromere close to one end, so that the chromosome has arms of very uneven length.

Adaptive landscape: A hypothetical three-dimensional representation of the interaction between genotype frequencies and fitness, in which the vertical axis is fitness. High points are called adaptive peaks and represent regions where all small changes in genotype result in lowered mean fitness of individuals, while valleys are regions of low fitness. Evolution is portrayed by the movement of a population uphill across an ever-changing landscape, searching for adaptive peaks, with its progress being barred by the valleys. An adaptive landscape is very much a conceptual tool rather than an attempt to model reality.

Adaptive radiation: The evolutionary diversification of a monophyletic lineage, leading to an array of types each adapted to particular environmental conditions.

Allele: When a gene or other DNA sequence exhibits variability, each alternative form is known as an allele.

Allopatric speciation: The genetic divergence of two or more geographically separated populations to such a point that they are considered to be taxonomically distinct.

Allopolyploid: Individuals having two chromosome sets, one each from two parent species, which subsequently double to produce an even number of chromosomes.

Altruistic: Any behaviour by which an individual increases the fitness of a conspecific without gaining a fitness benefit itself.

Anagenesis: Evolution as a result of cumulative changes occurring through time within a single evolutionary lineage, and in the absence of any lineage subdivision.

Aneuploidy: Differing by one or a few chromosomes from the number which is normal for the species.

Anisogamy: Anisogamous species are those in which the male and female gametes differ in size.

Arms race: A coevolutionary pattern involving two or more interacting species (e.g. predator and prey or parasite and host), where evolutionary change in one party leads to the evolution of a counter adaptation in the other.

Artificial selection: Human attempts to exaggerate natural traits by breeding selectively from those organisms which show these traits most strongly.

Autopolyploidy: Formation of individuals containing more than two sets of chromosomes of the same species, usually by a doubling of the chromosome number.

Balanced polymorphism: A genetic polymorphism in which the various forms are maintained in the population at constant frequencies, or their frequencies cycle in a more or less regular manner, and both alleles are present above a frequency that could be maintained by recurrent mutation.

Balanced view of the genome: View that most sexual species contain many polymorphic loci, that their genomes exhibit high levels of heterozygosity, and that much of this variation is maintained by a balance of selective advantage and disadvantage. This view was held by many selectionists during the first half of the twentieth century.

Band-sharing coefficient: A pairwise measure of similarity which is often used as a means to estimate relatedness using multibanded DNA fingerprint profiles. It is calculated simply as twice the number of bands in common divided by the total number of bands scored.

Batesian mimicry: The resemblance shown by one species (the mimic) to another (the model) that is better protected by poisonous or distasteful qualities or active defence (sting, bite, etc.).

B-chromosome: Any chromosome occurring in animals or plants that differs from normal chromosomes (autosomes and sex chromosomes) in respect of morphology, number, meiotic behaviour, mitotic behaviour and/or genetic effectiveness.

Biological species concept: A definition of a species as comprising a group of actually or potentially interbreeding natural populations which are reproductively isolated from other such groups.

Biparental inheritance: Genes and genetic elements in sexual organisms which are inherited from both parents (e.g. most nuclear genes) (*cf.* uniparental inheritance).

Centromere: The region on a chromosome that becomes attached to the spindle during cell division.

Character displacement: The divergence of two closely related species for a given trait when they live sympatrically, driven by selection to lower the intensity of interspecific competition.

Chiasma (plural: **chiasmata**): Regions of contact between homologous chromosomes during pairing in meiosis (from late prophase to the start of anaphase I) at which crossing over takes place to effect the exchange of homologous sections of non-sister chromatids.

Chromosomal fission: The splitting into two parts of a single chromosome. If both parts are to be maintained, a new centromere must arise.

Chromosomal fusion: The unification of two chromosomes into a single chromosome with the loss of a centromere.

Chromosome: Individual DNA molecules which act as the largest discrete unit of inheritance. Eukaryotes carry several different linear chromosomes whereas bacteria carry a single circular form.

Cladogenesis: The formation of independently evolving lineages from a single ancestral line through speciation.

Classical view of the genome: View that the genomes of most sexual species are composed of monomorphic loci, with polymorphic loci being relatively rare, these polymorphic loci being the result of occasional mutations which are either being eliminated from the population, or are spreading through the population as a result of selection.

Codon: A sequence of three nucleotide bases along a DNA or RNA chain which represents the code of a single amino acid.

Competitive exclusion principle: The principle that ecologically identical species cannot exist in the same habitat for an indefinite period.

Concerted evolution: Individual repeat units of DNA sequences which are present in multiple copies in the genome are often found to be more similar within a species or population than they are between species or populations. This pattern implies non-independence and is referred to as concerted evolution.

Condition-dependent handicaps: Secondary sexual traits which are potentially disadvantageous to their bearers due to reduced survival chances, but which may confer a reproductive advantage by being the subject of mating preferences. The development of these handicaps is positively correlated to the condition of the individual bearing them, thereby having potential to act as fitness indicators.

Co-operative behaviour: interactions between two individuals in which the inclusive fitness of both is increased.

Cryptic simplicity: A repetitive structure in DNA sequences which is imperfect and hence not obvious. Cryptic simplicity is quantified in terms of the overall excess of particular di-, tri- or tetranucleotide motifs relative to expectations based on random nucleotide order.

Cytoplasmic genes: Genes located in the cytoplasm of cells rather than in the nucleus, usually applied to those associated with mitochondria and chloroplasts.

Cytoplasmic male sterility: The phenomenon in some higher plants whereby a maternally inherited factor prevents the production of viable pollen.

Degeneracy: In the genetic code, since there are 64 possible codons and only 20 amino acids, some amino acids have two or more codons. Degeneracy refers to this lack of a one-to-one correspondence.

Denaturing gradient gel electrophoresis (DGGE): A method for separating DNA molecules by sequence rather than length. In DGGE, DNA fragments are caused to move into an ever greater concentration of chemical denaturing agents. Each fragment effectively stops moving when it begins to 'fray', and the point at which it does this depends on the proportion and order of the four nucleotide bases it contains.

Deoxyribonucleic acid: The principal heritable material of all cells. Chemically it is a polymer of nucleotides, each nucleotide subunit consisting of the pentose sugar 2-deoxy-D-ribose, phosphoric acid and one of the four nitrogenous bases adenine, cytosine, guanine or thymine.

DGGE: *See* denaturing gradient gel electrophoresis.

Diphyletic: A taxonomic group containing two distinct phylogenetic lineages is said to be diphyletic. The better we become at basing taxonomy on accurate phylogenies, the more diphyletic groups will be split into their two constituent, monophyletic groups.

Direct selection: In sexual selection when the mating choice made by an individual (usually the female) leads to an increase in the number of progeny it raises during its lifetime.

Directional selection: Selection which acts on individuals showing a phenotypic distribution in such a way that those individuals towards, or at one end of, the distribution are favoured.

DNA: *See* deoxyribonucleic acid.

DNA fingerprinting: The first genetic method capable of identifying individuals uniquely, based on minisatellite DNA sequences and invented by Prof. Alec Jeffreys. Often used more loosely to refer to any genetic method which identifies individuals with high confidence.

Dominant: Any allele whose phenotypic effect is the same whether present in a heterozygote or a homozygote. By extension, the effect of a dominant allele supplants that of its homologue, which must be recessive.

Effective population size: The effective size of the population is the size of an ideal population which would exhibit equivalent properties with respect to genetic drift.

End labelling: The labelling of a DNA strand by the addition of a radioactively labelled phosphate group to one end.

Endosymbiotic: Organisms which live within the cell or body of another.

Epc: *See* extra-pair copulation.

Epistasis: In terms of phenotype, epistasis is an interaction between two different genes such that the epistatic gene interferes with or blocks the expression of the other (which is referred to as the hypostatic gene). Also used in connection with fitness, where the relative fitnesses of the different genotypes at one locus depend upon the background at another.

Eukaryotes: Single or multicellular organisms in which the cell(s) contains a well-defined nucleus containing multiple linear chromosomes and which is enclosed by a nuclear membrane.

Eusocial: Colonially living organisms in which a proportion of individuals make no personal reproductive contribution, instead helping to raise sibling individuals.

Exon: Most eukaryotic genes are interrupted, comprising alternating coding and non-coding regions. These are called exons and introns respectively.

Extra-pair copulation: Copulation by one of an established pair of individuals with an individual that is not their normal partner.

F statistics: Equations which estimate inbreeding at various levels from the degree of homozygotic excess over and above the expectations of random mating.

Female choice: One of Darwin's two mechanisms of sexual selection. The phenomenon whereby females exercise choice over which of two or more available males they mate with.

Fisherian mechanism of sexual selection: The spread of a female choice gene because of genetic correlation with the preferred male trait, the spread initially owing to a naturally selected advantage, and later being associated with a sexually selected advantage.

Fisherian runaway: The theoretical outcome of Fisher's mechanism of the evolution of female choice, involving a selective feedback loop between a female preference for a male trait and the trait itself. Used as an explanation of extreme male ornamentation, the process runs until limited by the fitness costs of the extreme trait.

Fitness peaks: Genotype–environment combinations in which fitness is maximised such that any small genetic perturbation results in selection for a return to the original genotype (*cf.* adaptive landscape).

Fixation index (F_{ST}): A measure of population subdivision based on the excess of homozygotes in each population sampled over expectations based on a single panmictic population.

Founder effect: Any change in the genetic constitution of a population which results entirely from the process of sampling. Being strongest when the sample size is small, founder effects are usually associated with colonising or founding populations.

Frequency-dependent selection: Selection when the fitness of a genotype or phenotype varies depending on the frequency of that genotype or phenotype in the population.

Gametogenesis: Formation of male and female gametes or sex cells.

Gel electrophoresis: A technique used to detect differences in proteins and polypeptide chains based on differences in the size and electrical charges of the molecules.

Gene conversion: A non-reciprocal interaction between homologous chromosomes which has the effect of replacing a sequence on one homologue with a similar sequence from the other.

Gene flow: Genetic exchange between populations resulting from the dispersal of gametes, zygotes or individuals.

Genetic bottleneck: The reduction in genetic variability in a population resulting from a period of low population size.

Genetic linkage: Genes are said to be linked when they lie on the same chromosome. The term usually refers to an association between two or more genes or alleles caused by co-segregation during meiosis.

Genetic load: The degree to which the average fitness of individuals in a population is decreased, compared with the potential optimal genotype, by the presence and expression of genes that lower fitness.

Genetic markers: Any allele (or sequence) that is used experimentally to identify a sequence, gene, chromosome or individual.

Genetic polymorphism: The occurrence together in space and time of two or more discontinuous forms of a heritable character.

Genome: The entire complement of genetic material in a cell. In eukaryotes this word is sometimes used in referring to the material in just a single set of chromosomes.

Genotype: The genetic make-up of an individual in respect of one genetic locus, a group of loci or even its total genetic complement.

'Good genes' model of sexual selection: The theory whereby a female preference is selected because the preferred trait is carried by males whose genes confer greater than average fitness. Consequently, the gene for the preference tends to be inherited by progeny who will also carry 'good genes' and will spread.

Gradualists: Evolutionists who took the view that most evolutionary change resulted from the action of selection on heritable continuous variation exhibited by individuals in populations (*cf.* mutationists).

Group selection: Selection in which a group of individuals, rather than the individuals themselves, are the unit of selection. This type of selection has been proposed to explain certain types of behaviour such as altruism.

Handicap principle: In sexual selection, the principle that any character chosen by a female as an indicator of quality (genetic or phenotypic) can only be a good indicator of fitness if it can be expressed only by males of high quality (*see also* condition-dependent handicaps).

Haploid: Of cells (particularly gametes) or individuals having a single set of chromosomes.

Haplotype: The genotype of a haploid organism or chromosome. Often used to refer to mitochondrial genotypes.

Haplo-diploid: Life cycles where one sex is haploid and the other diploid.

Haplo-diplontic: Life cycle where haploid and diploid phases are of roughly equal duration.

Hardy–Weinberg equilibrium: The state in which genotype frequencies in a population are in accord with expectations of the H–W law.

Hardy–Weinberg law: A basic principle of population genetics which allows genotypic frequencies to be predicted from gene frequencies. The law states that in an effectively infinite population of sexually reproducing, randomly mating, diploid organisms containing a gene with two alleles having frequencies p and q, the frequencies of the homozygote genotypes will be p^2 and q^2 and that of the heterozygote will be

2pq within a single generation, and will not change thereafter in the absence of mutation, migration and selection.

Hermaphrodite: An individual that bears some tissues that are identifiably male and others that are female, and that can produce mature male and female gametes.

Heterogametic sex: The sex that carries sex chromosomes of different types and thus produces gametes of two types with respect to the sex chromosome they carry. In humans, males are the heterogametic sex, carrying one X and one Y chromosome.

Heterozygote: An individual that bears different alleles of a particular gene, one from each parent.

Heterozygote advantage: The situation in which the fitness of heterozygote individuals exceeds that of either type of homozygote.

Hitch-hiking (of genes): Any gene that increases in frequency as a consequence of being in linkage disequilibrium with a selectively advantageous gene, rather than as a consequence of its own fitness effect on its bearer.

Homogametic sex: The sex which carries sex chromosomes that are the same. In humans, females are the homogametic sex, carrying two X chromosomes.

Homologous: Genes or other sequences are said to be homologous if the similarities they share result from sharing a common ancestor.

Homozygote: An individual that bears two copies of the same allele at a given gene locus.

Hybrid dysgenesis: Germ-line abnormalities which are seen in crosses between different strains or populations. The best examples are found in *Drosophila*, where the phenotype is caused by transposable elements such as the P-element becoming active and jumping in the germ line, causing widespread mutations.

Ideal population: A population in which there is random union of gametes (effectively equivalent to random mating), in which all individuals contribute equally to the next generation and which is of constant size.

Inbreeding coefficient (F_{IS}): A measure of the reduction of heterozygosity in an individual which can be ascribed to inbreeding.

Inclusive fitness: The fitness of an individual plus the effect that individual has on the fitness of other individuals, weighted by their relatedness to it.

Indirect selection: In sexual selection when the mating choice made by an individual (usually the female) increases the fitness of its progeny rather than increasing the individual's own lifetime reproductive success.

Intersexual selection: Selection resulting from variation in the competitive ability of individuals of one sex to attract and be mated by members of the opposite sex.

Interspecific competition: Competition occurring between individuals of different species.

Intragenomic conflict: The situation where the action of one gene in increasing its probability of transmission produces selection for repression of its action.

Intrasexual selection: Selection resulting from variation in the competitive ability of members of one sex (usually males) to gain access to the opposite sex.

Intron: *See* exon.

Inversion: Of a chromosome, in which a segment of the chromosome has been reversed.

Isochore: Eukaryotic chromosomes tend to comprise large domains, several hundred kilobases in length, each with a characteristic nucleotide base composition. Each domain is termed an isochore.

Isogamy: Where gametes which fuse in fertilisation are morphologically identical.

Jumping genes: *See* mobile element.

Karyotype: The chromosome complement, in terms of number, size and constitution, of a cell or an organism.

Kin selection: Selection acting on an individual in favour of the survival, not of that individual, but of its relatives.

Lek: A site where males of a species congregate in a dense group, and which is visited by females for the sole purpose of mating.

Limiting similarity: The degree to which the niches of two different species may overlap without competitive exclusion.

Linkage disequilibrium: When genotypes at two or more loci differ in frequency from random expectations the loci are said to be in linkage disequilibrium. Such a state can arise through the action of natural selection, the existence of population substructure, mutation and non-random mating. Loci may be in linkage disequilibrium whether or not they are on the same chromosome.

Male competition: One of Darwin's two mechanisms of sexual selection. Males compete among themselves (e.g. by aggressive displays or fighting) to gain access to females.

Maximum likelihood: A way of determining which of two or more competing hypotheses (such as alternative phylogenetic trees) yields best fits to the data.

Meiosis: The process of two nuclear divisions and a single replication of the chromosome complement of a cell by which the number of chromosomes in resultant daughter cells is reduced to one-half during gamete production.

Meiotic drive: Any type of abnormal meiosis that results in a deviation from expected Mendelian segregation ratios in heterozygotes.

Meiotic spindle: A collection of microtubular fibres to which chromatids attach (at the centromere) during meiosis and which are involved in the segregation of chromatids into daughter cells.

Metacentric chromosome: A chromosome with a centromere in the middle region.

Metaphase plate: The equatorial region of a dividing cell where the chromosomes line up during metaphase.

Microsatellite: A block of short tandemly repeated motifs in which the basic repeat unit is five or fewer bases in length.

Minisatellite: Any of a group of many dispersed arrays of short tandemly repeated motifs. The family is to an extent unified by the presence of a so-called core sequence, possibly related to the bacterial recombination signal chi, in every repeat unit. Minisatellites can exhibit extreme length due to the frequent gain and loss of repeat units.

Mitochondrial DNA: The circular, double-stranded genome of eukaryotic mitochondria.

Mitosis: The normal process of cell division in growth, involving the replication of chromosomes, and the division of the nucleus into two, each with the identical complement of chromosomes to the original cell.

Mobile element: A DNA sequence which is capable of moving from one position in the genome to another via some form of precise recombination event.

Modifier gene: A gene that interacts with another, modifying its phenotypic expression.

Molecular clock: A theoretical clock based on the premise that the rates at which nucleotide (or amino acid) substitutions become fixed in evolutionary lineages is approximately constant for a given DNA sequence (or polypeptide chain) and reflects the time since taxa diverged.

Molecular drive: The theoretical mechanism by which multigene families evolve in a concerted manner, involving any of a variety of processes of genomic turnover (e.g. transposition, unequal exchange, gene conversion). Molecular drive is thought to operate independently of both selection and drift (*cf.* concerted evolution).

Monogamy: Situation in which an individual has a single mating partner over a set time period.

Monomorphic: The opposite of polymorphic, a sequence, gene or organism in which all individuals in a given sample are indistinguishable.

Monophyletic: A taxonomic group whose members have all evolved from a single ancestor.

mtDNA: *See* mitochondrial DNA.

'Muller's ratchet': The theoretical proposition that populations without recombination will tend to accumulate slightly deleterious mutations.

Mutation: Any change which alters the identity or order of nucleotide bases within a chromosome.

Mutational load: The load of selective deaths in a population resulting from the eradication from the population of deleterious mutations by selection.

Mutationists: Evolutionists who took the view that most adaptations and species arose as the result of major mutations (*cf.* selectionists, gradualists).

Mutualistic interactions: Interaction between two organisms in which the lifetime reproductive success of both is increased. Often applied to interactions between members of different species.

Natural selection: According to Charles Darwin, the main mechanism giving rise to evolution. The mechanism by which heritable traits which increase an organism's chances of survival and reproduction are more likely to be passed on to the next generation than less advantageous traits.

Neutral mutation theory of molecular evolution: The theory, associated with Mitoo Kimura, whereby the majority of mutations that occur are selectively neutral or effectively so, and evolution at the molecular level consists mainly of the chance fixation of neutral mutations through random genetic drift.

Neutral theory: *See* neutral mutation theory of molecular evolution.

Neo-Darwinian synthesis: A development of Darwin's evolutionary theory refined by incorporating modern biological knowledge, particularly Mendelian genetics, during the mid-twentieth century.

Nucleotide base: The structural unit of a nucleic acid. The major nucleotide bases in DNA are the purines adenine and guanine and the pyrimidines cytosine and thymine, the last of which is replaced by uracil in RNA.

Null alleles: Any allele which causes the absence of the normal gene product at the molecular level, or which causes the absence of a normal trait at the phenotypic level.

Operational sex ratio: The ratio of sexually active males to sexually active females in a population at any one moment in time.

Overdominance: *See* heterozygote advantage.

Panmictic population: A population in which individuals mate at random.

Paracentric inversion: A chromosomal inversion which does not include a centromere in the inverted segment.

Parapatric speciation: Speciation that occurs as a result of the divergence of two or more populations that occupy adjacent areas.

Parsimony: The use of the smallest number of evolutionary steps in the construction of a phylogenetic tree.

Parthenogenesis: Reproduction in which eggs develop without fertilisation by a male gamete.

PCR: *See* polymerase chain reaction.

Pericentric inversion: A chromosomal inversion which includes a centromere in the inverted segment.

Phenotype: The observable properties of an organism, resulting from the interaction between the organism's genotype and the environment in which it develops.

Phylogenetic tree: A diagrammatic representation of genetic distances between populations, species or higher taxa, the branching of which is said to resemble a tree.

Physical linkage: An association between two or more genes due to their being on the same chromosome.

Plasmid: An autonomously replicating, double-stranded DNA molecule found usually in bacterial cells. Most plasmids are circular, although a few linear plasmids are known. Some plasmids are essential to their host cells.

Plastid: Any of the specialised pigment-containing organelles of plant cells that can synthesise and store substances.

Pleiotropy: The situation in which a single gene effects two or more apparently unrelated phenotypic traits of an organism.

Polyandry: Situation in which males have a single mate while individual females may have more than one mate within a breeding season.

Polygamy: When organisms of either or both sexes have more than one mating partner within a single breeding season.

Polygenic: Of traits, whose phenotypic expression is controlled by many genes, none of which has an over-riding effect.

Polygyny: Situation in which females have a single mate while individual males may have more than one mate within a breeding season.

Polymerase chain reaction: A method of amplifying specific DNA sequences by means of repeated rounds of primer-directed DNA synthesis.

Polymorphism: The presence of several forms of a trait, gene or DNA sequence in a population. *See* genetic polymorphism.

Polyploid: Of individuals or cells, having more than two complete chromosome sets (e.g. triploid = three sets, tetraploid = four sets, hexaploid = six sets, etc.).

Polytene: Of interphase chromosomes occurring in various cell types of some organisms, which have replicated repeatedly and are paired lengthways to produce a thick chromosome with bands which are visible under the light microscope.

Post-zygotic reproductive isolation mechanism: Any characteristic of an organism that prevents gene flow between it and members of other species by causing sterility, decreased viability, and/or decreased fitness in zygotes formed as a result of fertilisation between gametes of the organism and those of another species.

Pre-zygotic reproductive isolation mechanism: Any characteristic of an organism that prevents gene flow between it and members of other species by reducing the likelihood of zygote formation, usually by decreasing the likelihood of copulation between the organism and a member of another species.

Profile: Usually used to describe the complex pattern of electrophoretic bands in a DNA fingerprint. More recently usage has spread to include the outputs from any DNA typing which uses highly variable markers.

Prokaryotes: Single-celled organisms which lack a membrane-bound nucleus or any membrane-bound organelles.

Protein isozymes: Electrophoretically distinguishable forms of a single protein, coded for by a single genetic locus.

Pseudoautosomal region: Of sex chromosomes, the region of homology between the X and Y chromosomes. Genes in this region behave in the same way as autosomal genes.

Pseudogene: A gene showing significant sequence homology (75+ per cent) to a functional gene, but which has lost any normal function, often through gaining internal stop codons.

Random amplified polymorphic DNA: Use of non-specific PCR reactions to uncover genetic variability.

Random genetic drift: Fluctuation in the frequencies of neutral genes and neutral alleles in a population due the fact that each generation is only a sample of the one it replaces.

RAPD: *See* random amplified polymorphic DNA.

Recessive: A recessive allele is not expressed phenotypically when present in a heterozygote, but only when in a homozygote. The opposite of dominant.

Recombination: The generation of new combinations of chromosomes and parts of chromosomes during meiosis through both crossing-over and the independent segregation of chromosomes.

Red Queen: In *Alice in Wonderland* the Red Queen was famous for stating that one had to keep running in order to remain in the same place. Ecological arms races and other examples of coevolution are regarded as being similar, in that there is no obvious destination, yet there is constant change.

Reinforcement: The evolution of reproductive isolating mechanisms (usually prezygotic) between two partially reproductively isolated populations through selection to lower the frequency of hybrid matings.

Reproductive isolation: The complete cessation of gene flow between two groups of organisms. Used in sexually reproducing organisms as a measure of whether speciation has occurred.

Restriction enzymes: Any enzyme which recognises a specific DNA sequence, usually four or six nucleotides in length, and cleaves the DNA at or near the recognition site.

Restriction fragment length polymorphism: Detection of DNA sequence variability by the use of one or more restriction enzymes followed by electrophoresis. Any mutations which change the spacing of cleavage sites will be detected.

RFLP: *See* restriction fragment length polymorphism.

Ribonucleic acid: A polynucleotide consisting of a chain of sugar and phosphate units to which are attached various nitrogenous bases (adenine, cytosine, guanine and uracil). This macromolecule has many diverse functions: where DNA is the reference library, RNA includes a range of individual books, summaries, pamphlets and photocopied sheets.

RNA: *See* ribonucleic acid.

RNA mediated events: Any of a poorly understood range of mechanisms by which information from an RNA sequence becomes transcribed back into the DNA of the genome.

Robertsonian change: An alteration in chromosomal structure resulting from chromosomal fusion or fission.

Secondary sexual characters: Characters which differ between the sexes, excluding the primary sex organs.

Segregation distorter: Any genetic mechanism that leads to a systematic bias in the representation of one or other allele of a heterozygote in the functional gametes produced by an individual.

Segregational load: The load imposed on a sexual population as a result of overdominance, where segregation among favoured heterozygotes provides a continual source of less-fit homozygotes.

Selectionists: Evolutionists who took the view that most evolutionary change resulted from the action of natural selection on small heritable differences between the individuals in populations (*cf.* mutationists).

Sensory exploitation hypothesis: The hypothesis that mate choice (particularly female choice) may evolve out of a pre-existing bias in the sensory biology of the organism.

Sex limited: Of genetically controlled traits which have their phenotypic expression limited to one sex, irrespective of whether the gene(s) concerned are situated on autosomes or sex chromosomes.

Sex role reversal: Of species in which the operational sex ratio is female biased.

Sexual dimorphism: The existence, within a species, of morphological differences between the sexes.

Sexual selection: Selection which promotes traits that will increase an organism's success in mating and ensuring that its gametes are successful in fertilisation. This is distinct from natural selection which acts simply on traits which influence fecundity and survival.

'Sexy sons': According to Fisher's theory of the evolution of female choice, females who choose to mate with males carrying a particular trait will produce 'sexy sons', progeny who tend to carry the preferred trait and so be attractive to other females with the same mating preference.

Single-strand conformation polymorphism: A method for detecting small differences between otherwise identical fragments of DNA, usually fragments which are less than 300 base pairs long. The method relies on allowing single-stranded DNA to assume three-dimensional secondary structures whose form depends critically on the precise sequence. Conformational changes alter electrophoretic mobility and can therefore be quantified.

Slippage: Misalignment of the two DNA strands in a helix associated with short, tandemly repeated sequences, rather like when a shirt is buttoned up incorrectly.

Sperm competition: The competition between sperm from two or more males, within a single female, for fertilisation of the ova.

SSCP: *See* single-strand conformation polymorphism.

Stable polymorphism: *See* balanced polymorphism.

Stringency: Of complementary DNA strands given an opportunity to pair. High stringency implies that the conditions are at or near the point at which only completely complementary strands will pair. Low stringency conditions (usually associated with a lower temperature) are those where mismatching base pairs do not prevent pairing.

Substitution: A mutation involving the replacement of one nucleotide base in a DNA sequence by another.

Substitutional load: The load of selective deaths imposed upon a population during the replacement of an older allele by a novel advantageous allele.

Super-gene: When a complex set of related traits are controlled by genes which lie so close to each other on the same chromosome that they behave as a single unit.

Switch gene: Any gene which controls the expression of one or more other genes in an on or off manner.

Sympatric speciation: The splitting of one lineage into two where the ranges of the two species overlap at the point of formation.

Synapsis: Chromosome pairing during meiosis.

Tangled bank: A theory of the evolution of sex based on the advantages of diversity of offspring in an environment with many niches and strong competition for resources.

Telomere: The structure at the ends of linear chromosomes that confers structural stability on the chromosome, generally reducing the incidence of terminal loss of material or fusion with other chromosomes or chromosomal fragments.

Trait groups: The situation where a population is structured into small, integral, short-lived groups of organisms.

Transition: A nucleotide base substitution involving the replacement of a purine by a purine or a pyrimidine by a pyrimidine.

Translocation: A chromosomal mutation characterised by the change in position of chromosome segments within the chromosome complement.

Transposition: A change in the position of a chromosomal segment without reciprocal exchange.

Transversion: A nucleotide base substitution mutation involving the replacement of a purine by a pyrimidine, or the reverse.

Twofold cost of sex: The difference in growth rate between a sexual population where one sex does not provision offspring and a comparable asexual population.

Ultra-selfish genes: Genes whose spread and maintenance occurs despite and because they cause damage to the individual in which they occur.

Unequal exchange: The effect of a crossing-over between misaligned chromosomes, usually between tandemly repeated elements. The result is two chromosomes, one carrying a duplication and the other having this section deleted.

Uniparental inheritance: In eukaryotes, the transmission of genetic elements (particularly those of organelles such as mitochondria, ribosomes and chloroplasts) from only one parent (*cf.* biparental inheritance).

Variable number of tandem repeats (VNTR): Short tandemly repeated sequences are prone to slippage, which can lead to high frequencies of gain and loss of repeat units. Such arrays show high levels of length polymorphism. VNTR was coined as an all-embracing term for sequences of this type (*cf.* minisatellite, microsatellite, slippage).

VNTRs: *See* variable number of tandem repeats.

Wahlund effect: The deficiency of heterozygotes in subdivided populations relative to Hardy–Weinberg expectations based on a single large population.

Zygote: The first cell of a new organism resulting from the fusion of two gametes.

References

ADACHI, J. and HASEGAWA, M. (1995). Phylogeny of whales: dependence of the inference on species sampling. *Molecular Biology and Evolution*, **12**, 177–179.

AGULNIK, S. I., AGULNIK, A. I. and RUVINSKY, A. O. (1991). Meiotic drive in female mice heterozygous for hsr inserts on chromosome 1. *Genetic Research*, **55**, 97–100.

ALMEIDA, J., CARPENTER, R., ROBBINS, T. P., MARTIN, C. and COEN, E. S. (1989). Genetic interactions underlying flower colour patterns in *Antirrhinum majus*. *Genes and Development*, **3**, 1758–1767.

ALTMANN, J., HAUSFATER, G. and ALTMANN, S. A. (1988). Determinants of reproductive success in savannah baboons. In Clutton-Brock, T. H. (ed.) *Reproductive success*, pp. 403–418. University of Chicago Press, Chicago.

AMOS, W. (1992). Analysis of polygamous systems using DNA fingerprinting. *Symposium of the Zoological Society of London*, **64**, 151–165.

AMOS, W. (1993). Use of molecular probes to analyse pilot whale pod structure: two novel analytical approaches. *Symposium of the Zoological Society of London*, **66**, 33–48.

AMOS, W. and PEMBERTON, J. P. (1992). DNA fingerprinting in non-human populations. *Current Opinions in Genetics and Development*, **2**, 857–860.

AMOS, W., BARRETT, J. A. and DOVER, G. A. (1991). Breeding behaviour of pilot whales revealed by DNA fingerprinting. *Heredity*, **67**, 49–55.

AMOS, W., SCHLÖTTERER, C. and TAUTZ, D. (1993a). Social structure of pilot whales revealed by analytical DNA typing. *Science*, **260**, 670–672.

AMOS, W., TWISS, S. S., POMEROY, P. P. and ANDERSON, S. S. (1993b). Male mating success in the grey seal, *Halichoerus grypus*: a study using DNA fingerprinting. *Proceedings of the Royal Society of London, Series B*, **252**, 199–207.

ANTONOVICS, J. (1968). Evolution in closely adjacent plant populations. V. Evolution of self fertility. *Heredity*, **23**, 219–238.

ARAK, A. and ENQUIST, M. (1993). Hidden preferences and the evolution of signals. *Philosophical Transactions of the Royal Society of London B*, **340**, 207–213.

ÁRNASON, Ú., ALLERDICE, P. W., LIEN, J. and WIDEGREN, B. (1988). Highly repetitive DNA in the baleen whale genera *Balaenoptera* and *Megaptera*. *Journal of Molecular Evolution*, **27**, 217–221.

ÁRNASON, Ú. and GULLBERG, A. (1994). Relationship of baleen whales established by cytochrome b gene sequence comparison. *Nature*, **367**, 726–728.

AVISE, J. C., BOWEN, B. W. and LAMB, T. (1989). DNA fingerprints from hypervariable mitochondrial genotypes. *Molecular Biology and Evolution*, **6**, 258–269.

AVISE, J. C., BOWEN, B. W., LAMB, T., MEYLAN, B. and BERMINGHAM, E. (1992). Mitochondrial DNA evolution at a turtle's pace: evidence for low genetic variability and reduced microevolutionary rate in the Testudinates. *Molecular Biology and Evolution*, **9**, 457–473.

AXELROD, R. (1984). *The evolution of cooperation*. Basic Books, New York.

AXELROD, R. and HAMILTON, W. D. (1981). The evolution of cooperation. *Science*, **211**, 1390–1396.

AYALA, F. J. and J. A. KIGER (1980). *Modern Genetics*. Benjamin Cummings, Menlo Park, CA.

AYLIFFE, M. A., LAWRENCE, G. J., ELLIS, J. G. and PRYOR, A. J. (1994). Heteroduplex molecules formed between allelic sequences cause non-parental RAPD bands. *Nucleic Acids Research*, **22**, 1632–1636.

BAILEY, W. J., HAYASAKA, K., SKINNER, C. G. *et al.* (1992). Reexamination of the African hominid trichotomy with additional sequences from the primate β-globin gene cluster. *Molecular Phylogenetics and Evolution*, **1**, 97–135.

BAKER, C. S., PALUMBI, S. R., LAMBERTSEN, R. H., WEINRICH, M. T., CALAMBOKIDIS, J. and O'BRIEN, S. J. (1990). Influence of seasonal migration on geographic distribution of mitochondrial DNA haplotypes in humpback whales. *Nature*, **344**, 238–240.

BAKER, H. G. (1959). Reproductive methods as factors in speciation in flowering plants. *Cold Spring Harbor Symposia on Quantitative Biology*, **24**, 177–191.

BAKKER, T. C. M. and POMIANKOWSKI, A. (1995). The genetic basis of female mate preferences. *Journal of Evolutionary Biology*, **8**, 129–171.

BALLARD, J. W. O. and KREITMAN, M. (1994). Unravelling selection in the mitochondrial genome of *Drosophila*. *Genetics*, **138**, 757–772.

BALMFORD, A., ALBON, S. and BLAKEMAN, S. (1992). Correlates of male mating success and female choice in a lek-breeding antelope. *Behavioural Ecology*, **3**, 112–123.

BANCHS, I., BOSCH, A., GUIMERÀ, J., LÁZARO, C., PUIG, A. and ESTIVILL, X. (1994). New alleles at microsatellite loci in CEPH families mainly arise from somatic mutations in the lymphoblastoid cell lines. *Human Mutation*, **3**, 365–372.

BARLOW, D. P., STÖGER, R., HERRMANN, B. G., SAITO, K. and SCHWEIFER, N. (1991). The mouse insulin-like growth-factor type-2 receptor is imprinted and closely linked to the *tme* locus. *Nature*, **349**, 84–87.

BARRETT, J. A. (1988). Frequency-dependent selection in plant–fungal interactions. *Philosophical Transactions of the Royal Society of London, Series B*, **319**, 473–483.

BASOLO, A. L. (1990). Female preference predates the evolution of the sword in swordtail fish. *Science*, **250**, 808–810.

BASOLO, A. L. (1995). Phylogenetic evidence for the role of a pre-existing bias in sexual selection. *Proceedings of the Royal Society of London B*, **259**, 307–311.

BATES, H. W. (1862). Contributions to an insect fauna of the Amazonian Valley. Lepidoptera: Heliconidae. *Transactions of the Linnaean Society of London*, **23**, 495–566.

BATESON, W. (1894). *Materials for the study of variation*. Macmillan, London.

BAUR, E. (1911). *Einfuhrung in die Vererbungslehre*. Borntrager, Berlin.

BECKMANN, J. S. and WEBER, J. L. (1992). Survey or rat and human microsatellites. *Genomics*, **12**, 627–631.

BEEMAN, R. W., FRIESEN, K. S. and DENELL, R. E. (1992). Maternal-effect selfish genes in flour beetles. *Science*, **256**, 89–92.

BELL, G. (1982). *The Masterpiece of Nature: the evolution and genetics of sexuality*. University of California Press, San Francisco.

BELYAEV, D. K. (1979). Destabilizing selection as a factor in domestication. *Journal of Heredity*, **70**, 301–308.

BERITASHVILI, D. R., SYROKVASHEVA, E. Y., MAZURCHUK, S. A. and RYSKOV, A. P. (1989). DNA fingerprinting of individual yeast chromosomes. *Dokl. Akad. Nauk USSR*, **305**, 457–459 (in Russian).

BERNARDI, G. (1993). The vertebrate genome: isochores and evolution. *Molecular Biology and Evolution*, **10**, 186–204.

BERNARDI, G., OLOFSSON, B., FILIPSKI, J. *et al.* (1985). The mosaic genome of warm-blooded vertebrates. *Science*, **228**, 953–957.

BERTA, P., HAWKINS, J. R., SINCLAIR, A. H. *et al.* (1990). Genetic evidence equating *SRY* and the testis determining factor. *Nature*, **348**, 448–450.

BEUKEBOOM, L. W. (1994). Bewildering Bs: an impression of the 1st B-chromosome conference. *Heredity*, **73**, 328–336.

BIRKHEAD, T. R. and MOLLER, A. (1992). *Sperm competition in birds: evolutionary causes and consequences*. Academic Press, London.

BLACK, D. M., JACKSON, M. S., KIDWELL, M. G. and DOVER, G. A. (1987). KP elements repress P-induced hybrid dysgenesis in *Drosophila melanogaster*. *EMBO Journal*, **6**, 4125–4135.

BLAIR, W. F. (1995). Mating call and stage of speciation in *Microhyla olivacea–M. carolinensis* complex. *Evolution*, **9**, 469–480.

BLAKESLEE, A. F. (1934). New Jimson weeds from old chromosomes. *Journal of Heredity*, **25**, 80–108.

BLANCHETOT, A. (1992). DNA fingerprinting analysis in the solitary bee *Megachile rotundata* – variability and nest mate genetic relationships. *Genome*, **35**, 681–688.

BLOUIN, M. S., DAME, J. B., TARRANT, C. A. and COURTNEY, C. H. (1992). Unusual population genetics of a parasitic nematode: mtDNA variation within and among populations. *Evolution*, **46**, 470–476.

BODMER, M. and ASHBURNER, M. (1984). Conservation and change in the DNA sequences coding for alcohol dehydrogenase in sibling species of *Drosophila*. *Nature*, **309**, 425–430.

BORGIA, G. and COLLIS, K. (1989). Female choice for parasite-free male satin bowerbirds and the evolution of bright male plumage. *Behavioural Ecology and Sociobiology*, **25**, 445–459.

BOWCOCK, A. M., RUIZ-LINARES, A., TOMFOHRDE, J., MINCH, E., KIDD, J. R. and CAVALLI-SFORZA, L. L. (1994). High resolution trees with polymorphic microsatellites. *Nature*, **368**, 455–457.

BOWERS, M. A. and BROWN, J. H. (1982). Body size and coexistence in desert rodents: chance or community structure? *Ecology*, **63**, 391–400.

BRAKEFIELD, P. M. (1987). Industrial melanism: do we have the answers? *Trends in Ecology and Evolution*, **2**, 117–122.

BRIDGES, C. B. (1936). The "bar" gene, a duplication. *Science*, **83**, 210–211.

BRONOWSKI, J. (1973). *The ascent of man*. British Broadcasting Corporation, London.

BROWN, W. M., GEORGE, M. J. and WILSON, A. C. (1979). Rapid evolution of animal mitochondrial DNA. *Proceedings of the National Academy of Sciences USA*, **76**, 1967–1971.

BROWN, W. M., PRAGER, E. M., WANG, A. and WILSON, A. C. (1982). Mitochondrial DNA sequences in primates: tempo and mode of evolution. *Journal of Molecular Evolution*, **18**, 225–239.

BRUFORD, M. W. and WAYNE, R. K. (1993). Microsatellites and their application to population genetic studies. *Current Opinions in Genetics and Development*, **3**, 939–943.

BULL, J. J., MOLINEUX, I. J. and WERREN, J. H. (1992). Selfish genes. *Science*, **256**, 65.

BURKE, T. (1989). DNA fingerprinting and other methods for the study of mating success. *Trends in Ecology and Evolution*, **4**, 139–144.

BURKE, T., DAVIES, N. B., BRUFORD, M. W. and HATCHWELL, B. J. (1989). Parental care and mating behaviour of polyandrous dunnocks *Prunella modularis* related to paternity by DNA fingerprinting. *Nature*, **338**, 249-251.

BUSH, G. L. (1969). Sympatric host race formation and speciation in frugivorous flies of the genus *Rhagoletis* (Diptera: Tephritidae). *American Naturalist*, **103**, 669–672.

BUSH, G. L. (1994). Sympatric speciation in animals: new wine in old bottles. *Trends in Ecology and Evolution*, **9**, 285–288.

BUTLIN, R. K. (1987). Speciation by reinforcement. *Trends in Ecology and Evolution*, **2**, 8–13.

BUTLIN, R. K. (1989). Reinforcement of premating isolation. In Otte, D. and Endler, J. A. (eds) *Speciation and its consequences*, pp. 158–179. Sinauer, Sunderland, MA.

CAIN, A. J. and CURREY, J. D. (1968). Climate and selection of banding morphs in *Cepaea* from the climatic optimum to the present day. *Philosophical Transactions of the Royal Society of London B*, **253**, 483–498.

CAIN, A. J. and SHEPPARD, P. M. (1950). Selection in the polymorphic land snail *Cepaea nemoralis* (L.). *Heredity*, **4**, 275–294.

CAIN, A. J. and SHEPPARD, P. M. (1952). The effects of natural selection on body colour in the land snail *Cepaea nemoralis*. *Heredity*, **6**, 217–231.

CAIN, A. J. and SHEPPARD, P. M. (1954). Natural selection in *Cepaea*. *Genetics*, **39**, 89–116.

CAIRNS, J., OVERBAUGH, J. and MILLER, S. (1988). The origin of mutants. *Science*, **335**, 142–145.

CALLEN, D. F., THOMPSON, A. D., SHEN, Y. *et al.* (1993). Incidence and origins of 'null' alleles in (AC)n microsatellite markers. *American Journal of Human Genetics*, **52**, 922–927.

CAMMARANO, P., PALM, P., CRETI, R., CECCARELLI, E., SANANGELANTONI, A. M. and TIBONI, O. (1992). Early evolutionary relationships among known life forms inferred from elongation factor EF-2/EF-G sequences: phylogenetic coherence and structure of the archaeal domain. *Journal of Molecular Evolution*, **34**, 396–405.

CANN, R. L., STONEKING, M. and WILSON, A. C. (1987). Mitochondrial DNA and human evolution. *Nature*, **325**, 31–36.

CARAYON, J. (1974). Insémination traumatique heterosexuelle et homosexuelle chez *Xylocoris maculipennis* (Hem. Anthocoridae). *C. R. Acad. Sci. Paris D*, **278**, 2803–2806.

CARPENTER, G. D. H. (1913). The inheritance of small variations in the pattern of *Papilio dardanus* Brown. *Transactions of the Entomological Society of London*, **1913**, 656–666.

CAVALIER-SMITH, T. (1978). Nuclear volume control by neuroskeletal DNA, selection for cell volume and the solution of the DNA C-value paradox. *Journal of Cell Science*, **34**, 247–278.

CAVALIER-SMITH, T. (1987). The origin of eukaryote and archaebacterial cells. *Annals of the New York Academy of Science*, **503**, 17–54.

CHAPMAN, T., LIDDLE, L. F., KALB, J. M., WOLFNER, M. F. and PARTRIDGE, L. (1995). Cost of mating in *Drosophila melanogaster* females is mediated by male accessory gland products. *Nature*, **373**, 241–244.

CHARLESWORTH, B. (1978). Model for evolution of Y chromosomes and dosage compensation. *Proceedings of the National Academy of Sciences USA*, **75**, 5618–5622.

CHARLESWORTH, B. (1979). Evidence against Fisher's theory of dominance. *Nature*, **278**, 848–849.

CHARLESWORTH, B. (1991). The evolution of sex chromosomes. *Science*, **251**, 1030–1033.

CHARLESWORTH, B. and GANDERS, F. R. (1979). The population genetics of gynodioecy with cytoplasmic-genic male-sterility. *Heredity*, **43**, 213–218.

CHESLEY, P. and DUNN, L. C. (1936). The inheritance of taillessness (anury) in the house mouse. *Genetics*, **21**, 525–536.

CLARKE, C. A. (1963). Interactions between major genes and polygenes in the determination of mimetic patterns of *Papilio dardanus*. *Evolution*, **17**, 404–413.

CLARKE, C. A. and SHEPPARD, P. M. (1959a). The genetics of some mimetic forms of *Papilio dardanus* Brown and *Papilio glaucus* Linn. *Journal of Genetics*, **56**, 236–260.

CLARKE, C. A. and SHEPPARD, P. M. (1959b). The genetics of *Papilio dardanus* Brown. I. Race *cenea* from South Africa. *Genetics*, **44**, 1347–1358.

CLARKE, C. A. and SHEPPARD, P. M. (1960a). The genetics of *Papilio dardanus* Brown. II. Races *dardanus, polytrophus, meseres* and *tibullus*. *Genetics*, **45**, 439–457.

CLARKE, C. A. and SHEPPARD, P. M. (1960b). The genetics of *Papilio dardanus* Brown. III. Race *antimorii* from Abyssinia and race *meriones* from Madagascar. *Genetics*, **45**, 683–697.

CLARKE, C. A. and SHEPPARD, P. M. (1960c). The evolution of dominance under disruptive selection. *Heredity*, **14**, 73–87.

CLARKE, C. A. and SHEPPARD, P. M. (1962). Disruptive selection and its effect on a metrical character in the butterfly *Papilio dardanus*. *Evolution*, **16**, 214–226.

CLARKE, C. A. and SHEPPARD, P. M. (1971). Further studies on the genetics of the mimetic butterfly *Papilio memnon*. *Philosophical Transactions of the Royal Society B*, **263**, 35–70.

CLARKE, C. A., SHEPPARD, P. M. and THORNTON, I. W. B. (1968). The genetics of the mimetic butterfly *Papilio memnon*. *Philosophical Transactions of the Royal Society B*, **254**, 37–89.

CLARKE, S. C. A., MANI, G. S. and WYNNE, G. (1985). Evolution in reverse: clean air and the peppered moth. *Biological Journal of the Linnaean Society*, **26**, 189–199.

CLUTTON-BROCK, T. H. and PARKER, G. A. (1995). Punishment in animal societies. *Nature*, **373**, 209–216.

CLUTTON-BROCK, T. H., HARVEY, P. H. and RUDDER, B. (1977). Sexual dimorphism, socionomic sex ratio and body weight in primates. *Nature*, **269**, 797–800.

CLUTTON-BROCK, T. H., ALBON, S. D., and GUINNESS, F. E. (1981). Parental investment in the male and female offspring in polygynous mammals. *Nature*, **289**, 487–489.

CLUTTON-BROCK, T. H., ALBON, S. D. and GUINNESS, F. E. (1984). Maternal dominance, breeding success, and birth sex ratios in red deer. *Nature*, **308**, 358–360.

COEN, E. S. and CARPENTER, R. (1986). Transposable elements in *Antirrhinum majus*: generators of genetic diversity. *Trends in Genetics*, **2**, 292–296.

COLLICK, A. and JEFFREYS, A. J. (1990). Detection of a novel minisatellite-specific DNA-binding protein. *Nucleic Acids Research*, **18**, 625–629.

COLLICK, A., DUNN, M. G. and JEFFREYS, A. J. (1991). Minisatellite binding protein MSBP-1 is a sequence-specific single-stranded DNA-binding protein. *Nucleic Acids Research*, **19**, 6399–6404.

CONDIT, R. and HUBBELL, S. P. (1991). Abundance and DNA sequence of two-base repeat regions in tropical plant genomes. *Genome*, **34**, 66–71.

CONNELL, J. H. (1961). The influence of interspecific competition and other factors on the distribution of the barnacle *Chthamalus stellatus*. *Ecology*, **42**, 710–723.

CONNELL, J. H. (1980). Diversity and the coevolution of competitors, or the ghost of competition past. *Oikos*, **35**, 131–138.

COOK, L. M., LEFÈBVRE, C. and MCNEILLY, T. (1972). Competition between metal tolerant and normal plant populations on normal soil. *Evolution*, **26**, 366–372.

COOLEY, L., KELLEY, R. and SPRADLING, A. (1988). Insertional mutagenesis of the *Drosophila* genome with single P elements. *Science*, **239**, 1121–1128.

COOPER, G. (1995). *Analysis of genetic variation and sperm competition in dragonflies.* DPhil Thesis, Oxford University.

COOPER, K. W. (1950). Normal spermatogenesis in *Drosophila*. In Demerec, M. (ed.) *Biology of Drosophila*. Wiley, New York.

CORRENS, C. (1909). Vererbunsversuche mit blass (gelb) grunen und buntblättrigen sippen bei *Mirabilis jalapa, Urtica pilulifera* und *Lunaria annua. Z. Indukt. Abstamm. Verebuns*, **1**, 291–329.

COSMIDES, L. and TOOBY, J. (1981). Cytoplasmic inheritance and intragenomic conflict. *Journal of Theoretical Biology*, **89**, 83–129.

COWIE, R. H. and JONES, J. S. (1985). Climatic selection on body colour in *Cepaea. Heredity*, **55**, 261–267.

COYNE, J. A. (1992). Genetics and speciation. *Nature*, **355**, 511–515.

COYNE, J. A. and ORR, H. A. (1989). Patterns of speciation in *Drosphila. Evolution*, **43**, 362–381.

CROW, J. F. (1988). The ultraselfish gene. *Genetics*, **118**, 389–391.

CROW, J. F. and KIMURA, M. (1965). Evolution in sexual and asexual populations. *American Naturalist*, **99**, 430–450.

DARWIN, C. R. (1859). *On the Origin of Species by Means of Natural Selection, or the Preservation of Favoured Races in the Struggle for Life*. John Murray, London.

DARWIN, C. (1862). *On the various contrivances by which British and foreign orchids are fertilized by insects*. John Murray, London.

DARWIN, C. R. (1871). *The Descent of Man, and Selection in Relation to Sex*. John Murray, London.

DAVIES, N. B. and BROOKE, M. D. L. (1989a). An experimental study of co-evolution between the cuckoo *Cuculus canorus* and its hosts. I. Host egg discrimination. *Journal of Animal Ecology*, **58**, 207–224.

DAVIES, N. B. and BROOKE, M. D. L. (1989b). An experimental study of co-evolution between the cuckoo *Cuculus canorus* and its hosts. II. Host egg markings, chick discrimination and general discussion. *Journal of Animal Ecology*, **58**, 225–236.

DAWKINS, R. (1982). *The Extended Phenotype*. Oxford University Press, Oxford.

DAWKINS, R. (1990). Parasites, Desiderata lists and the paradox of the organism. *Parasitology*, **100**, S63–S73.

DAWKINS, R. and KREBS, J. R. (1979). Arms races between and within species. *Proceedings of the Royal Society of London, Series B*, **205**, 489–511.

DE CARVALHO, A. B. and KLACZKO, L. B. (1993). Autosomal supressors of sex-ratio in *Drosophila mediopunctata. Heredity*, **71**, 546–551.

DE CARVALHO, A. B., PIEXOTO, A. A. and KLACZKO, L. B. (1989). Sex-ratio in *Drosphila mediopunctata. Heredity*, **62**, 425–428.

DE CHIARA, T. M., EFSTRATIADIS, A. and ROBERTSON, E. J. (1990). A growth deficiency phenotype in heterozygous mice carrying an insulin-like growth factor-II gene disrupted by targetting. *Nature*, **345**, 78–80.

DE CHIARA, T. M., ROBERTSON, E. J. and EFSTRATIADIS, A. (1991). Parental imprinting of the mouse insulin-like growth factor-II gene. *Cell*, **64**, 849–859.

DEBRY, R. W., ABELE, L. G., WEISS, S. H. *et al.* (1993). Dental HIV transmission? *Nature*, **361**, 691.

DEKA, R., SHRIVER, M. D., YU, L. M., ASTON, C. E., CHAKRABORTY, R. and FERRELL, R. (1994). Conservation of human chromosome 13 polymorphic microsatellite (CA)n repeats in chimpanzees. *Genomics*, **22**, 226–230.

DIAMOND, J. M. (1975). Assembly of species communities. In Cody, J. L. and Diamond, J. M. (eds) *Ecology and evolution of communities*. Belknap, Cambridge, MA.

DIGBY, L. (1912). The cytology of *Primula kewensis* and other related *Primula* hybrids. *Annals of Botany*, **26**, 357–388.

DIONNE, F. T., TURCOTTE, L., THIBAULT, M.-C., BOULAY, M. R., SKINNER, J. S. and BOUCHARD, C. (1991). Mitochondrial DNA sequence polymorphism, VO2max, and response to endurance training. *Medicine and Science in Sports and Exercise*, **23**, 177–185.

DI RIENZO, A., PETERSON, A. C., GARZA, J. C., VALDES, A. M. and SLATKIN, M. (1994). Mutational processes of simple sequence repeat loci in human populations. *Proceedings of the National Academy of Sciences USA*, **91**, 3166–3170.

DIVER, C. (1929). Fossil records of Mendelian mutants. *Nature*, **124**, 183.

DOBZHANSKY, T. (1937). *Genetics and the origin of species*. Columbia University Press, New York.

DOBZHANSKY, T. (1947). Genetics of natural populations: XIV. *Genetics*, **32**, 142–160.

DOBZHANSKY, T. (1951). *Genetics and the origin of species*. Columbia University Press, New York.

DOBZHANSKY, T. (1956). Genetics of natural populations, XXV. *Evolution*, **10**, 82–92.

DOBZHANSKY, T. (1961). On the dynamics of chromosomal polymorphism in *Drosophila*. *Symposia of the Royal Entomological Society of London*, **1**, 30–42.

DOBZHANSKY, T. and PAVLOVSKY, O. (1957). An experimental study of interaction between genetic drift and natural selection. *Evolution*, **11**, 311–319.

DOBZHANSKY, T. and SPASSKY, N. (1954). Environmental modification of heterosis in *Drosophila pseudoobscura*. *Proceedings of the National Academy of Science*, **40**, 407–415.

DORIT, R. L., AKASHI, H. and GILBERT, W. (1995). Absence of polymorphism at the *ZFY* locus on the human Y chromosome. *Science*, **268**, 1183–1186.

DOUGLAS, A. E. and SMITH, D. C. (1989). Are endosymbioses mutualistic? *Trends in Ecology and Evolution*, **4**, 350–352.

DOVER, G. A. (1982). Molecular drive: a cohesive mode of species evolution. *Nature*, **299**, 111–117.

DUNFORD, C. (1977). Kin selection for ground squirrel alarm calls. *American Naturalist*, **111**, 782–785.

DUYAO, M., ABROSE, C., MYERS, R. *et al.* (1993). Trinucleotide repeat length instability and age of onset in Huntington's disease. *Nature Genetics*, **4**, 387–392.

EANES, W. F., GAFFNEY, P. M., KOEHN, R. K. and SIMON, C. M. (1977). A study of sexual selection in natural populations of the milkweed beetle, *Tetraopes tetraophthalmus*. In Christiansen, F. B. and Fenchel, T. M. (eds) *Lecture Notes in Biomathematics: 19 Measuring Selection in Natural Populations*. Springer-Verlag, Berlin.

EBERHARD, W. G. (1985). *Sexual selection and animal genitalia*. Harvard University Press, Cambridge, MA.

EDELSTON, R. S. (1864). *The Entomologist*, **2**, 150.

EDWARDS, S. V. (1993). Long-distance gene flow in a cooperative breeder detected in genealogies of mitochondrial DNA. *Proceedings of the Royal Society of London, Series B*, **252**, 177–185.

EHRLICH, A. H. and EHRLICH, P. R. (1978). Reproductive strategies in the butterflies. I. Mating frequency, plugging and egg numbers. *Journal of the Kansas Entomological Society*, **51**, 666–697.

EMLEN, S. T. and ORING, L .W. (1977). Ecology, sexual selection and the evolution of mating systems. *Science*, **197**, 215–223.

ENDLER, J. A. (1977). *Geographic variation, speciation and clines*. Princeton University Press, Princeton, New Jersey.

ENGELS, W. R. (1992). The origin of P elements in *Drosophila melanogaster*. *BioEssays*, **14**, 681–686.

ENQUIST, M. and ARAK, A. (1993). Selection of exaggerated male traits by female aesthetic senses. *Nature*, **361**, 446–448.

ENQUIST, M. and ARAK, A. (1994). Symmetry, beauty and evolution. *Nature*, **372**, 169–172.

EPPLEN, J. T., AMMER, H, EPPLEN, C. *et al.* (1991). Oligonucleotide fingerprinting using simple repeat motifs: a convenient, ubiquitously applicable method to detect hypervariability for multiple purposes. In Burke, T., Dolf, G., Jeffreys, A. J. and Wolff, R. (eds) *DNA Fingerprinting: Approaches and Applications*, pp. 50–69. Birkhauser Verlag, Basel.

ESHEL, I. and FELDMAN, M. W. (1970). On the evolutionary effects of recombination. *Theoretical Population Biology*, **1**, 88–100.

ESTOUP, A., SOLIGNAC, M., HARRY, M. and CORNUET, J.-M. (1993). Characterization of (GT)n and (CT)n microsatellites in two insect species: *Apis mellifera* and *Bombus terrestris*. *Nucleic Acids Research*, **21**, 1427–1431.

EYRE-WALKER, A. (1993). Recombination and mammalian genome evolution. *Proceedings of the Royal Society of London, Series B*, **252**, 237–243.

EYRE-WALKER, A. and BULMER, M. (1993). Reduced synonymous substitution rate at the start of eubacterial genes. *Nucleic Acids Research*, **21**, 4599–4603.

FEDER, J. L., CHILCOTE, C. A. and BUSH, G. L. (1990). The geographic patterns of genetic differentiation between host associated populations of *Rhagoletis pomonella* (Diptera: Tephritidae) in the Eastern United States and Canada. *Evolution*, **44**, 570–594.

FEDER, J. L., OPP, S. B., WLAZLO, B., REYNOLDS, K., GO, W. and SPISAK, S. (1994). Host fidelity is an effective premating barrier between sympatric races of the apple maggot fly. *Proceedings of the National Academy of Sciences USA*, **91**, 7990–7994.

FELSENSTEIN, J. (1974). The evolutionary advantage of recombination. *Genetics*, **78**, 737–756.

FELSENSTEIN, J. (1985a). Confidence limits on phylogenies: an approach using the bootstrap. *Evolution*, **39**, 783–791.

FELSENSTEIN, J. (1985b). Recombination and sex: is Maynard Smith necessary. In Greenwood, H. and Slatkin, M. (eds) *Evolution: Essays in honour of John Maynard Smith*, pp. 209–220. Cambridge University Press, Cambridge.

FINDLAY, C. S., ROCKWELL, R. F., SMITH, J. A. and COOKE, F. (1985). Life history studies of the lesser snow goose (*Anser caerulescens caerulescens*). VI Plumage polymorphism, assortative mating and fitness. *Evolution*, **39**, 904–914.

FISCHER, E. A. (1980). The relationship between mating system and simultaneous hermaphroditism in the coral reef fish. *Animal Behavior*, **28**, 620–633.

FISCHER, S. G. and LERMAN, L. S. (1979). Length-independent separation of DNA restriction fragments in two-dimensional gel electrophoresis. *Cell*, **16**, 191–200.

FISCHER, S. G. and LERMAN, L. S. (1983). DNA fragments differing by single base-pair substitutions are separated in denaturing gradient gels: correspondence with melting theory. *Proceedings of the National Academy of Sciences USA*, **80**, 1579–1583.

FISHER, R. A. (1918). The correlation between relatives on the supposition of Mendelian inheritance. *Transactions of the Royal Society of Edinburgh*, **52**, 399–433.

FISHER, R. A. (1922). On the dominance ratio. *Proceedings of the Royal Society of Edinburgh*, **42**, 321–341.

FISHER, R. A. (1927). On some objections to mimicry theory – statistical and genetic. *Transactions of the Royal Entomological Society of London*, **75**, 269–278.

FISHER, R. A. (1928). The possible modification of the response of the wild type to recurrent mutations. *American Naturalist*, **62**, 115–126.

FISHER, R. A. (1930). *The genetical theory of natural selection*. Dover, New York.

FISHER, R. A. and FORD, E. B. (1947). The spread of a gene in natural conditions in a colony of the moth *Panaxia dominula* L. *Heredity*, **1**, 143–174.

FISHER, R. A. and FORD, E. B. (1950). The 'Sewall Wright' effect. *Heredity*, **4**, 117–119.

FITCH, D. H. A., BUGAJ-GAWEDA, B. and EMMONS, S. W. (1995). 18S ribosomal RNA gene phylogeny for some rhabditidae related to *Caenorhabditis*. *Molecular Biology and Evolution*, **12**, 346–358.

FORD, E. B. (1937). Problems of heredity in the Lepidoptera. *Biological Reviews*, **12**, 461–503.

FORD, E. B. (1940). Polymorphism and taxonomy. In Huxley, J. S. (ed.) *The New Systematics*, pp. 493–513. Clarendon Press, Oxford.

FORD, E. B. (1953). The genetics of polymorphism in the Lepidoptera. *Advances in Genetics*, **5**, 43–87.

FORD, E. B. (1964). *Ecological Genetics*. Methuen, London.

FORD, E. B. (1975). *Ecological Genetics*. Chapman and Hall, London.

FORTERRE, P., NADIA, B.-L. and LABEDAN, B. (1993). Universal tree of life. *Nature*, **362**, 795.

FRANK, S. A. (1991). Divergence of meiotic drive-suppression systems as an explanation for sex biased hybrid sterility and inviability. *Evolution*, **45**, 262–267.

FRYER, J. C. F. (1913). An investigation by pedigree breeding into the polymorphism of *Papilio polytes* Linn. *Philosophical Transactions of the Royal Society of London B*, **204**, 227–254.

GAUSE, G. F. (1934). *The struggle for existence*. Williams and Wilkins, Baltimore.

GEORGHIOU, G. P. and TAYLOR, C. E. (1976). Pesticide resistance as an evolutionary phenomenon. *Proceedings of the 15th International Congress of Entomology*, Washington DC, 759–785.

GERSHENSON, S. (1928). A new sex ratio abnormality in *Drosophila obscura*. *Genetics*, **13**, 488–507.

GIBBS, H. L., WEATHERHEAD, P. J., BOAG, P. T., WHITE, B. N., TABAK, L. M. and HOYSAK, D. J. (1990). Realized reproductive success of polygynous red-winged blackbirds revealed by DNA markers. *Science*, **250**, 1394–1397.

GILBERT, L. E. (1976). Postmating female odor in *Heliconius* butterflies: A male contributed aphrodisiac. *Science*, **193**, 419–420.

GILBURN, A. S., FOSTER, S. P. and DAY, T. H. (1992). Female mating preference for large size in *Coelopa frigida* (seaweed fly). *Heredity*, **69**, 209–216.

GILBURN, A. S., FOSTER, S. P. and DAY, T. H. (1993). Genetic correlation between a female mating preference and the preferred male character in seaweed flies (*Coelopa frigida*). *Evolution*, **47**, 1788–1795.

GLESENER, R. R. and TILMAN, D. (1978). Sexuality and the components of environmental uncertainty: clues from geographic parthenogenesis in terrestrial animals. *American Naturalist*, **112**, 659–673.

GOJOBORI, T., LI, W.-H. and GRAUR, D. (1982). Patterns of nucleotide substitution in pseudo-genes and functional genes. *Journal of Molecular Evolution*, **18**, 360–369.

GOLDSTEIN, D. B., LINARES, A. R., CAVALLI-SFORZA, L. L. and FELDMAN, M. W. (1995). An evaluation of genetic distances for use with microsatellite loci. *Genetics*, **139**, 463–471.

GRANT, P. R. (1984). Recent research on the evolution of land birds of the Galapagos. *Biological Journal of the Linnaean Society of London*, **21**, 113–136.

GRAVES, J., HAY, R. T., SCALLAN, M. and ROWE, S. (1992). Extra-pair paternity in the shag, *Phalacrocorax aristotelis* as determined by DNA fingerprinting. *Journal of the Zoological Society of London*, **226**, 399–408.

GRAY, I. C. and JEFFREYS, A. J. (1991). Evolutionary transience of hypervariable minisatellites in man and the primates. *Proceedings of the Royal Society of London, Series B*, **243**, 241–253.

GRAY, M. W., SANKOFF, D. and CEDERGREN, R. J. (1984). On the evolutionary descent of organisms and organelles: a global phylogeny based on a highly conserved structural core in small subunit ribosomal RNA. *Nucleic Acids Research*, **12**, 5837–5852.

GREIG, J. C. (1979). Principles of genetic conservation in relation to wildlife management in Southern Africa. *South Africa Wildlife Research*, **9**, 57–78.

GRENFELL, B. T., PRICE, O. F., ALBON, S. D. and CLUTTON-BROCK, T. H. (1992). Overcompensation and population cycles in an ungulate. *Nature*, **355**, 823–826.

HADRYS, H., BALICK, M. and SCHIERWATER, B. (1992). Applications of random amplified polymorphic DNA (RAPD) in molecular ecology. *Molecular Ecology*, **1**, 55–63.

HAFNER, M. S. and NADLER, S. A. (1988). Phylogenetic trees support the coevolution of parasites and their hosts. *Nature*, **332**, 258–259.

HAFNER, M. S., SUDMAN, P. D., VILLABLANCA, F. X., SPRADLING, T. A., DEMASTES, J. W. and NADLER, S. A. (1994). Disparate rates of molecular evolution in cospeciating hosts and parasites. *Science*, **265**, 1087–1090.

HAGELBERG, E. (1994). Ancient DNA studies. *Evolutionary Anthropology*, **2(6)**, 199–207.

HAIG, D. and GRAHAM, C. (1991). Genomic imprinting and the strange case of the Insulin-like growth factor II receptor. *Cell*, **64**, 1045–1046.

HANSON, M. R. (1991). Plant mitochondrial mutations and male sterility. *Annual Review of Genetics*, **25**, 461–486.

HALANYCH, K. M. (1991). 5S ribosomal RNA sequences inappropriate for phylogenetic reconstruction. *Molecular Biology and Evolution*, **8**, 249–253.

HALDANE, J. B. S. (1922). Sex ratio and unisexual sterility in hybrid animals. *Journal of Genetics*, **12**, 101–109.

HALDANE, J. B. S. (1932). *The causes of evolution*. Harper, New York.

HALDANE, J. B. S. (1937). The effect of variation on fitness. *American Naturalist*, **71**, 337–349.

HALDANE, J. B. S. (1955). Population genetics. *New Biology*, **18**, 34–51.

HALDANE, J. B. S. (1957). The cost of natural selection. *Journal of Genetics*, **55**, 511–524.

HALL, J. C. and ROSBASH, M. (1988). Genetics and molecular biology of rhythms. *BioEssays*, **7**, 108–112.

HAMILTON, W. D. (1964). The genetical evolution of social behaviour I, II. *Journal of Theoretical Biology*, **7**, 1–52.

HAMILTON, W. D. (1967). Extraordinary sex ratios. *Science*, **156**, 477–488.

HAMILTON, W. D. and ZUK, M. (1982). Heritable true fitness and bright birds: a role for parasites? *Science*, 218, 384–387.

HAMILTON, W. D., AXELROD, R. and TANESE, R. (1990). Sexual reproduction as an adaptation to resist parasites (a review). *Proceedings of the National Academy of Sciences USA*, **87**, 3566–3573.

HANCOCK, J. M. and DOVER, G. A. (1988). Molecular coevolution among cryptically simple expansion segments of eukaryotic 26S/28S rRNAs. *Molecular Biology and Evolution*, **5**, 377–391.

HANCOCK, J. M. and DOVER, G. A. (1990). 'Compensatory slippage' in the evolution of ribosomal RNA genes. *Nucleic Acids Research*, **18**, 5949–5954.

HANCOCK, J. M., TAUTZ, D. and DOVER, G. A. (1988). Evolution of the secondary structures and compensatory mutations of the ribosomal RNAs of *Drosophila melanogaster*. *Molecular Biology and Evolution*, **5**, 393–414.

HARCOURT, A. H., HARVEY, P. H., LARSON, S. G. and SHORT, R. V. (1981). Testis weight, body weight and breeding systems in primates. *Nature*, **293**, 55–57.

HARRISON, J. W. H. (1928). A further induction of melanism in the lepidopterous insect, *Selenia bilunaria* Esp. and its inheritance. *Proceedings of the Royal Society of London B*, **102**, 338–347.

HARSHMAN, L. G. and PROUT, T. (1994). Sperm displacement without sperm transfer in *Drosophila melanogaster*. *Evolution*, **48**, 758–766.

HARVEY, P. H., KAVANAGH, M. and CLUTTON-BROCK. T. H. (1978). Sexual dimorphism in primate teeth. *Journal of Zoology*, **186**, 475–485.

HARVEY, P. H., MAY, R. M. and NEE, S. (1994). Phylogenies without fossils. *Evolution*, **48**, 523–529.

HASEGAWA, M. and HASHIMOTO, T. (1993). Ribosomal RNA trees misleading. *Nature*, **361**, 23.

HASTINGS, I. M. (1992). Population genetic aspects of deleterious cytoplasmic genomes and their effect on the evolution of sexual reproduction. *Genetic Research*, **59**, 215–225.

HEDGES, S. B., KUMAR S., TAMURA, K. and STONEKING, M. (1991). Human origins and analysis of mitochondrial DNA sequences. *Science*, **255**, 737–739.

HENDRICKS, L. E., HUYSMANS, E., VANDENBERGHE, A. and DE WACHTER, R. (1986). Primary structure of the 5S ribosomal RNAs of 11 arthropods and applicability of 5S RNA to the study of metazoan evolution. *Journal of Molecular Evolution*, **24**, 103–109.

HICKEY, W. A. and CRAIG, G. B. J. (1966). Genetic distortion of sex ratio in a mosquito, *Aedes aegypti*. *Genetics*, **53**, 1177–1196.

HILLIS, D. M. and MORITZ, C. (1990). An overview of applications of molecular systematics. In Hillis, D. M. and Moritz, C. (eds) *Molecular systematics*, pp. 508–515. Sinauer, Sunderland, MA.

HILLIS, D. M., HUELSENBECK, J. P. and CUNNINGHAM, C. W. (1994). Application and accuracy of molecular phylogenies. *Science*, **264**, 671–676.

HOELZEL, A. R., HALLEY, J. and O'BRIEN, S. J. (1993). Elephant seal genetic variation and the use of simulation models to investigate historical population bottlenecks. *Journal of Heredity*, **84**, 443–449.

HOLMES, E. C. and GARNETT, G. P. (1994). Genes, trees and infections: molecular evidence in epidemiology. *Trends in Ecology and Evolution*, **9**, 256–260.

HOLMES, W. G. and SHERMAN, P. W. (1982). The ontogeny of kin recognition in two species of ground squirrel. *American Zoologist*, **22**, 491–517.

HONMA, M, YOSHII, T., ISHIYAMA, I., MITANI, K., KOMINAMI, R. and MURAMATSU, M. (1989). Individual identification from semen by the deoxyribonucleic acid (DNA) fingerprint technique. *Journal of Forensic Science*. **34**, 222–227.

HÖSS, M., KOHN, M., PÄÄBO, S., KNAUER, F. and SCHRÖDER, W. (1992). Excrement analysis by PCR. *Nature*, **359**, 199.

HOUCK, M. A., CLARK, J. B., PETERSON, K. R. and KIDWELL, M. G. (1991). Possible horizontal transfer of *Drosophila* genes by the mite *Proctolaelaps regalis. Science*, **253**, 1125–1129.

HOWARD, R. S. and LIVELY, C. M. (1994). Parasitism, mutation accumulation and the maintenance of sex. *Nature*, **367**, 554–557.

HOWLETT, R. J. and MAJERUS, M. E. N. (1987). The understanding of industrial melanism in the peppered moth (*Biston betularia*) (Lepidoptera: Geometridae). *Biological Journal of the Linnaean Society*, **30**, 31–44.

HRDY, S. B. (1977). *The langurs of Abu: female and male strategies of reproduction.* Harvard University Press, Cambridge, MA.

HURST, G. D. D. and MAJERUS, M. E. N. (1993). Why do maternally inherited microorganisms kill males? *Heredity*, **71**, 81–95.

HURST, G. D. D., MAJERUS, M. E. N. and WALKER, L. E. (1992). Cytoplasmic male killing elements in *Adalia bipunctata* (Linnaeus) (Coleoptera: Coccinellidae). *Heredity*, **69**, 84–91.

HURST, G. D. D., MAJERUS, M. E. N. and WALKER, L. E. (1993). The importance of cytoplasmic male killing elements in natural populations of the two spot ladybird, *Adalia bipunctata* (Linnaeus) (Coleoptera: Coccinellidae). *Biological Journal of the Linnaean Society of London*, **49**, 195–202.

HURST, L. D. (1990). Parasite diversity and the evolution of diploidy, multicellularity and anisogamy. *Journal of Theoretical Biology*, **144**, 429–443.

HURST, L. D. (1991). The incidences and evolution of cytoplasmic male killers. *Proceedings of the Royal Society of London, Series B*, **244**, 91–99.

HURST, L. D. (1993). *scat*$^+$ is a selfish gene analogous to *medea* in *Tribolium castaneum. Cell*, **75**, 407–408.

HURST, L. D. and HAMILTON, W. D. (1992). Cytoplasmic fusion and the nature of sexes. *Proceedings of the Royal Society of London, Series B*, **247**, 189–194.

HURST, L. D. and POMIANKOWSKI, A. N. (1991). Causes of sex ratio bias may account for unisexual sterility in hybrids: a new explanation for Haldane's rule and related phenomena. *Genetics*, **128**, 841–858.

HUTCHINSON, G. E. (1959). A homage to Santa Rosalia, or why are there so many different kinds of animals? *American Naturalist*, **93**, 145–159.

INOUYE, D. W. (1980). The terminology of floral larceny. *Ecology*, **61**, 1251–1253.

IRELAND, H., KEARNS, P. W. E. and MAJERUS, M. E. N. (1986). Interspecific hybridisation in the Coccinellidae: some observations on an old controversy. *Entomologists' Record and Journal of Variation*, **98**, 181–185.

IWASA, Y., POMIANKOWSKI, A. and NEE, S. (1991). The evolution of costly mate preferences. II. The 'Handicap principle'. *Evolution*, **45**, 1431–1442.

JACOBSON, E. (1909). Beobachtungen uber den Polymorphismus von *Papilio memnon. Tijdschr. Ent.*, **52**, 125–157.

JAENIKE, J. (1993). Rapid evolution of host-specificity in a parasitic nematode. *Evolutionary Ecology*, **7**, 103–108.

JAMES, A. C. and JAENIKE, J. (1990). 'Sex ratio' meiotic drive in *Drosophila testacea. Genetics*, **126**, 651–656.

JANZEN, D. H. (1966). Coevolution of mutualism between ants and acacias in Central America. *Evolution*, **23**, 1–27.

JANZEN, D. H. (1967). Interaction of the bull's horn acacia (*Acacia cornigera* L.) with an ant inhabitant (*Pseudomyrmex ferruginea* F. Smith) in eastern Mexico. *University of Kansas Science Bulletin*, **47**, 315–558.

JANZEN, D. (1980). When is it coevolution? *Evolution*, **34**, 611–612.

JEFFREYS, A. J., BROOKFIELD, J. F. Y. and SEMEONOFF, R. (1985b). Positive identification of an immigration test-case using DNA fingerprints. *Nature*, **317**, 818–819.

JEFFREYS, A. J., WILSON, V. and THEIN, S. L. (1985a). Hypervariable 'minisatellite' regions in human DNA. *Nature*, **314**, 67–73.

JEFFREYS, A. J., ROYLE, N. J., WILSON, V. and WONG, Z. (1988a). Spontaneous mutation rates to new length alleles at tandem-repetitive hypervariable loci in human DNA. *Nature*, **332**, 278–281.

JEFFREYS, A. J., WILSON, V., NEUMANN, R. and KEYTE, J. (1988b). Amplification of human minisatellites by the polymerase chain reaction: towards DNA fingerprinting of single cells. *Nucleic Acids Research*, **16**, 10953–10971.

JEFFREYS, A. J., MACLEOD, A., TAMAKI, K., NEIL, D. L. and MONCKTON, D. G. (1991). Minisatellite repeat coding as a digital approach to DNA typing. *Nature*, **354**, 204–209.

JEFFREYS, A. J., TAMAKI, K., MACLEOD, A., MONCKTON, D. G., NEIL, D. L. and ARMOUR, J. A. L. (1994). Complex gene conversion events in germline mutation at human minisatellites. *Nature Genetics*, **6**, 136–145.

JERMANN, T. M., OPITZ, J. G., STACKHOUSE, J. and BENNER, S. A. (1995). Reconstructing evolutionary history of the artiodactyl ribonuclease. *Nature*, **374**, 57–59.

JOBLING, M. A., FRETWELL, N., DOVER, G. A. and JEFFREYS, A. J. (1994). Digital coding of human Y-chromosomes – MVR-PCR at Y-specific minisatellites. *Cytogenetics and Cell Genetics*, **67**, 390.

JOHNSTONE, R. A. (1994). Female preference for symmetrical males as a by-product of selection for mate recognition. *Nature*, **372**, 172–175.

JONES, C. S., LESSELLS, C. M. and KREBS, J. R. (1991). Helpers-at-the-nest in European bee-eaters (*Merops apiaster*): a genetic analysis. In Burke, T., Dolf, G., Jeffreys, A. J. and Wolff, R. (eds) *DNA fingerprinting: approaches and applications*, pp. 169–192. Birkhauser Verlag, Basel.

JONES, I. L. and HUNTER, F. M. (1993). Mutual sexual selection in a monogamous seabird. *Nature*, **362**, 238–239.

JONES, J. S. (1973). The genetic structure of a southern peripheral population of the snail *Cepaea nemoralis* (L.). *Proceedings of the Royal Society of London B*, **183**, 371–384.

JONES, J. S. (1982). Genetic differences in individual behaviour associated with shell polymorphism in the snail *Cepaea nemoralis*. *Nature*, **298**, 749–750.

JONES, S. J. M. (1995). An update and lessons from whole genome sequencing projects. *Current Opinions in Genetics and Development*, **5**, 349–353.

JUKES, T. H. and CANTOR, C. R. (1969). Evolution of protein molecules. In Munro H. N. (ed.) *Mammalian protein metabolism*, pp. 21–132. Academic Press, New York.

KANEDA, H., HAYASHI, J.-I., TAKAHAMA, S., TAYA, C., LINDAHL, K. F. and YONEKAWA, H. (1995). Elimination of paternal mitochondrial DNA in intraspecific crosses during early mouse embryogenesis. *Proceedings of the National Academy of Sciences USA*, **92**, 4542–4546.

KANESHIRO, K. Y. and BOAKE, C. R. B. (1987). Sexual selection and speciation: issues raised by Hawaiian *Drosophila*. *Trends in Ecology and Evolution*, **2**, 207–212.

KANG, S., JAWORSKI, A., OHSHIMA, K. and WELLS, R. D. (1995). Expansion and deletion of CTG repeats from human disease genes are determined by the direction of replication in *E. coli. Nature Genetics*, **10**, 213–218.

KERNAGHAN, R. P. and EHRMAN, L. (1970). Antimycoplasmal antibiotics and hybrid sterility in *Drosophila paulistorum*. *Science*, **169**, 63–64.

KETTLEWELL, H. B. D. (1955). Selection experiments on industrial melanism in the Lepidoptera. *Heredity*, **9**, 323–342.

KETTLEWELL, H. B. D. (1956). Further selection experiments on industrial melanism in the Lepidoptera. *Heredity*, **10**, 287–301.

KETTLEWELL, H. B. D. (1973). *The evolution of melanism*. Clarendon Press, Oxford.

KHUSH, G. S. (1973). *Cytogenetics of aneuploids*. Academic Press, New York.

KIDWELL, M. G. (1994). The evolutionary history of the P family of transposable elements. *Journal of Heredity*, **85**, 339–346.

KIMURA, M. (1968). Evolutionary rate at the molecular level. *Nature*, **217**, 624–626.

KIMURA, M. (1983). *The neutral theory of molecular evolution*. Cambridge University Press, Cambridge.

KIPLING, D. and COOKE, H. J. (1990). Hypervariable ultra-long telomeres in mice. *Nature*, **347**, 400–402.

KIRKPATRICK, M. (1983). Sexual selection and the evolution of female choice. *Evolution*, **36**, 1–12.

KIRKPATRICK, M. and RYAN, M. J. (1991). The evolution of mating preferences and the paradox of the lek. *Nature*, **350**, 33–38.

KOCHER, T. D., THOMAS, W. K., MEYER, A. *et al.* (1989). Dynamics of mitochondrial DNA evolution in animals: amplification and sequencing with conserved primers. *Proceedings of the National Academy of Sciences USA*, **86**, 6196–6200.

KONDRASHOV, A. S. (1982). Selection against deleterious mutations in large sexual and asexual populations. *Genetic Research*, **40**, 325–332.

KONDRASHOV, A. S. (1988). Deleterious mutations and the evolution of sexual reproduction. *Nature*, **336**, 435–440.

KONDRASHOV, A. S. (1992). Species and speciation. *Nature*, **356**, 752.

KONDRASHOV, A. S. (1993). Classification of hypotheses on the advantage of amphimixis. *Journal of Heredity*, **84**, 372–387.

KONDRASHOV, A. S. (1994). The asexual ploidy cycle and the origin of sex. *Nature*, **370**, 213–216.

KONDRASHOV, A. S. and CROW, J. F. (1991). Haploidy or diploidy: which is better? *Nature*, **351**, 314–315.

KORENBERG, J. R. and ENGELS, W. R. (1978). Base ratio, DNA content, and quinacrine brightness of human chromosomes. *Proceedings of the National Academy of Sciences USA*, **75**, 3382–3386.

LABANDEIRA, C. C., DILCHER, D. L., DAVIS, D. R. and WAGNER, D. L. (1994). Ninety-seven million years of angiosperm–insect association: paleobiological insights into the meaning of coevolution. *Proceedings of the National Academy of Sciences USA*, **91**, 12 278–12 282.

LACK, D. (1947). *Darwin's finches*. Cambridge University Press, Cambridge.

LACK, D. (1966). *Population studies of birds*. Clarendon Press, Oxford.

LAMBORN, W. A. (1912). Three families of *P. dardanus* Brown, bred from *hippocoon* F. females in the Lagos district. *Proceedings of the Entomological Society of London*, **1912**, xii–xvii.

LANDE, R. (1980). Sexual dimorphism, sexual selection and adaptation in polygenic characters. *Evolution*, **34**, 292–305.

LANDE, R. (1981). Models of speciation by sexual selection on polygenic traits. *Proceedings of the National Academy of Sciences USA*, **78**, 3721–3725.

LAW, R. and HUTSON, V. (1992). Intracellular symbionts and the evolution of uniparental cytoplasmic inheritance. *Proceedings of the Royal Society, Series B*, **248**, 69–77.

LAWTON, J. H. (1984). Non-competitive populations, non-convergent communities, and vacant niches: the herbivores of bracken. In Strong, D. R., Simberloff, D., Abele, L. G. and Thistle, A. B. (eds) *Ecological communities: conceptual issues and the evidence.* Princeton University Press, Princeton, New Jersey.

LEIGH, E. G. (1971). *Adaptation and diversity.* Freeman Cooper and Co, San Francisco.

LEIGH, G. F. (1904). Synepigonic series of *Papilio cenea* (1902–1903) and *Hypolimnas misippus* (1904) together with observations on the life history of the former. *Transactions of the Entomological Society of London*, **1904**, 677–694.

LEIGH, G. F. (1912). Families of *Papilio dardanus* Brown, bred in Natal from female parents of the *trophonius* Westw. form. *Proceedings of the Entomological Society of London*, **1912**, cxxxv–cxxxvi.

LENINGTON, S. (1983). Social preferences for partners carrying 'good genes' in wild house mice. *Animal Behavior*, **31**, 325–333.

LENINGTON, S. (1994). Of mice, men and the MHC. *Trends in Ecology and Evolution, 9,* 455–456.

LEVENE, H., PAVLOVSKY, O. and DOBZHANSKY, T. (1954). Interaction of the adaptive values in polymorphic experimental populations of *Drosophila pseudoobscura. Evolution,* **8,** 335–349.

LEVIN, D. A. (1975). Pest pressure and recombination systems in plants. *American Naturalist*, **109**, 437–451.

LEVINGS, C. S. III and PRING, D. R. (1976). Restriction endonuclease analysis of mitochondrial DNA from normal and Texas cytoplasmic male sterile maize. *Science*, **193**, 158–160.

LEWIS, W. M. J. (1985). Nutrient scarcity as an evolutionary cause of haploidy. *American Naturalist*, **125**, 692–701.

LEWONTIN, R. C. and HUBBY, J. L. (1966). A molecular approach to the study of genic heterozygosity in natural populations. II Amount of variation and degree of hetero-zygosity in natural populations of *Drosophila pseudoobscura. Genetics*, **54**, 595–609.

LITT, M. and LUTY, J. A. (1989). A hypervariable microsatellite revealed by in vitro ampli-fication of a dinucleotide repeat within the cardiac muscle actin gene. *American Journal of Human Genetics*, **44**, 397–401.

LIVELY, C. M. (1987). Evidence of a New Zealand snail for the maintenace of sex by parasitism. *Nature*, **328**, 519–521.

LLOYD, D. G. (1974). Theoretical sex ratios of dioecious and gynodioecious angiosperms. *Heredity*, **32**, 11–31.

LLOYD, D. G. (1975). The maintenance of gynodioecy and androdioecy in angiosperms. *Genetica*, **45**, 325–339.

LLOYD, D. G. (1976). The transmission of genes via pollen and ovules in gynodioecious angiosperms. *Theoretical Population Biology*, 9, 299–316.

LOVE, J. M., KNIGHT, A. M., MCALEER, M. and TODD, J. A. (1990). Towards construction of a high resolution map of the mouse genome using PRC-analysed microsatellites. *Nucleic Acids Research*, **18**, 4123–4130.

LYNCH, M. (1990). The similarity index and DNA fingerprinting. *Molecular Biology and Evolution*, **7**, 478–484.

LYNCH, M. and JARRELL, P. E. (1993). A method for calibrating molecular clocks and its application to animal mitochondrial DNA. *Genetics*, **135**, 1197–1208.

LYNCH, M. and MILLIGAN, B. G. (1994). Analysis of population genetic structure with RAPD markers. *Molecular Ecology*, **3**, 91–99.

LYTTLE, T. W. (1991). Segregation distorters. *Annual Review of Genetics*, **25**, 511–557.

MADSEN, T., SHINE R., LOMAN, J. and HAKANSSON, T. (1992). Why do female adders copulate so frequently? *Nature*, **355**, 440–441.

MAJERUS, M. E. N. (1986). The genetics and evolution of female choice. *Trends in Ecology and Evolution*, **1**, 1–7.

MAJERUS, M. E. N. (1989). Melanic polymorphism in the peppered moth, *Biston betularia*, and other Lepidoptera. *Journal of Biological Education*, **23**, 267–284.

MAJERUS, M. E. N. (1994). *Ladybirds*. Harper Collins, London.

MAJERUS, M. E. N., O'DONALD, P. and WIER, J. (1982). Female mating preference is genetic. *Nature*, **300**, 521–523.

MAJERUS, M. E. N., O'DONALD, P., KEARNS, P. W. E. and IRELAND, H. (1986). The genetics and evolution of female choice. *Nature*, **321**, 164–167.

MANDEL, J.-L. (1994). Trinucleotide diseases on the rise. *Nature Genetics*, **7**, 453–455.

MARGULIS, L. (1970). *Origin of eukaryotic cells*. Yale University Press.

MARTIN, A. P. and PALUMBI, S. R. (1993). Body size, metabolic rate, generation time, and the molecular clock. *Proceedings of the National Academy of Sciences USA*, **90**, 4087–4091.

MARTIN, A. P., NAYLOR, G. J. P. and PALUMBI, S. R. (1992). Rates of mitochondrial DNA evolution in sharks are slow compared with mammals. *Nature*, **357**, 153–155.

MAY, R. M. and ANDERSON, R. M. (1983). Epidemiology and genetics in the coevolution of parasites and hosts. *Proceedings of the Royal Society of London, Series B*, **219**, 281–313.

MAYNARD SMITH, J. (1965). Group selection and kin selection. *Nature*, **201**, 1145–1147.

MAYNARD SMITH, J. (1976a). A short-term advantage for sex and recombination through sib-competition. *Journal of Theoretical Biology*, **63**, 245–258.

MAYNARD SMITH, J. (1976b). Sexual selection and the handicap principle. *Journal of Theoretical Biology*, **57**, 239–242.

MAYNARD SMITH, J. (1978). *The evolution of sex*. Cambridge University Press, Cambridge.

MAYNARD SMITH, J. (1986). Contemplating life without sex. *Nature*, **324**, 300–301.

MAYNARD SMITH, J. (1991). Theories of sexual selection. *Trends in Ecology and Evolution*, **6**, 146–151.

MAYR, E. (1942). *Systematics and the origin of species*. Columbia University Press, New York.

MAYR, E. (1963). *Animal species and evolution*. Belknap, Cambridge, MA.

MCALLISTER, M. K. and ROITBERG, B. P. (1987). Adaptive suicidal behaviour of pea aphids. *Nature*, **328**, 797–799.

MCCLINTOCK, B. (1951). Chromosome organization and genic expression. *Cold Spring Harbor Symposia on Quantitative Biology*, **16**, 13–47.

MCNEILLY, T. (1968). Evolution in closely adjacent plant populations. III. *Agrostis tenuis* on a small copper mine. *Heredity*, **23**, 99–108.

MCNEILLY, T. and ANTONOVICS, J. (1968). Evolution in closely adjacent plant populations. IV. Barriers to gene flow. *Heredity*, **23**, 2205–2218.

MEUSEL, M. S. and MORITZ, R. F. A. (1993). Transfer of mitochondrial DNA during fertilization of honeybee (*Apis mellifera*) eggs. *Current Genetics*. **24**, 539–543.

MEYER, A., MORRISSEY, J. and SCHARTL, M. (1994). Molecular phylogeny of fishes of the genus *Xiphophorus* suggests repeated evolution of a sexually selected trait. *Nature*, **368**, 539–542.

MIKKOLA, K. (1979). Resting site selection of *Oligia* and *Biston* moths (Lepidoptera: Noctuidae and Geometridae). *Annales Entomologici Fennici*, **45**, 81–87.

MILINKOVITCH, M. C. (1995). Molecular phylogeny of cetaceans prompts revision of morphological transformations. *Trends in Genetics*, **10**, 328–334.

MILINKOVITCH, M. C., ORTÍ, G. and MEYER, A. (1993). Revised phylogeny of whales suggested by mitochondrial ribosomal DNA sequences. *Nature*, **361**, 346–348.

MILINKOVITCH, M. C., ORTÍ, G. and MEYER, A. (1995). Novel phylogeny of whales revisited but not revised. *Molecular Biology and Evolution*, **12**, 518–520.

MØLLER (1990). Effects of a haematophagous mite on the barn swallow (*Hirundo rustica*): a test of the Hamilton and Zuk hypothesis. *Evolution*, **44**, 771–784.

MONATH, T. P. (1994). Dengue: the risk to developed and developing countries. *Proceedings of the National Academy of Sciences USA*, **91**, 2395–2400.

MONCKTON, D. G., NEUMANN, R., GURAM, T. *et al.* (1994). Minisatellite mutation rate variation associated with a flanking DNA sequence polymorphism. *Nature Genetics*, **8**, 162–170.

MOORE, S. S., SARGEANT, L. L., KING, T. J., MATTICK, J. S., GEORGES, M. and HETZEL, D. J. S. (1991). The conservation of dinucleotide microsatellites among mammalian genomes allows use of heterologous PCR primer pairs in closely related species. *Genomics*, **10**, 654–660.

MORAN, N. A., MUNSON, M. A., BAUMANN, P. and ISHIKAWA, H. (1993). A molecular clock in ·endosymbiotic bacteria is calibrated using the insect hosts. *Proceedings of the Royal Society of London B*, **253**, 167–171.

MORGAN, T. H., BRIDGES, C. B. and STURTEVANT, A. H. (1925). The genetics of *Drosophila*. *Bibliogr. Genet.*, **2**, 1–262.

MORIN, P. A. and WOODRUFF, D. S. (1992). Paternity exclusion using multiple hypervariable microsatellite loci amplified from nuclear DNA of hair cells. In Martin, R. D., Dixson, A. F. and Wickings, E. J. (eds) *Paternity in primates: genetic tests and theories*, pp. 63–81. Karger, Basel.

MORITZ, C. (1994). Applications of mitochondrial DNA analysis in conservation: a critical review. *Molecular Ecology*, **3**, 401–411.

MULLER, H. J. (1932). Some genetic aspects of sex. *American Naturalist*, **66**, 118–138.

MULLER, H. J. (1933). Further studies on the causes of gene mutations. *Proceedings of the Sixth International Congress on Genetics*, **1**, 213–255.

MULLER, H. J. (1942). Isolating mechanisms, evolution and temperature. *Biological Symposia*, **6**, 71–125.

MULLER, H. J. (1950). Our load of mutations. *American Journal of Human Genetics*, **2**, 111–176.

MULLER, H. J. (1964). The relation of recombination to mutational advance. *Mutation Research*, **1**, 2–9.

MYLVAGANAM, S. and DENNIS, P. P. (1992). Sequence heterogeneity between two genes encoding 16S rRNA from the halophilic archaebacterium *Haloarcula marismortui*. *Genetics*, **130**, 399–410.

NAKAMURA, H. (1990). Brood parasitism by the cuckoo *Cuculus canorus* in Japan and the start of new parasitism on the azure-winged magpie *Cyanopica cyana*. *Japanese Journal of Ornithology*, **39**, 1–18.

NEE, S., HOLMES, E. C., RAMBAUT, A. and HARVEY, P. H. (1995). Inferring population histories from molecular phylogenies, *Philosophical Transactions of the Royal Society of London, Series B* (in press).

NEWTON, W. C. F. and PELLEW, C. (1929). *Primula kewensis* and its derivatives. *Journal of Genetics*, **20**, 405–466.

NICHOLLS, M. (1979). Ecological genetics of copper tolerant *Agrostis tenuis* Sibth. Upublished PhD thesis, University of Liverpool.

NIGRO, V., POLITANO, L., NIGRO, G., ROMANO S. C., MOLINARI, A. M. and PUCA, G. A. (1992). Detection of nonsense mutation in the dystrophin gene by multiple SSCP. *Human Molecular Genetics*, **1**, 517–520.

NILSSON, L. A. (1988). The evolution of flowers with deep corolla tubes. *Nature*, **334**, 147–149.

NILSSON, L. A., JONSSON, L., RASON, L. and RANDRIANJOHANY, E. (1985). Monophily and pollination mechanisms in *Angracum* in a guild of long-tongued hawkmoths. *Biological Journal of the Linnaean Society of London*, **26**, 1–19.

NISBET, I. C. T. (1977). Courtship feeding and clutch size in common terns *Sterna hirundo*. In Stonehouse, B. and Perrins, C. M. (eds) *Evolutionary Ecology*, pp. 101–109. Macmillan, London.

NOOR, M. A. (1995). Speciation driven by natural selection in *Drosophila*. *Nature*. **375**, 674–675.

NUR, U., WERREN, J. H., EICKBUSH, D. G., BURKE, W. D. and EICKBUSH, T. H. (1988). A 'selfish' B-chromosome that enhances its transmission by eliminating the paternal genome. *Science*, **240**, 512–514.

NYBOM, H. and SCHAAL, B. A. (1990). DNA 'fingerprints' reveal genotypic distributions in natural populations of blackberries and raspberries (Rubus, Rosaceae). *American Journal of Botany*, **77**, 883–888.

O'DONALD, P. (1967). A general model of sexual and natural selection. *Heredity*, **31**, 145–156.

O'DONALD, P. (1969). 'Haldane's Dilemma' and the rate of natural selection. *Nature*, **221**, 815–816.

O'DONALD, P. (1980). *Genetic models of sexual selection*. Cambridge University Press, Cambridge, UK.

O'DONALD, P. (1983). *The arctic skua*. Cambridge University Press, Cambridge.

OHNO, S. (1970). *Evolution by gene duplication*. Springer-Verlag, New York.

O'NEILL, S., GIORDANO, R., COLBERT, A. M. E., KARR, T. L. and ROBERTSON, H. M. (1992). 16S rRNA phylogenetic analysis of the bacterial endosymbionts associated with cytoplasmic incompatibility in insects. *Proceedings of the National Academy of Sciences USA*, **89**, 2694–2702.

ORGEL, L. E. and CRICK, F. H. C. (1980). Selfish DNA: The ultimate parasite. *Nature*, **284**, 604–607.

ORR, H. A. (1995). Somatic mutation favours the evolution of diploidy. *Genetics*, **139**, 1441–1447.

OU, C. Y., CIESIELSKI, C. A., MYERS, G. *et al.* (1992). Molecular epidemiology of HIV transmission in a dental practice. *Science*, **256**, 1165–1171.

OWEN, D. F. (1965). Change in sex ratio in an African butterfly. *Nature*, **205**, 744.

OWEN, D. F., OWEN, J. and CHANTER, D. O. (1973). Low mating frequencies in an African butterfly. *Nature*, **244**, 116–117.

PACKER, C. A., GILBERT, D. A., PUSEY, A. E. and O'BRIEN, S. J. (1991). A molecular genetic analysis of kinship and cooperation in African lions. *Nature*, **351**, 562–565.

PALSBØLL, P. J., VADER, A., BAKKER, I. and ELGEWELY, M. R. (1992). Determination of gender in cetaceans by the polymerase chain-reaction. *Canadian Journal of Zoology*, **70**, 2166–2170.

PALUMBI, S. R. and BAKER, C. S. (1994). Contrasting population structure from nuclear intron sequences and mtDNA of humpback whales. *Molecular Biology and Evolution*, **11**, 426–435.

PARKER, G. A. (1970). Sperm competition and its evolutionary consequences in insects. *Biological Reviews*, **45**, 525–567.

PARKER, G. A. and SMITH, J. L. (1975). Sperm competition and the evolution of the pre-copulatory passive phase behaviour in *Locusta migratoria migratorioides*. *Journal of Entomology (A)*, **49**, 155–171.

PARKER, G. A. and SIMMONS, L. W. (1994). Evolution of phenotypic optima and copula duration in dungflies. *Nature*, **370**, 53–56.

PARKER, G. A., BAKER, R. R. and SMITH, V. G. F. (1972). The origin and evolution of gamete dimorphism and the male–female phenomenon. *Journal of Theoretical Biology*, **36**, 529–553.

PARKER, M. A. (1994). Pathogens and sex in plants. *Evolutionary Ecology*, **8**, 560–584.

PELLMYR, O. and THOMPSON, J. N. (1992). Multiple occurrences of mutualism in the yucca moth lineage. *Proceedings of the National Academy of Sciences USA*, **89**, 2927–2929.

PEMBERTON, J. M., ALBON, S. D., GUINNESS, F. E. and CLUTTON-BROCK, T. H. (1991). Counterveiling selection in different fitness components in female red deer. *Evolution*, **45**, 93–103.

PEMBERTON, J. M., ALBON, S. D., GUINNESS, F. E., CLUTTON-BROCK, T. H. and DOVER, G. A. (1992). Behavioral estimates of male mating success tested by DNA fingerprinting in a poly-gynous mammal. *Behavioural Ecology*, **3**, 66–75.

PEMBERTON, J. M., SLATE J., BANCROFT, D. R. and BARRETT, J. A. (1995). Non-amplifying alleles at microsatellite loci. *Molecular Ecology*, **4**, 249–252.

PERRINS, C. M. (1965). Population fluctuations and clutch size in the great tit *Parus major* L. *Journal of Animal Ecology*, **34**, 601–647.

PERROT, V., RICHERD, S. and VALÉRO, M. (1991). Transition from haploidy to diploidy. *Nature*, **351**, 315–317.

PETERS, L. L. and BARKER, J. E. (1993). Novel inheritance of the murine severe combined anemia and thrombocytopenia (scat) phenotype. *Cell*, **74**, 135–142.

PETRIE, M. (1994). Improved growth and survival of offspring of peacocks with more elaborate trains. *Nature*, **371**, 598–599.

PETRIE, M., HALLIDAY, T. and SANDERS, C. (1991). Peahens prefer peacocks with elaborate trains. *Animal Behaviour*, **41**, 323–331.

PITNICK, S., SPICER, G. S. and MARHOW, T. A. (1995). How long is a giant sperm? *Nature*, **375**, 109.

PLOWMAN, A. B. and BOUGOURD, S. M. (1993). Selectively advantageous effects of B-chromo-somes on germination behaviour in *Allium schoenoprasum* L. *Heredity*, **72**, 587–593.

POLICANSKY, D. and DEMPSEY, B. (1978). Modifiers and 'sex ratio' in *Drosophila pseudoobs-cura*. *Evolution*, **32**, 922–924.

POLICANSKY, D. and ELLISON, J. (1970). Sex ratio in *Drosophila pseudoobscura*: spermiogenic failure. *Science*, **169**, 888–889.

POTTS, W. K. and WAKELAND, E. K. (1993). Evolution of MHC genetics diversity: a tale of incest, pestilence and sexual preference. *Trends in Genetics*, **9**, 406–412.

POULTON, E. B. (1906). Mimetic forms of *Papilio dardanus* (*merope*) and *Acraea johstonii*. *Transactions of the Entomological Society of London*, **1906**, 281–321.

POULTON, E. B. (1908). Heredity in six families of *Papilio dardanus* Brown, subspecies *cenea* Stoll., bred at Durban by Mr. G. F. Leigh. *Transactions of the Entomological Society of London*, **1908**, 427–445.

PROCTOR, H. C. (1991). Courtship in the water mite *Neumania papillator*: males capitalize on female adaptations for predation. *Animal Behaviour*, **42**, 589–598.

PROCTOR, H. C. (1992). Sensory exploitation and the evolution of male mating behaviour: a cladistic test using water mites (Acari: Parasitengona). *Animal Behaviour*, **44**, 745–752.

PUNNETT, R. C. (1915). *Mimicry in butterflies*. Cambridge University Press, Cambridge.

PUNNETT, R. C. (1927). *Mendelism*. Macmillan, London.

RABENOLD, P. P., RABENOLD, K. N., PIPER, W. H., HAYDOCK, J. and ZACK, S. W. (1990). Shared paternity revealed by genetic analysis in cooperatively breeding tropical wrens. *Nature*, **348**, 538–540.

RAND, D. M. (1994). Thermal habit, metabolic rate and the evolution of mitochondrial DNA. *Trends in Ecology and Evolution*, **9**, 125–131.

RASMUSSON, K. E., RAYMOND, J. D. and SIMMONS, M. J. (1993). Repression of hybrid dysgenesis in *Drosophila melanogaster* by individual naturally occurring *P* elements. *Genetics*, **133**, 605–622.

REDFIELD, R. J. (1994). Male mutation rates and the cost of sex to females. *Nature*, **369**, 145–147.

REHR, S. S., FEENY, P. P. and JANZEN, D. H. (1973). Chemical defences in Central American non-ant-acacias. *Journal of Animal Ecology*, **42**, 405–416.

RICE, W. R. (1987). The accumulation of sexually antagonistic genes as a selective agent promoting the evolution of reduced recombination between primitive sex chromosomes. *Evolution*, **41**, 911–914.

RICE, W. R. (1992). Sexually antagonistic genes: experimental evidence. *Science*, **256**, 1436–1439.

RICE, W. R. (1994). Degeneration of a nonrecombining chromosome. *Science*, **263**, 230–232.

RICE, W. R. and HOSTERT, E. E. (1993). Laboratory experiments on speciation: what have we learned in 40 years? *Evolution*, **47**, 1637–1653.

RICHARDS, R. I. and SUTHERLAND, G. R. (1994). Simple repeats not replicated simply. *Nature Genetics*, **6**, 114–116.

RICHERD, S., COUVET, D. and VALÉRO, M. (1993). Evolution of haploid and diploid phases in life cycles. II. Maintenance of haplo-diplontic cycle. *Journal of Evolutionary Biology*, **6**, 253–280.

RICO, C., KUHNLEIN, U. and FITZGERALD, G. J. (1991). Spawning patterns in the 3-spined stickleback (*Gasterosteus aculeatus* L.) – an evaluation by DNA fingerprinting. *Journal of Fish Biology*, **39**, 151–158.

RIDLEY, M. (1993). *The Red Queen: sex and the evolution of human nature*. Viking, London.

RIEDY, M. F., HAMILTON III, W. J., and AQUADRO, C. F. (1992). Excess of non-parental bands in offspring from known primate pedigrees assayed using RAPD PCR. *Nucleic Acids Research*, **20**, 918.

RIGAUD, T. and JUCHAULT, P. (1992). Genetic control of the vertical transmission of a cytoplasmic sex factor in *Armadillidium vulgare*. *Heredity*, **68**, 47–52.

ROUSSET, F., BOUCHON, D., PINTUREAU, B., JUCHAULT, P. and SOLIGNAC, M. (1992). *Wolbachia* endosymbionts responsible for various alterations of sexuality in arthropods. *Proceedings of the Royal Society of London, Series B*, **250**, 91–98.

ROY, M. S., GEFFEN, E., SMITH, D., OSTRANDER, E. A. and WAYNE, R. K. (1994). Patterns of differentiation and hybridization in North American wolflike canids, revealed by analysis of microsatellite loci. *Molecular Biology and Evolution*, **11**, 553–570.

RUBINSZTEIN, D. C., AMOS, W., LEGGO, J. *et al.* (1994). Mutational bias provides a model for the evolution of Huntington's disease and predicts a general increase in disease prevalence. *Nature Genetics*, **7**, 525–530.

RUBINSZTEIN, D. C., AMOS, W., LEGGO, J. *et al.* (1995). Microsatellites are generally longer in humans compared to their homologues in non-human primates: evidence for directional evolution at microsatellite loci. *Nature Genetics* (in press).

RYAN, M. J. (1983). Sexual selection and communication in a neotropical frog, *Physlaemus pustulosus*. *Evolution*, **37**, 261–272.

RYAN, M. J. (1985). *The Tungara Frog, a study in sexual selection and communication.* Chicago University Press, Chicago.

RYAN, M. J., FOX, J. H., WILCZYNSKI, W. and RAND, A. S. (1990). Sexual selection for sensory exploitation in the frog *Physalemus pustulosus*. *Nature*. **343**, 66–67.

RYDER, O. A., KUMAMOTO, A. T., DURRANT, B. S. and BENIRSCHKE, K. (1989). Chromosomal divergence and reproductive isolation in dik-diks. In Otte, D. and Endler, J. A. (eds) *Speciation and its consequences*, pp. 208–228. Columbia University Press, New York.

SAITOU, N. and NEI, M. (1987). The neighbour-joining method: a new method for reconstructing phylogenetic trees. *Molecular Biology and Evolution*, **4**, 406–425.

SANDLER, L. and NOVITSKI, E. (1958). Meiotic drive as an evolutionary force. *American Naturalist*, **91**, 105–110.

SCHIERWATER, B. and ENDER, A. (1993). Different thermostable DNA polymerases may amplify different RAPD products. *Nucleic Acids Research*, **21**, 4647–4648.

SCHLIEWEN, U. K., TAUTZ, D. and PÄÄBO, S. (1994). Sympatric speciation suggested by monophyly of Crater Lake cichlids. *Nature*, **368**, 629–632.

SCHLÖTTERER, C. and TAUTZ, D. (1992). Slippage synthesis of simple sequence DNA. *Nucleic Acids Research*, **20**, 211–215.

SCHLÖTTERER, C., AMOS, W. and TAUTZ, D. (1991). Conservation of polymorphic simple sequence loci in cetacean species. *Nature*, **354**, 63–64.

SCHLUTER, D. (1994). Experimental evidence that competition promotes divergence in adaptive radiation. *Science*, **266**, 798–801.

SCHLUTER, D. and MCPHAIL, J. D. (1992). Ecological character displacement and speciation in sticklebacks. *American Naturalist*, **140**, 85–108.

SCOTT, D. K. (1988). Reproductive success in Bewick's swans. In Clutton-Brock, T. H. (ed.) *Reproductive success*, pp. 220–236. Chicago University Press, Chicago.

SEDLMAIR, H. (1956). Verhaltens-, Resistenz- und Gehause Unterschiede bei den polymorphen Banderschnecken *Cepaea hortensis* (Mull.) und *Cepaea nemoralis* (L.). *Biol. Zentbl.*, **75**, 281–313.

SHAW, D. (1995). When DNA turns. *New Scientist*, 25 March, 28–33.

SHAW, M. W. (1984). The population genetics of the B-chromosome of *Myrmeleotettix maculatus* (Thumb.) (Orthoptera: Acrididae). *Heredity*, **54**, 187–194.

SHAW, M. W. and HEWITT, G. M. (1990). B chromosomes, selfish DNA and theoretical models: where next? In Futuyama, D. and Antonovics, J. (eds) *Oxford surveys in evolutionary biology*, pp. 197–223. Oxford University Press, Oxford.

SHEFFIELD, V. C., COX, D. R. and MYERS, R. M. (1989). Attachment of a 40-base-pair G + C rich sequence to genomic DNA fragments by the polymerase chain reaction results in improved detection of single base changes. *Proceedings of the National Academy of Sciences USA*, **86**, 232–236.

SHEFFIELD, V. C., BECK, J. S., KWITEK, A. E., SANDSTROM, D. W. and STONE, E. M. (1993). The sensitivity of single-strand conformation polymorphism analysis for the detection of single base substitutions. *Genomics*, **16**, 325–332.

SHEPPARD, P. M. (1952a). Natural selection in two colonies of the polymorphic land snail *Cepaea nemoralis*. *Heredity*, **6**, 233–238.

SHEPPARD, P. M. (1952b). A note on non-random mating in the moth *Panaxia dominula* (L.). *Heredity*, **6**, 239–241.

SHEPPARD, P. M. and COOK, L. M. (1962). The manifold effects of the *medionigra* gene in the moth *Panaxia dominula* and the maintenance of a polymorphism. *Heredity*, **17**, 415–426.

SHERMAN, P. W. (1977). Nepotism and the evolution of alarm calls. *Science*, **197**, 1246–1253.

SHERMAN, P. W. (1981). Kin selection, demography, and Belding's ground squirrel nepotism. *Behavioural Ecology and Sociobiology*, **8**, 251–259.

SHRIVER, M. D., JIN, L., CHAKRABORTY, R. and BOERWINKLE, E. (1993). VNTR allele frequency distributions under the stepwise mutation model: a computer simulation approach. *Genetics*, **134**, 983–993.

SIMBERLOFF, D. S. and BOKECLEN, W. (1981). Santa Rosalia reconsidered: size ratios and competition. *Evolution*, **35**, 1206–1228.

SKIBINSKI, D. O. F., GALLAGHER, C. and BEYNON, C. M. (1994). Mitochondrial DNA inheritance. *Nature*, **368**, 817–818.

SKINNER, S. W. (1982). Maternally inherited sex ratio in the parasitoid wasp *Nasonia vitripennis*. *Science*, **215**, 1133–1134.

SLATKIN, M. (1995). A measure of population subdivision based on microsatellite allele frequencies. *Genetics*, **139**, 457–462.

SMITH, M. F. and PATTON, J. L. (1991). Variation in mitochondrial cytochrome *b* sequence in natural populations of South American akodontine rodents (Muridae: Sigmodontidae). *Molecular Biology and Evolution*, **8**, 85–103.

SMITH, M. F., THOMAS, W. K. and PATTON, J. P. (1992). Mitochondrial-DNA like sequences in the genome of an akodontine rodent. *Molecular Biology and Evolution*, **9**, 204–215.

SNELL, R. G., MACMILLAN, J. C., CHEADLE, J. P., *et al.* (1993). Relationship between trinucleotide repeat expansion and phenotypic variation in Huntington's disease. *Nature Genetics*, **4**, 393–397.

SOGIN, M. L., HINKLE, G. and LELPE, D. D. (1993). Universal tree of life. *Nature*, **362**, 795.

SOLER, M. and MØLLER, A. P. (1990). Duration of sympatry and co-evolution between the great spotted cuckoo and its magpie host. *Nature*, **343**, 748–750.

SOMERSON, N. L., EHRMAN, L., KOCKA, J. P. and GOTTLIEB, F. J. (1984). Streptococcal L-forms from *Drosophila paulistorum* semispecies cause sterility in male progeny. *Proceedings of the National Academy of Sciences USA*, **81**, 282–285.

STALKER, H. D. (1961). The genetic systems modifying meiotic drive in *Drosophila paramelanica*. *Genetics*, **46**, 177–202.

STEBBINS, G. L. and HILL, G. J. C. (1980). Did multicellular plants invade the land? *American Naturalist*, **115**, 342–353.

STEEL, M. A., LOCKHART, P. J. and PENNY, D. (1993). Confidence in evolutionary trees from biological sequence data. *Nature*, **364**, 440–442.

STEINEMANN, M. and STEINEMANN, S. (1992). Degenerating Y chromosome of *Drosophila miranda*: a trap for retrotransposons. *Proceedings of the National Academy of Sciences USA*, **89**, 7591–7595.

STEINEMANN, M., STEINEMANN, S. and LOTTSPEICH, F. (1993). How Y chromosomes become genetically inert. *Proceedings of the National Academy of Sciences USA*, **90**, 5737–5741.

STERN, D. (1994). A phylogenetic analysis of soldier evolution in the aphid family Hormaphididae. *Proceedings of the Royal Society of London, Series B*, **256**, 203–209.

STEWART, K. J. and HARCOURT, A. H. (1987). Gorillas: variation in female relationships. In Smuts, B. B., Cheney, D. L., Seyfarth, R. M., Wrangham, R. W. and Struhsaker, T. T. (eds) *Primate societies*, pp. 155–164. University of Chicago Press, Chicago.

STRONG, D. R., LAWTON, J. H. and SOUTHWOOD, T. R. E. (1984). *Insects on plants: community patterns and mechanisms*. Blackwell, Oxford.

STURTEVANT, A. H. and DOBZHANSKY, T. (1936). Geographical distribution and cytology of 'sex ratio' in *Drosophila pseudoobscura* and related species. *Genetics*, **21**, 473–490.

SWYNNERTON, G. F. M. (1919). Evidence of Mendelian heredity in *Papilio dardanus* Brown. *Proceedings of the Entomological Society of London*, **1919**, xxx–xxxiii.

TABERLET, P. and BOUVERT, J. (1992). Bear conservation genetics. *Nature*, **358**, 197.

TAKAHATA, N. (1991). Statistical models of the overdispersed molecular clock. *Theoretical Population Biology*, **39**, 329–344.

TAUBER, C. A. and TAUBER, M. J. (1989). Sympatric speciation in insects: perception and perspective. In Otte, D. and Endler J. A. (eds) *Speciation and its consequences*, pp. 307–344. Sinauer, Sunderland, MA.

TAUTZ, D. (1989). Hypervariability of simple sequences as a general source of polymorphic DNA markers. *Nucleic Acids Research*, **17**, 6462–6471.

TAUTZ, D. and RENZ, M. (1984). Simple sequences are ubiquitous components of eukaryotic genomes. *Nucleic Acids Research*, **12**, 4127–4138.

TAUTZ, D., TRICK, M. and DOVER, G. A. (1986). Cryptic simplicity as a major source of genetic variation. *Nature*, **322**, 652–656.

TAUTZ, D., TAUTZ, C., WEBB, D. A. and DOVER, G. A. (1987). Evolutionary divergence of promoters and spacers in the rDNA family of four *Drosophila* species: implications for molecular coevolution in multigene families. *Journal of Molecular Biology*, **195**, 525–542.

TAYLOR, D. R. (1994). The genetic basis of sex ratio in *Silene alba* (= *S.latifoloia*). *Genetics*, **136**, 641–651.

TEMPLETON, A. R. (1989). The meaning of species and speciation: a genetic perspective. In Otte, D. and Endler, J. A. (eds) *Speciation and its consequences*. Sinauer, Sunderland, MA.

TEMPLETON, A. R. (1991). Human origins and analysis of mitochondrial DNA sequences. *Science*, **255**, 737.

TEMPLETON, A. R., HOLLICHER, H., LAWLER, S. and JOHNSTON, J. S. (1989). Natural selection and ribosomal DNA in *Drosophila*. *Genome*, **31**, 296–303.

THOMAS, J. H. (1995). Genomic imprinting proposed as a surveillance mechanism for chromosome loss. *Proceedings of the National Academy of Sciences USA*, **92**, 480–482.

THOMAS, R. H. (1994). What is a guinea-pig? *Trends in Ecology and Evolution*, **9**, 159–160.

THOMPSON, J. N. (1994). *The coevolutionary process*. University of Chicago Press, Chicago.

THORNHILL, R. (1976). Sexual selection and nuptial feeding behaviour in *Bittacus apicalis* (Insecta: Mecoptera). *American Naturalist*, **110**, 529–548.

TRIVERS, R. L. (1971). The evolution of reciprocal altruism. *Quarterly Review of Biology*, **46**, 35–57.

330 References

TURNER, J. R. G. (1985). Fisher's evolutionary faith and the challenge of mimicry. In Dawkins, R. and Ridley, M. (eds) *Oxford surveys in evolutionary biology*, pp. 159–196. Oxford University Press, Oxford.

TUTT, J. W. (1896). *British moths*. George Routledge, London.

VAN VALEN, L. M. (1973). A new evolutionary law. *Evolutionary Theory*, **1**, 1–30.

VAN VALEN, L. M. (1983). How pervasive is coevolution? In Nitecki, M. H. (ed.) *Coevolution*, pp. 1–20. University of Chicago Press, Chicago.

VERGNAUD, G., MARIAT, D., APION, F., AURIAS, A., LATHROP, M. and LAUTHIER, V. (1991). The use of synthetic tandem repeats to isolate new VNTR loci – cloning of a human hypermutable sequence. *Genomics*, **11**, 135–144.

VIGILANT, L., STONEKING, M., HARPENDING, H., HAWKES, K. and WILSON, A. C. (1991). African populations and the evolution of human mitochondrial DNA. *Science*, **253**, 1503–1507.

VOELKER, R. A. (1972). Preliminary characterization of 'sex ratio' and rediscovery and reinterpretation of 'male sex ratio' in *Drosphila affinis*. *Genetics*. **71**, 597–606.

VOSSBRINK, C. R. and WOESE, C. R. (1986). Eukaryotic ribosomes that lack 5.8S RNA. *Nature*, **320**, 287–288.

VOSSBRINK, C. R., MADDOX, J. V., FRIEDMAN, S., DEBRUNNER-VOSSBRINK, B. A. and WOESE, C. R. (1987). Ribosomal RNA sequences suggest microsporidia are extremely ancient eukaryotes. *Nature*, **326**, 411–414.

WADDINGTON, C. H. (1957). *The strategy of the genes*. Allen and Unwin, London.

WALLACE, A. R. (1858). On the tendency of varieties to depart indefinitely from the original type. *Journal of the Linnean Society of London (Zoological)*, **3**, 45.

WARBURTON, P. E. and WILLARD, H. F. (1990). Genomic analysis of sequence variation in tandemly repeated DNA. *Journal of Molecular Biology*, **216**, 3–16.

WASHBURN, J. O., MERCER, D. R. and ANDERSON, J. R. (1991). Regulatory role of parasites: impact on host population shifts with resource availability. *Science*, **253**, 185–188.

WAYNE, R. K. and JENKS, S. M. (1991). Mitochondrial DNA analysis implying extensive hybridisation of the endangered red wolf *Canis rufus*. *Nature*, **351**, 565–568.

WEBER, J. L. (1990). Informativeness of human (dC-dA)n.(dG-dT)n polymorphisms. *Genomics*, **7**, 524–530.

WEBER, J. L. and MAY, P. E. (1989). Abundant class of human DNA polymorphisms which can be typed using the polymerase chain reaction. *American Journal of Human Genetics*, **44**, 338–396.

WEBER, J. L. and WONG, C. (1993). Mutation of human short tandem repeats. *Human Molecular Genetics*, **2**, 1123–1128.

WEISSENBACH, J., GYAPAY, G., DIB, C. et al. (1992). A second-generation linkage map of the human genome. *Nature*, **359**, 794–801.

WEISSMANN, A. (1887). On the signification of polar globules. *Nature*, **36**, 607–609.

WELLER, P., JEFFREYS, A. J., WILSON, V. and BLANCHETOT, A. (1984). Organisation of the human myoglobin gene. *EMBO Journal*, **3**, 439–446.

WERREN, J. H. (1991). The paternal sex ratio chromosome of *Nasonia*. *American Naturalist* **137**, 392–402.

WERREN, J. H., SKINNER, S. W. and CHARNOV, E. L. (1981). Paternal inheritance of a daughter's SR factor. *Nature*, **293**, 467–468.

WERREN, J. H., HURST, G. D. D., ZHANG, W., BREEUWER, J. A. J., STOUTHAMER, R. and MAJERUS, M. E. N. (1994). Rickettsial relative associated with male killing in the ladybird beetle (*Adalia bipunctata*). *Journal of Bacteriology*, **176**, 388–394.

WILKINSON, G. S. (1984). Reciprocal food sharing in the vampire bat. *Nature*, **308**, 181–184.

WILLEMS, P. J. (1994). Dynamic mutations hit double figures. *Nature Genetics*, **8**, 213–215.

WILLIAMS, G. C. (1975). *Sex and evolution*. Princeton University Press, Princeton, New Jersey.

WILLIAMS, J. G. K., KUBELIK, A. R., LIVAK, K. J. and TINGEY, S. V. (1990). DNA polymorphisms amplified by arbitrary primers are useful as genetic markers. *Nucleic Acids Research*, **18**, 6531–6535.

WILSON, A. C., CANN, R. L., CARR, S. M. *et al.* (1985). Mitochondrial DNA and two perspectives on evolutionary genetics. *Biological Journal of the Linnaean Society of London*, **26**, 375–400.

WILSON, C., PEARSON, R. K., BELLEN, H. J., O'KANE, C. J., GROSSNIKLAUS, U. and GEHRING, W. J. (1989). P-element-mediated enhancer detection: an efficient method for isolating and characterising developmentally regulated genes in *Drosophila*. *Genes and Development*, **3**, 1301–1313.

WILSON, D. S. (1975). A theory of group selection. *Proceedings of the National Academy of Sciences USA*, **72**, 143–146.

WILSON, D. S. (1980). *The natural selection of populations and communities*. Benjamin/Cummings, Menlo Park, California.

WOESE, C. R., KANDLER, O. and WHEELIS, M. L. (1990). Towards a natural system of organisms: proposal for the domains Archaea, Bacteria and Eucarya. *Proceedings of the National Academy of Sciences USA*, **87**, 4576–4579.

WONG, Z., WILSON, V., JEFFREYS, A. J. and THEIN, S. L. (1986). Cloning a selected fragment from a human DNA 'fingerprint': isolation of an extremely polymorphic minisatellite. *Nucleic Acids Research*, **14**, 4605–4616.

WOOD, R. J. and NEWTON, M. E. (1991). Sex ratio distortion caused by meiotic drive in mosquitoes. *American Naturalist*, **137**, 379–391.

WOOLFENDEN, G. E. and FITZPATRICK, J. W. (1984). *The Florida scrub jay*. Princeton University Press, Princeton, New Jersey.

WRANGHAM, R. W. (1986). Ecology and social relationships in two species of chimpanzee. In Rubensztein, D. I. and Wrangham, R. W. (eds) *Ecological aspects of social evolution*, pp. 352–378. Princeton University Press, Princeton, New Jersey.

WRIGHT, S. (1932). The roles of mutation, inbreeding, crossbreeding and selection in evolution. *Proceedings of the VIth International Congress of Genetics*, **1**, 356–366.

WRIGHT, S. (1934). Physiological and evolutionary theories of dominance. *American Naturalist*, **68**, 25–53.

WRIGHT, S. (1935). The analysis of variance and the correlations between relatives with respect to deviations from an optimum. *Journal of Genetics*, **30**, 243–256.

WRIGHT, S. (1948). On the roles of directed and random changes in gene frequencies in the genetics of populations. *Evolution*, **2**, 279–294.

WRIGHT, S. (1951). The genetic structure of populations. *Annals of Eugenics*, **15**, 323–354.

WRIGHT, S. (1977). *Evolution and the genetics of populations. Vol. 3. Experimental results and evolutionary deductions*. University of Chicago Press, Chicago.

WU, C. I. and PALOPOLI, M. F. (1994). Genetics of postmating reproductive isolation in animals. *Annual Review of Genetics*, **28**, 283–308.

WYNNE-EDWARDS, V. C. (1962). *Animal dispersion in relation to social behaviour*. Oliver & Boyd, Edinburgh.

YAMAGISHI, S. and FUJIOKA, M. (1986). Heavy brood parasitism by the common cuckoo *Cuculus canorus* on the azure winged magpie *Cyanopica cyana*. *Tori*, **34**, 91–96.

332   References

YANG., Z, GOLDMAN, N. and FRIDAY, A. E. (1994). Comparison of models for nucleotide substitution used in maximum-likelihood phylogenetic estimation. *Molecular Biology and Evolution*, **11**, 316–324.

YEN, J. H. and BARR, A. R. (1973). The etiologic agent of cytoplasmic incompatibility in *Culex pipiens*. *Journal of Invertebrate Pathology*, **22**, 242–250.

ZAHAVI, A. (1975). Mate selection – a selection for a handicap. *Journal of Theoretical Biology*, **53**, 205–214.

ZAHAVI, A. (1977). The cost of honesty (further remarks on the handicap principle). *Journal of Theoretical Biology*, **67**, 603–605.

ZOUROS, E., FREEMAN, K. R., BALL, A. O. and POGSON, G. H. (1992). Direct evidence for extensive paternal mitochondrial DNA inheritance in the marine mussel *Mytilus*. *Nature*, **359**, 412–414.

Index